Air Force SMARTBOOK

Second Edition
(AFOPS2)

OPERATIONS & PLANNING

Guide to Curtis E. LeMay Center & Joint Air Operations Doctrine

The Lightning Press
Norman M Wade

The Lightning Press

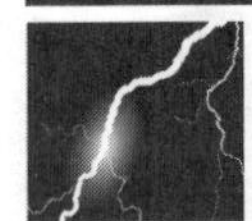

2227 Arrowhead Blvd.
Lakeland, FL 33813
24-hour Voicemail/Fax/Order: 1-800-997-8827
E-mail: SMARTbooks@TheLightningPress.com

www.TheLightningPress.com

(AFOPS2) The Air Force Operations & Planning SMARTbook, 2nd Ed.

Guide to Curtis E. LeMay Center & Joint Air Operations Doctrine

AFOPS2: The Air Force Operations & Planning SMARTbook, 2nd Ed. (Guide to Curtis E. LeMay Center & Joint Air Operations Doctrine) is the second edition of our Air Force SMARTbook. Topics and references of the 376-pg AFOPS2 include airpower fundamental and principles (Volume 1), command and organizing (Volume 3); command and control (Annex 3-30/3-52), airpower (doctrine annexes), operations and planning (Annex 3-0), planning for joint air operations (JP 3-30/3-60), targeting (Annex 3-60), and combat suppo (Annex 4-0, 4-02, 3-10, and 3-34).

ISBN: 978-1-935886-75-4

About our cover photo: An F-22 Raptor pilot lines up his aircraft March 31, 2011, to be refueled by a KC-135 Stratotanker. (U.S. Air Force photo/TSgt Kendra M. Owenby)

Printed and bound in the United States of America.

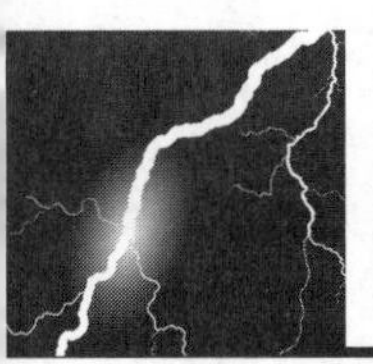

[AFOPS2] Notes to Reader

As the nation's most comprehensive provider of military airpower, the **Air Force** conducts continuous and concurrent air, space, and cyberspace operations.

Airpower exploits the third dimension of the operational environment; the electromagnetic spectrum; and time to leverage speed, range, flexibility, precision, tempo, and lethality to create effects from and within the air, space, and cyberspace domains. From this multi-dimensional perspective, Airmen can apply military power against an enemy's entire array of diplomatic, informational, military, and economic instruments of power, at long ranges and on short notice.

The Air Force conducts operations along a varying scale of military involvement and violence, referred to as the **range of military operations (ROMO)**. They range from continuous and recurring operations such as military engagement, security cooperation, and deterrence; through smaller-scale contingencies and crisis response operations, as well as irregular warfare; to major operations and campaigns such as declared wars.

The Air Force designs, plans, conducts, and assesses operations according to an **effects-based approach (EBAO)**. There are some significant differences between the focus of strategy during steady-state conditions and the focus during contingencies and major operations. Contingency planning and steady-state planning employ a common logical approach and process referred to as the **common framework for operation**s, which helps to foster coherence in Air Force strategy creation.

The JFC's estimate of the operational environment and articulation of the objectives needed to accomplish the mission form the basis for determining components' objectives. The JFACC uses the JFC's mission, commander's estimate and objectives, commander's intent, CONOPS, and the components' objectives to develop a course of action (COA). When the JFC approves the JFACC's COA, it becomes the basis for more detailed **joint air operations planning**—expressing what, where, and how joint air operations will affect the adversary or current situation.

Targeting is the process of selecting and prioritizing targets and matching the appropriate response to them, considering operational requirements and capabilities. The targeting cycle supports the joint force commander's (JFC) joint operation planning and execution with a comprehensive, iterative and logical methodology for employing ways and means to create desired effects that support achievement of objectives.

(AFOPS2) The Air Force Operations & Planning SMARTbook, 2nd Ed.

Core Doctrine (Volumes 1-3)

Volume 1

Volume 2

Volume 3

Chap 1: Airpower Fundamentals & Principles (Volume 1)

Air Force Doctrine Volume 1, Air Force Basic Doctrine, is the senior statement of Air Force doctrine. It discusses the fundamental beliefs that underpin the application of Air Force capabilities across the range of military operations. It provides guidance on the proper employment of airpower, sets the foundation for educating Airmen on airpower, guides the development of all other doctrine, and provides insight where personal experience may be lacking.

Chap 2: Commanding & Organizing (Volume 3)

Organization is critically important to effective and efficient operations. Service and joint force organization and command relationships—literally, who owns what, and who can do what with whom, and when—easily create the most friction within any operation. Air Force organization and preferred command arrangements are designed to address unity of command, a key principle of war.

Joint Air Operations Doctrine (Joint Pubs)

JP 3-30

JP 3-52

JP 3-60

Chap 6: Joint Air Operations Planning (JP 3-30, 3-52, 3-60)

The JFACC uses the JFC's mission, commander's estimate and objectives, commander's intent, CONOPS, and the components' objectives to develop a course of action (COA). When the JFC approves the JFACC's COA, it becomes the basis for more detailed joint air operations planning—expressing what, where, and how joint air operations will affect the adversary or current situation.

The JFACC is responsible for planning joint air operations and uses the joint operation planning process for air (JOPPA) to develop a JAOP that guides employment of the air capabilities and forces made available to accomplish missions assigned by the JFC. The JFC will normally delegate the authority to conduct execution planning, coordination, and deconfliction associated with joint air targeting to the JFACC and will ensure that this process is a joint effort. The joint air tasking cycle process provides an iterative, cyclic process for the planning, apportionment, allocation, coordination, and tasking of joint air missions and sorties within the guidance of the JFC.

Doctrine Annexes (& Air Force Instructions)

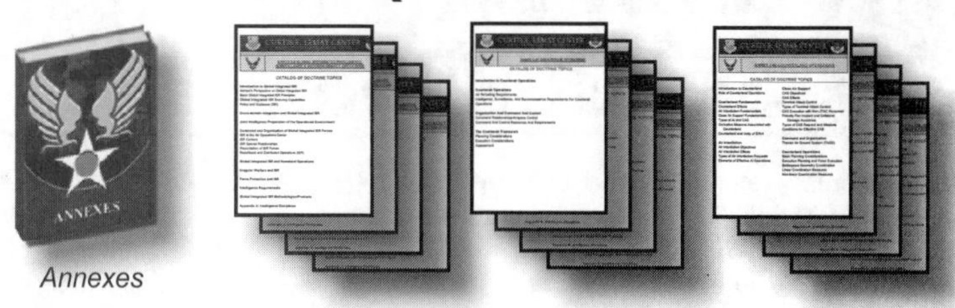

Annexes

Chap 3: Command & Control (Annexes 3-30 & 3-52)

Command and control (C2) and organization are inextricably linked. Forces should be organized around the principle of unity of command. Clear lines of authority, with clearly identified commanders at appropriate echelons exercising appropriate control, are essential to achieving unity of effort, reducing confusion, and maintaining priorities. Airspace control is defined as "capabilities and procedures used to increase operational effectiveness by promoting the safe, efficient, and flexible use of airspace.

Chap 4: Airpower (Annexes 2-0, 3-01, 3-03, 3-04, 3-05, 3-12, 3-13, 3-14, 3-17, 3-50, 3-51, 3-61, 3-17, 3-72)

Airpower entails the use of military power and influence to achieve objectives at all levels by controlling and exploiting air, space, and cyberspace. Airpower is a vital component of successful military operations and can often provide for decisive, rapid, and more efficient attainment of enduring advantage. It has been an asymmetric advantage for the United States in many operations. Defeating enemy forces has traditionally been the most important of the tasks assigned to the military, and while that remains vitally important, national strategic guidance increasingly emphasizes the importance of preventing conflict, deterring adversaries, and shaping the operational environment so as to obtain continuing strategic advantage for the US and its allies.

Chap 5: Operations & Planning (Annex 3-0, AFI 10-401/421)

Air Force Doctrine Annex 3-0 is the Air Force's foundational doctrine publication on strategy and operational design, planning, employment, and assessment of airpower. It presents the Air Force's most extensive explanation of the effects-based approach to operations (EBAO) and contains the Air Force's doctrinal discussion of operational design and some practical considerations for designing operations to coerce or influence adversaries. It presents doctrine on cross-domain integration and steady-state operations–emerging, but validated concepts that are integral to and fully complement EBAO. A common framework of processes helps to foster coherence in Air Force strategy creation through a process of operational design, effects-based approach to planning operations, execution, and assessment.

Chap 7: Targeting (Annex 3-60)

Targeting is the process of selecting and prioritizing targets and matching the appropriate response to them, considering operational requirements and capabilities. This process is systematic, comprehensive, and continuous. Combined with a clear understanding of operational requirements, capabilities, and limitations, the targeting process identifies, selects, and exploits critical vulnerabilities of target systems and their associated targets to achieve the commanders' objectives and desired end state.

Chap 8: Combat Support (Annexes 4-0, 4-02, 3-34, 3-10)

The Air Force defines combat support (CS) as the foundational and crosscutting capability to field, base, protect, support, and sustain Air Force forces across the range of military operations. CS enables airpower through the integration of its functional communities to provide the core effects, core processes, and core capabilities required to execute the Air Force mission.

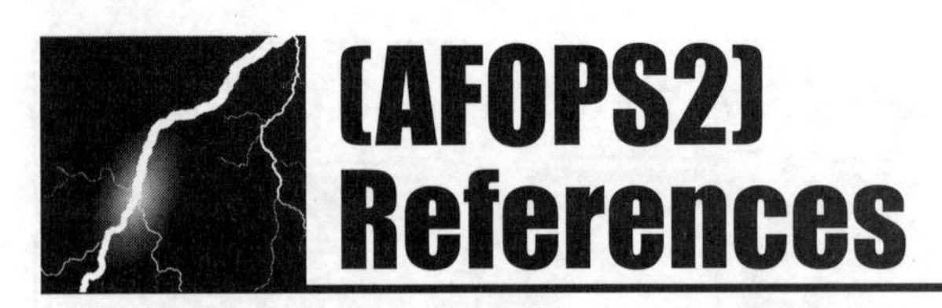

[AFOPS2] References

The following primary references were used to compile *AFOPS2: The Air Force Operations & Planning SMARTbook, 2nd Ed.* All references are open-source, public domain, available to the general public, and/or designated as "approved for public release; distribution is unlimited." *AFOPS2: The Air Force Operations & Planning SMARTbook, 2nd Ed.* does not contain classified or sensitive material restricted from public release.

Air Force Core Doctrine

Volume 1, Basic Doctrine (27 Feb 15)

Volume 2, Leadership (08 Aug 15)

Volume 3, Command (22 Nov 16)

Air Force Doctrine Annexes

Annex 2-0, Global Integrated ISR Ops (29 Jan 15)

Annex 3-0, Operations and Planning (04 Nov 16)

Annex 3-01, Counterair Ops (01 Feb 16)

Annex 3-2, Irregular Warfare (12 Jul 16)

Annex 3-03, Counterland Ops (17 Mar 17)

Annex 3-04, Countersea Ops (07 Nov 14)

Annex 3-05, Special Ops (09 Feb 17)

Annex 3-10, Force Protection (17 Apr 17)

Annex 3-12, Cyberspace Ops (30 Nov 11)

Annex 3-13, Information Ops (28 Apr 16)

Annex 3-14, Counterspace Ops (27 Aug 18)

Annex 3-17, Air Mobility Ops (05 Apr 16)

Annex 3-22, Foreign Internal Defense (10 Jul 15)

Annex 3-27, Homeland Ops (28 Apr 16)

Annex 3-30, Command and Control (7 Nov 14)

Annex 3-34, Engineer Ops (15 Aug 17)

Annex 3-40, Counter-WMD Ops (05 Apr 16)

Annex 3-50, Personnel Recovery (23 Oct 17)

Annex 3-51, Electronic Warfare Ops (10 Oct 14)

Annex 3-52, Airspace Control (23 Aug 17)

Annex 3-59, Weather Ops (27 May 15)

Annex 3-60, Targeting (14 Feb 17)

Annex 3-61, Public Affairs (28 Jul 17)

Annex 3-70, Strategic Attack (25 May 17)

Annex 3-72, Nuclear Ops (19 May 15)

Annex 4-0, Combat Support (21 Dec 15)

Annex 4-02, Medical Ops (29 Sep 15)

Air Force Instructions

AFI 10-401, Air Force Operations Planning & Execution (w/Chg 4, 13 Mar 12)

AFI 10-421, Operations Planning for the Steady-State (25 Jun 15)

AFI 13-1AOC, Volume 3, Operational Procedures-Air Operations Center (AOC) (w/ Chg 1, 18 May 12)

AFI 13-103, AFFOR Staff Operations, Readiness and Structures (19 Aug 14)

Joint Publications

JP 3-01, Countering Air and Missile Threats (May '18)

JP 3-08, Interorganizational Cooperation (12 Oct 16).

JP 3-14, Space Operations (10 Apr 18)

JP 3-16, Multinational Operations (16 Jul 13)

JP 3-17, Air Mobility Operations (20 Sept 13)

JP 3-30, Command and Control of Joint Air Operations (10 Feb 2014)

JP 3-52, Joint Airspace Control (13 Nov 2014)

JP 3-60, Joint Targeting (31 Jan 2013)

(AFOPS2) Table of Contents

Chap 1

Airpower Fundamentals & Principles

Chap 2

Commanding & Organizing

Chap 3

Command & Control

Chap 4

Airpower

Chap 5

Operations & Planning

Chap 6

Planning for Joint Air Operations

Chap 7

Targeting

Chap 8 Combat Support

Chap 1

I. Introduction to Air Force Basic Doctrine

Ref: Volume 1, Air Force Basic Doctrine (27 Feb 15), introduction.

Air Force Doctrine Volume 1, Air Force Basic Doctrine, is the senior statement of Air Force doctrine. It discusses the fundamental beliefs that underpin the application of Air Force capabilities across the range of military operations. It provides guidance on the proper employment of airpower, sets the foundation for educating Airmen on airpower, guides the development of all other doctrine, and provides insight where personal experience may be lacking.

As a whole, Air Force doctrine describes the various operations and activities that underpin the Service's ability to provide global vigilance, global reach, and global power, which allows us to anticipate threats and provide strategic reach to curb crises with overwhelming power to prevail.

Global Vigilance

Global Vigilance is the ability to gain and maintain awareness – to keep an unblinking eye on any entity – anywhere in the world; to provide warning and to determine intent, opportunity, capability, or vulnerability; then to fuse this information with data received from other Services or agencies and use and share relevant information with the joint force commander.

Global Reach

Global Reach is the ability to project military capability responsively – with unrivaled velocity and precision – to any point on or above the earth, and provide mobility to rapidly supply, position, or reposition joint forces.

Global Power

Global Power is the ability to hold at risk or strike any target anywhere in the world, assert national sovereignty, safeguard joint freedom of action, and achieve swift, decisive, precise effects.

The global context in which Airmen must anticipate and plan will remain ambiguous; unlike the Cold War era, there is no single, clearly defined opponent against which we can design forces and anticipate strategy. Air Force studies of the likely future operating environment, such as the Air Force Strategic Environment Assessment, provide a perspective on future trends and implications. Some key points are summarized as follows:

- Changes are leading to a shift in the balance of power, a more multi-polar world, and potentially adverse deviations to traditional US alliances and partnerships.
- The potential demand for certain types of operations—especially those associated with irregular warfare (IW), humanitarian operations, special operations, information gathering, and urban operations—will likely increase, and effective deterrence will likely become more challenging.
- Adversaries are gaining access to potential new and enhanced technologies and their associated capabilities. These capabilities, which will challenge Air Force operations include more lethal and precise weapon systems, enablers, and defenses; improved capabilities in space and cyberspace; weapons of mass destruction; and emerging and disruptive technology.

II. Air Force Doctrine

Ref: Volume 1, Air Force Basic Doctrine (27 Feb 15), pp. 4 to 21.

Doctrine Defined

Doctrine is defined as *"fundamental principles by which the military forces or elements thereof guide their actions in support of national objectives. It is authoritative but requires judgment in application"* (Joint Publication [JP] 1-02, Department of Defense Dictionary of Military and Associated Terms).

"... fundamental principles..."

Doctrine is a body of carefully developed, sanctioned ideas which has been officially approved or ratified corporately, and not dictated by any one individual. Doctrine establishes a common frame of reference including intellectual tools that commanders use to solve military problems. It is what we believe to be true about the best way to do things based on the evidence to date.

"...military forces..."

For the purposes of Air Force doctrine, this includes all Airmen, both uniformed and Department of the Air Force civilians. These constitute the uniformed warfighters, their commanders, and the capabilities and support that they employ. They operate across the range of military operations (ROMO) and can be task-organized into the "right force" for any particular joint contingency.

"...in support of national objectives..."

Military forces should always conduct operations in order to support objectives that create continuing advantage for our nation.

"...guide their actions... authoritative... judgment..."

Doctrine is a guide to action, not a set of fixed rules; it recommends, but does not mandate, particular courses of action.

Air Force doctrine describes and guides the proper use of airpower in military operations. It is what we have come to understand, based on our experience to date. The Air Force promulgates and teaches its doctrine as a common frame of reference on the best way to prepare and employ Air Force forces. Subsequently, doctrine shapes the manner in which the Air Force organizes, trains, equips, and sustains its forces. Doctrine prepares us for future uncertainties and provides a common set of understandings on which Airmen base their decisions. Doctrine consists of the fundamental principles by which military forces guide their actions in support of national objectives; it is the linchpin of successful military operations. It also provides us with common terminology, conveying precision in expressing our ideas. In application, doctrine should be used with judgment. It should never be dismissed out of hand or through ignorance of its principles, nor should it be employed blindly without due regard for the mission and situation at hand. On the other hand, following doctrine to the letter is not the fundamental intent. Rather, good doctrine is somewhat akin to a good "commander's intent:" it provides sufficient information on what to do, but does not specifically say how to do it. Airmen should strive above all else to be doctrinally sound, not doctrinally bound.

In the current turbulent environment of expeditionary operations and the arena of homeland security, doctrine provides an informed starting point for the many decisions Airmen make in what seems to be a continuous series of deployments. Airmen no longer face the challenge of starting with a blank sheet of paper; with doctrine, Airmen now have a good outline that helps answer several basic questions:

- What is my mission? How should I approach it?
- What should my organization look like, and why?
- What are my lines of authority within my organization and within the joint force?
- What degrees of control do I have over my forces?
- How am I supported? Who do I call for more support?
- How should I articulate what the Air Force provides to the joint force?

From one operation to the next, many things are actually constant. Doctrine, properly applied, often can provide a 70-, 80-, or even 90-percent solution to most questions, allowing leaders to focus on the remainder, which usually involves tailoring for the specific operation. Good doctrine informs, provides a sound departure point, and allows flexibility.

A study of airpower doctrine should draw a distinction between theory and practice. Theory is less constrained by limited empirical context, and designed to encourage debate and introspection with an eye towards improving military advantage. It is part of a vital, iterative investigation of what works under particular circumstances, and why. Theoretical discussion is critical to a successful military. This publication does not present a comprehensive theory for airpower. Instead, it focuses on those ideas and validated concepts, grounded in experience and Service consensus. This is the heart of doctrine.

Finally, a study of airpower doctrine should also distinguish between doctrine and public relations-like pronouncements concerning the Air Force's role. There have been many of the latter since the Air Force's inception. Some have been developed with an eye towards influencing public and congressional perception of the Air Force's role and value. Others have been made in a strategic planning context (e.g., a "vision-mission-goals" development process) that are a normal part of formal, long range corporate planning. Such statements are not enduring and not doctrine; they should be viewed in the context in which they were created.

Policy, Strategy, and Doctrine

The term "doctrine" is frequently (and incorrectly) used when referring to policy or strategy. These terms are not interchangeable; they are fundamentally different. Because policy and strategy may impact each other, it is important to first understand their differences before delving into a discussion of doctrine.

Policy

Policy is guidance that is directive or instructive, stating what is to be accomplished. It reflects a conscious choice to pursue certain avenues and not others. Thus, while doctrine is held to be relatively enduring, policy is more mutable and also directive. Policies may change due to changes in national leadership, political considerations, or for fiscal reasons. At the national level, policy may be expressed in such broad vehicles as the National Security Strategy or Presidential Executive Orders. Within military operations, policy may be expressed not only in terms of objectives, but also in rules of engagement (ROE)—what we may or may not strike, or under what circumstances we may strike particular targets.

Strategy

Strategy defines how operations should be conducted to accomplish national policy objectives. Strategy is the continuous process of matching ends, ways, and means to accomplish desired goals within acceptable levels of risk. Strategy originates in policy and addresses broad objectives, along with the designs and plans for achieving them.

Doctrine

Doctrine presents considerations on how to accomplish military goals and objectives. It is a storehouse of analyzed experience and wisdom. Military doctrine is authoritative, but unlike policy, is not directive.

- The proliferation of inexpensive technology enabled by globalization is greatly enhancing the ability of both state and non-state actors to challenge not only US military power and interests, but also international support for the United States, domestic US resolve, and the US economy and homeland security. In some cases, small numbers of sophisticated systems employed by non-state actors may deter US intervention.
- US advantages derived from space and cyberspace will decline relative to select potential adversaries who will approach parity with the Unites States in terms of their command and control and situational awareness capabilities. These and other adversaries will also be increasingly able to degrade US strengths in these areas.
- As an adversary's capabilities are brought to bear, portions of the operational environment can change from permissive to contested or highly contested.
- Strategic planners may need to rethink existing assumptions and force structures and develop new concepts that integrate nuclear, conventional, IW, and non-kinetic capabilities.
- There may be regions where many states possess nuclear weapons. These states may have conflicting doctrines and beliefs regarding their use. What may deter one actor may not deter another, and may even result in unintended negative consequences in other areas. Also, traditional deterrence models may not necessarily apply to rogue states and apply even less to non-state actors.
- The Cold War notion of controlling escalation may no longer be sufficient.

In summary, the United States will likely remain the world's single largest military power, but its relative advantage may shrink. Additionally, increasingly contested areas may reduce access, not only to the global commons, but to forward operating bases. The Air Force will likely face states and entities that have lower bars to entry to areas that can challenge existing US strengths. The need for IW capabilities will likely also continue, while strategic leverage such as effective deterrence may become more difficult and complex.

Against this backdrop, doctrine should be flexible enough to adapt and evolve to situations as they arise. Air Force doctrine should continually strive to provide a better, more relevant baseline for ongoing and future operations.

A note on terminology in Air Force doctrine: The Air Force prefers—and in fact, plans and trains—to employ in the joint fight through a commander, Air Force forces (COMAFFOR) who is normally also multi-hatted as joint force air component commander (JFACC), area air defense commander, airspace control authority, space coordinating authority, and electronic warfare control authority; when involved in multinational operations, the JFACC may become a combined force air component commander (CFACC).

To simplify nomenclature, Air Force doctrine simply uses the term "COMAFFOR," with the presumption that the COMAFFOR may also be designated with multiple hats.

Similarly, Air Force doctrine recognizes that the AOC, in joint or combined operations is correctly known as a joint AOC (JAOC) or combined AOC (CAOC). However, doctrine simply uses the term "AOC."

II. Airpower

Ref: Volume 1, Air Force Basic Doctrine (27 Feb 15), pp. 23 to 35.

Airpower is defined as "the ability to project military power or influence through the control and exploitation of air, space, and cyberspace to achieve strategic, operational, or tactical objectives." The proper application of airpower requires a comprehensive doctrine of employment and an Airman's perspective. As the nation's most comprehensive provider of military airpower, the Air Force conducts continuous and concurrent air, space, and cyberspace operations. The air, space, and cyberspace capabilities of the other Services serve primarily to support their organic maneuver paradigms; the Air Force employs air, space, and cyberspace capabilities with a broader focus on theater-wide and national-level objectives. Through airpower, the Air Force provides the versatile, wide-ranging means towards achieving national objectives with the ability to deter and respond immediately to crises anywhere in the world.

Airpower exploits the third dimension of the operational environment; the electromagnetic spectrum; and time to leverage speed, range, flexibility, precision, tempo, and lethality to create effects from and within the air, space, and cyberspace domains. From this multi-dimensional perspective, Airmen can apply military power against an enemy's entire array of diplomatic, informational, military, and economic instruments of power, at long ranges and on short notice. Airpower can be applied across the strategic, operational, and tactical levels of war simultaneously, significantly increasing the options available to national leadership. Due to its range, speed, and flexibility, airpower can compress time, controlling the tempo of operations in our favor. Airpower should be employed with appropriate consideration of land and maritime power, not just during operations against enemy forces, but when used as part of a team that protects and aids friendly forces as well.

Much of what airpower can accomplish from within these three domains is done to critically affect events in the land and maritime domains—this is the heart of joint-domain integration, a fundamental aspect of airpower's contribution to US national interests. Airmen integrate capabilities across air, space, and cyberspace domains to achieve effects across all domains in support of joint force commander objectives.

The Third Dimension

Airmen exploit the third dimension, which consists of the entire expanse above the earth's surface. Its lower limit is the earth's surface (land or water), and the upper limit reaches toward infinity. This third dimension consists of the air and space domains. From an operational perspective, the air domain can be described as that region above the earth's surface in which aerodynamics generally govern the planning and conduct of military operations, while the space domain can be described as that region above the earth's surface in which astrodynamics generally govern the planning and conduct of military operations.

Airmen also exploit operational capabilities in cyberspace. Cyberspace is "a global domain within the information environment consisting of the interdependent network of information technology infrastructures, including the Internet, telecommunications networks, computer systems, and embedded processors and controllers." In contrast to our surface-oriented sister Services, the Air Force uses air, space, and cyberspace capabilities to create effects, including many on land and in the maritime domains, that are ends unto themselves, not just in support of predominantly land or maritime force activities.

I. The Foundations of Airpower

Airpower stems from the use of lethal and nonlethal means by air forces to achieve strategic, operational, and tactical objectives. The Air Force can rapidly provide national leadership and joint commanders a wide range of military options for meeting national objectives and protecting national interests.

Elevation above the earth's surface provides relative advantages and has helped create a mindset that sees conflict more broadly than other forces. Broader perspective, greater potential speed and range, and three-dimensional movement fundamentally change the dynamics of conflict in ways not well understood by those bound to the surface. The result is inherent flexibility and versatility based on greater mobility and responsiveness.

With its speed, range, and three-dimensional perspective, **airpower operates in ways that are fundamentally different from other forms of military power.** Airpower has the ability to conduct operations and impose effects throughout an entire theater and across the range of military operations (ROMO), unlike surface forces that typically divide up the battlefield into individual operating areas. Airmen generally view the application of force more from a functional than geographic standpoint, and classify targets by generated effects rather than physical location.

By making effective use of the third dimension, the electromagnetic spectrum, and time, airpower can seize the initiative, set the terms of battle, establish a dominant tempo of operations, better anticipate the enemy through superior observation, and take advantage of tactical, operational, and strategic opportunities. Thus, airpower can simultaneously strike directly at the adversary's centers of gravity, vital centers, critical vulnerabilities, and strategy. Airpower's ability to strike the enemy rapidly and unexpectedly across all of these critical points adds a significant impact to an enemy's will in addition to the physical blow. This capability allows airpower to achieve effects well beyond the tactical effects of individual actions, at a tempo that disrupts the adversary's decision cycle.

Airpower can be used to rapidly express the national will wherever and whenever necessary. The world at large perceives American airpower to be a politically acceptable expression of national power which offers reasonable alternatives to long, bloody ground battles, while making an impact on the international situation. While a "boots-on-the-ground" presence may often be required, airpower makes that presence more effective, in less time, and often with fewer casualties.

The Air Force provides national leadership and joint commanders with options, the threat of which may accomplish political objectives without the application of lethal force. The means is embedded in the ability to respond rapidly to crises anywhere in the world and across the ROMO.

The Air Force provides the unique ability to hold at risk a wide range of an adversary's options and possible courses of action; this is increasingly the key to successful joint campaigns. Airpower is increasingly the first military instrument brought to bear against an enemy in order to favorably influence the overall campaign. Frequently, and especially during the opening days of a crisis, airpower may be the only military instrument available to use against an enemy; this may be especially true if friendly ground forces are not immediately present in a given region.

Air Force forces can respond rapidly to apply effects. The same spacecraft which Airmen employ to observe hostile territory prior to the outbreak of hostilities provide key intelligence to battle planners. The same aircraft which provide visible to battle planners. The same aircraft which provide visible to battle planners. The same aircraft which provide visible

Airpower is more than dropping bombs, strafing targets, firing missiles, providing precision navigation and timing, or protecting networks. It is also a way of influencing world situations in ways which support national objectives.

To most observers in the post-Cold War world, the use of military power is politically less acceptable than in previous times. This is true even if we act in a purely humanitarian endeavor or influence a given international political situation with a modest show of force. In international disasters, natural or man-made, from the Berlin Airlift to earthquake relief operations in Pakistan, the Air Force is the only military force in the world which has the airlift and air refueling capability to provide immediate relief supplies and personnel in response to global emergencies. Air Force aircraft delivering relief supplies serve not only to alleviate the immediate situation, but also to provide a visible symbol of the care, concern, and capability of the United States. Through careful building of partnerships, Air Force forces can favorably shape the strategic environment by assessing, advising, training, and assisting host nation air forces in their efforts to counter internal or external threats. The perception of credible US forces underpins many deterrence and assurance strategies. Such activities lead to greater regional stability and security.

Airpower's speed, range, flexibility, precision, and lethality provide a spectrum of employment options with effects that range from tactical to strategic. This range of effects is an important contribution. A surface-centric strategy often seeks its outcome through the destruction of hostile land forces and the occupation of territory. However, destruction of hostile land forces may be only a tactical or operational objective and may not achieve the desired strategic outcome. Further, territorial occupation, with its attendant large cultural footprint, may not be feasible or politically acceptable. Sea power, with its ability to project force and disrupt the economic lifeline of a maritime-capable adversary, also provides the potential for strategic results. However, slow surface speeds can constrain its capability to respond rapidly from one theater to another. In addition, it may be extremely vulnerable in littoral regions. Often, in such circumstances, the political risks outweigh the actual military risks.

Airpower has a degree of versatility not found in any other force. Many aircraft can be employed in a variety of roles and shift rapidly from defense to offense. Aircraft may conduct a close air support mission on one sortie, and then be rearmed and subsequently used to suppress enemy surface-to-surface missile attacks or to interdict enemy supply routes on the next. In time-sensitive scenarios, aircraft en route to one target, or air mobility aircraft in support of one mission, can be reassigned new targets or re-missioned as new opportunities emerge. Multi role manned and unmanned platforms may perform intelligence, surveillance, and reconnaissance, command and control, and attack functions all during the same mission, providing more potential versatility per sortie. Finally, aircraft can be repositioned within a theater to provide more responsiveness, while space and cyberspace capabilities can be reprioritized.

Joint campaigns rely upon this versatility. However, many airpower capabilities are limited in number; dividing or parceling out airpower into "penny-packets" violate the tenet of synergy and principle of mass. To preserve unity of effort, joint force commanders normally vest a single air commander with control of all airpower capabilities.

Historically, armies, navies, and air forces massed large numbers of troops, ships, or aircraft to create significant impact on the enemy. Today, the technological impact of precision guided munitions enables a relatively small number of aircraft to directly achieve national as well as military strategy objectives. When combined with stealth technologies, airpower today can provide shock and surprise without unnecessarily exposing friendly forces.

With those characteristics considered, one should remember that **air, space, and cyberspace superiority are the essential first ingredients in any successful modern military operation.** Military leaders recognize that successful military operations can be conducted only when they have gained the required level of control of the domains above the surface domains. Freedom to conduct land and naval operations is substantially enhanced when friendly forces are assured that the enemy cannot disrupt operations from above.

II. Airmindedness

Ref: Volume 1, Air Force Basic Doctrine (27 Feb 15), p. 33.

The perspective of Airmen is necessarily different; it reflects a unique appreciation of airpower's potential, as well as the threats and survival imperatives unique to Airmen. The study of airpower leads to a particular expertise and a distinctive point of view that General Henry H. "Hap" Arnold termed "airmindedness."

Airmen normally think of airpower and the application of force from a functional rather than geographical perspective. Airmen do not divide up the battlefield into operating areas as some surface forces do; airmindedness entails thinking beyond two dimensions, into the dimensions of the vertical and the dimension of time. Airmen think spatially, from the surface to geosynchronous orbit. Airmen typically classify targets by the effect their destruction would have on the adversary instead of where the targets are physically located. This approach normally leads to more inclusive and comprehensive perspectives that favor strategic solutions over tactical ones. Finally, Airmen also think of power projection from inside the US to anywhere on the globe in hours (for air operations) and even nanoseconds (for space and cyberspace operations).

Airmindedness impacts Airmen's thoughts throughout all phases of operations. It is neither platform- nor situation-specific. Airmindedness enables Airmen to think and act at the tactical, operational, and strategic levels of war, simultaneously if called for. Thus, the flexibility and utility of airpower is best fully exploited by an air-minded Airman.

The Airman's Perspective

The practical application of "airmindedness" results in the Airman's unique perspective, which can be summarized as follows.

- Control of the vertical dimension is generally a necessary precondition for control of the surface
- Airpower is an inherently strategic force
- Airpower can exploit the principles of mass and maneuver simultaneously to a far greater extent than surface forces
- Airpower can apply force against many facets of enemy power. Air Force-provided capabilities can be brought to bear against any lawful target within an enemy's diplomatic, informational, military, economic, and social structures simultaneously or separately.
- Air Force forces are less culturally intrusive in many scenarios
- Airpower's inherent speed, range, and flexibility combine to make it one of the most versatile components of military power
- Airpower results from the effective integration of capabilities, people, weapons, bases, logistics, and all supporting infrastructure
- The choice of appropriate capabilities is a key aspect in the realization of airpower. Weapons should be selected based on their ability to create desired effects on an adversary's capability and will. Achieving the full potential of airpower requires timely, actionable intelligence and sufficient command and control capabilities to permit commanders to exploit precision, speed, range, flexibility, and versatility.
- Supporting bases with their people, systems, and facilities are essential to launch, recovery, and sustainment of Air Force forces
- Airpower's unique characteristics necessitate that it be centrally controlled by Airmen. Airpower can quickly intervene anywhere, regardless of whether it is used for strategic or tactical purposes

Chap 1

III. The Range of Military Operations

Ref: Volume 1, Air Force Basic Doctrine (27 Feb 15), pp. 36 to 47.

I. The Range of Military Operations (ROMO)

Military operations slide along an imprecise scale of violence and scale of military involvement, from theater-wide major operations and campaigns; to smaller scale contingencies and crisis response operations; to engagement, security cooperation, and deterrence (see figure, "The Range of Military Operations"). No two operations are alike; scope, duration, tempo, and political context vary widely. Some operations may even change from one form to another, either escalating or de-escalating; several may exist simultaneously. Military leaders carefully assess the nature of the missions they may be assigned, not only to properly determine the appropriate mix of forces but also to discern implied requirements. Some operations involve open combat between regular forces; in others, combat may be tangential to the main effort. In some operations, the US military's contribution may not involve combat at all; simply providing an organizational framework for an interagency force and key elements of infrastructure may be all that's required.

Range of Military Operations (ROMO)

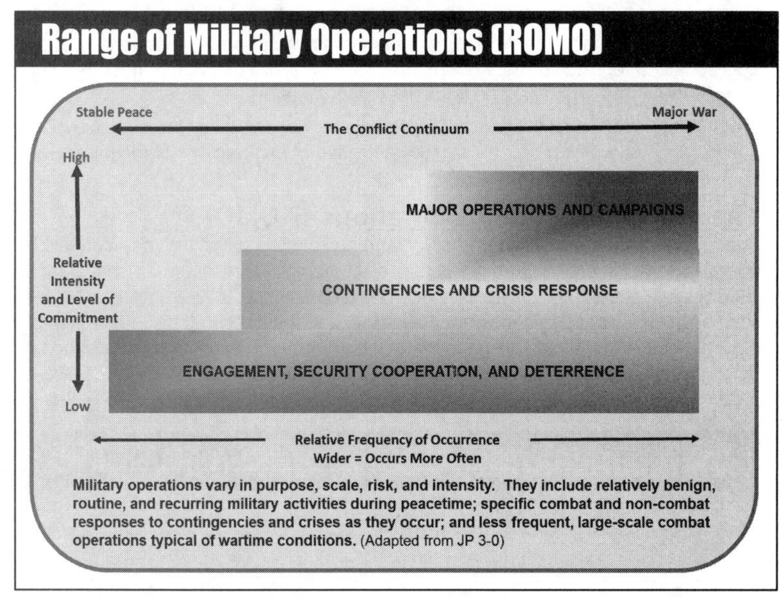

Military operations vary in purpose, scale, risk, and intensity. They include relatively benign, routine, and recurring military activities during peacetime; specific combat and non-combat responses to contingencies and crises as they occur; and less frequent, large-scale combat operations typical of wartime conditions. (Adapted from JP 3-0)

For further discussion on the ROMO from JP 3-0, see following pages (pp. 1-10 to 1-11). Refer also to Annex 3-0, Operations and Planning, and Joint Publication 1, Doctrine for the Armed Forces of the United States.

II. Steady-State Operations

The Department of Defense (DOD) and the Air Force have increased the emphasis on the military engagement, security cooperation, and deterrence portion of the range of military operations (ROMO). A key milestone was the 2008 release of the inaugural Guidance for Employment of the Force and complementary Joint Strategic

Joint Operations, Unified Action, & the Range of Military Operations (ROMO)

Ref: JP 3-0, Joint Operations (Jan '17) and Annex 3-0 (4 Nov 16), pp. 28 to 32.

Services may accomplish tasks and missions in support of Department of Defense (DOD) objectives. However, the DOD primarily employs two or more services in a single operation, particularly in combat, through joint operations. The general term, joint operations, describes military actions conducted by joint forces or by Service forces employed under command relationships. A joint force is one composed of significant elements, assigned or attached, of two or more military departments operating under a single joint force commander. Joint operations exploit the advantages of interdependent Service capabilities through unified action, and joint planning integrates military power with other instruments of national power to achieve a desired military end state.

Unified Action

Whereas the term joint operations focuses on the integrated actions of the Armed Forces of the United States in a unified effort, the term unified action has a broader connotation. JFCs are challenged to achieve and maintain operational coherence given the requirement to operate in conjunction with interorganizational partners. CCDRs play a pivotal role in unifying joint force actions, since all of the elements and actions that comprise unified action normally are present at the CCDR's level. However, subordinate JFCs also integrate and synchronize their operations directly with the operations of other military forces and the activities of nonmilitary organizations in the operational area to promote unified action.

Unified action is a comprehensive approach that synchronizes, coordinates, and when appropriate, integrates military operations with the activities of other governmental and nongovernmental organizations to achieve unity of effort.

When conducting operations for a joint force commander, Army forces achieve unified action by synchronizing actions with the activities of components of the joint force and unified action partners.

The Range of Military Operations (ROMO)

The range of military operations is a fundamental construct that provides context. Military operations vary in scope, purpose, and conflict intensity across a range that extends from military engagement, security cooperation, and deterrence activities to crisis response and limited contingency operations and, if necessary, to major operations and campaigns. Use of joint capabilities in military engagement, security cooperation, and deterrence activities helps shape the operational environment and keep the day-to-day tensions between nations or groups below the threshold of armed conflict while maintaining US global influence.

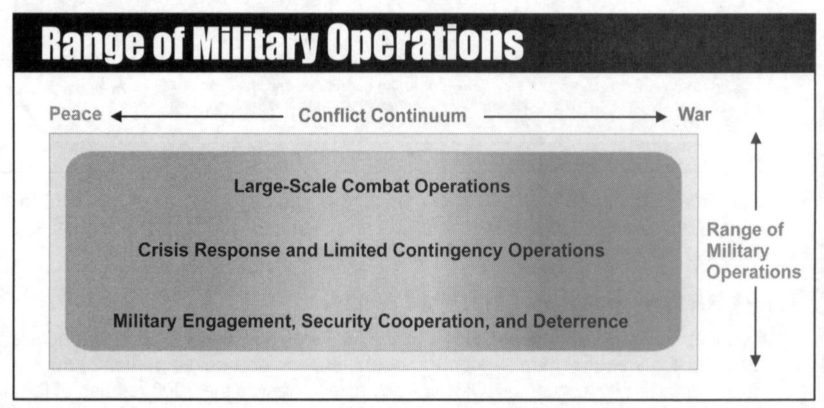

A. Military Engagement, Security Cooperation, and Deterrence

These ongoing activities establish, shape, maintain, and refine relations with other nations and domestic civil authorities (e.g., state governors or local law enforcement). The general strategic and operational objective is to protect US interests at home and abroad:

- Arms control operations
- Counterdrug operations
- Military-to-military contacts
- Unilateral and multilateral exercises
- Building partner capacity
- Senior leader engagements with international and domestic partners
- Security assistance
- Shows of force
- Demonstrations
- Theater security package-related operations
- National Guard Bureau State Partnership Program

B. Crisis Response & Limited Contingency Operations

A crisis response or limited contingency operation can be a single small-scale, limited-duration operation or a significant part of a major operation of extended duration involving combat. The associated general strategic and operational objectives are to protect US interests and/or prevent surprise attack or further conflict. Examples include:

- Combating terrorism
- Some types of counterproliferation operations, (when arms control operations are not successful)
- Consequence management (especially of weapons of mass destruction [WMD]-related events)
- Enforcement of sanctions and maritime intercept operations
- Enforcing exclusion zones
- Ensuring freedom of navigation and passage, in both maritime and aerial operations, including protection of shipping and overflight
- Ensuring freedom of action in air, space, and relevant portions of cyberspace
- Noncombatant evacuation operations
- Peace operations

C. Large-Scale Combat Operations

When required to achieve national strategic objectives or protect national interests, the US national leadership may decide to conduct a major operation or campaign normally involving large-scale combat. During **major operations**, joint force actions are conducted simultaneously or sequentially in accordance with a common plan and are controlled by a single commander. A **campaign** is a series of related major operations aimed at achieving strategic and operational objectives within a given time and space.

Refer to JFODS5: The Joint Forces Operations & Doctrine SMARTbook (Guide to Joint, Multinational & Interorganizational Operations) for further discussion. Topics and chapters include joint doctrine fundamentals (JP 1), joint operations (JP 3-0), joint planning (JP 5-0), joint logistics (JP 4-0), joint task forces (JP 3-33), information operations (JP 3-13), multinational operations (JP 3-16), interorganizational cooperation (JP 3-08), plus more!

Capabilities Plan, which introduced a campaign planning methodology for steady-state operations. As a result, combatant commands now develop and execute steady-state campaign plans and commanders, Air Force forces (COMAFFORs), develop and execute steady-state campaign support plans. Both plans "operationalize" a commander's strategy. Even though fighting and winning the nation's wars remains the primary justification for standing and capable military forces, these same forces share in the responsibility to shape the operational environment, deter aggression, and prevent conflict. Airmen should be as proficient in steady-state operations as they are in conducting contingency and crisis operations.

The military engagement, security cooperation, and deterrence portion of the ROMO is increasingly referred to in DOD publications as the "steady state." Although DOD has not formally defined this term, the Air Force describes it as "a stable condition involving continuous and recurring operations and activities with simultaneous absence of major military, crisis response, and contingency operations. The steady state is characterized by shaping operations and activities at a relatively low level of intensity, urgency, and commitment of military forces."

The steady state is synonymous with shaping, and is designed to influence the environment in order to prevent and deter future conflict; mitigate operational risks associated with combat; and strengthen United States and partner capabilities to respond to major operations, campaigns, and contingencies. From the Airman's perspective, the focus of steady-state operations is to support the combatant commander's (CCDR's) steady-state campaign plan.

The term "steady state" also provides context and relevance for the many Airmen who conduct operations on a daily basis, not just during crises. Such Airmen include air mobility personnel performing intertheater airlift; space and cyberspace operators in the performance of global requirements; missileers and bomber crews on nuclear alert; Air National Guardsmen performing air defense alert; and intelligence, surveillance, and reconnaissance operators maintaining worldwide situational awareness. The daily contributions of these and other Airmen are just as important as the contributions of Airmen in periods of crisis.

See chap. 5 for discussion of Air Force planning, execution, and assessment in support of steady-state operations from Annex 3-0, Operations and Planning.

III. The Levels of War

Ref: Volume 1, Air Force Basic Doctrine (27 Feb 15), pp. 44 to 45.

Warfare is typically divided into three levels: strategic, operational, and tactical. These divisions have arisen because traditional war constrained forces to engage force-on-force, on the surface, at the tactical level, allowing effects to aggregate up from that level to the level of campaigns and other major operations, and finally to the level directly affecting an adversary's ability to wage war altogether. A given aircraft, dropping a given weapon, could conduct a "tactical," "operational," or "strategic" mission, depending on the planned results.

Strategic Level of War

Effects at the strategic level of war impair the adversary's ability to carry out war or hostilities in general. Strategic effects should neutralize the adversary's centers of gravity. At this level, the United States determines national or multinational (alliance or coalition) security objectives and guidance and uses all national resources to achieve objectives and desired end states. These national objectives in turn provide the direction for developing overall military objectives, which in turn are used to develop the military objectives and strategy for each theater or operation. Strategy is aimed at outcomes, thus strategic ends define this level. In some circumstances, there may be value in distinguishing between the nation's strategy as a whole and what might be termed the "theater-strategic" level, at which particular combatant commanders determine and direct the overall outcomes of major operations (or "wars") taking place within their particular areas of responsibility, explicitly tying these "theater-strategic" aims to overarching national strategy and policy. In general terms, the strategic level of war addresses the issues of WHY and WITH WHAT we will fight and WHY the enemy fights against us.

Operational Level of War

The operational level of war lies between the strategic and tactical levels. At this level, campaigns and major operations are designed, planned, conducted, sustained, assessed, and adapted to accomplish strategic goals within theaters or areas of operations. These activities imply a broader dimension of time or space than do tactics; they orchestrate tactical successes to achieve objectives at higher levels. The decision-making products at this level of planning identify required forces and resources balanced against operational risk. Operational effects such as air superiority, space superiority, cyberspace superiority, defeat of enemy surface forces, isolation of enemy forces in the battlespace, and disruption or destruction of enemy leadership functions are the means with which the operational commander supports the overall strategy. Operations involve the integration of tactical military missions and engagements to achieve strategic ends. Planning at the operational level of war determines WHAT we will affect, with WHAT courses of action, in WHAT order, for WHAT duration, and with WHAT RESOURCES.

Tactical Level of War

At the lowest end of the spectrum lies the tactical level of war, where individual battles and engagements are fought. While resulting effects may be described as operational or strategic, military actions occur almost entirely at the tactical level. Thus, even a global strike mission intended to produce a direct strategic effect on an adversary COG is ultimately a tactical action. To the Airman, the distinction between this level and higher levels of war is fairly clear-cut; Airmen tend not to fight large-scale battles (as surface forces use the term) but focus at the tactical level on individual engagements and "missions." The tactical level of air, space, and cyberspace warfare deals with how forces are employed, and the specifics of how engagements are conducted. Tactics are concerned with the unique employment of force, so application defines this level. In short, the tactical level of war deals with HOW we fight.

IV. The Nature of War

Ref: Volume 1, Air Force Basic Doctrine (27 Feb 15), pp. 38 to 43.

Because war underpins the reason for the Air Force's existence, an understanding of doctrine should also include an understanding of war. The ultimate objective of peacetime preparation of forces is their employment as instruments of national power to deter or win wars. Therefore, Airmen should understand the nature and consequences of war.

War is a violent struggle between rival parties to attain competing objectives. War is just one means used by nation-states, sub-national groups, or supranational groups to achieve disputed objectives. War has been a basic aspect of human affairs throughout history. The modern Western tendency to view war as an aberration in human affairs, only occasionally necessary as an operation with limited aims or an all-out campaign to destroy a clearly recognized evil, often distorts our understanding of warfare and its purposes. Warfare is ingrained in the very nature of certain cultures. While for nation states, war is an instrument of policy aiming at political objectives, it is also, even within this context, a phenomenon involving the full range of human emotions and irrationalities. War has a dynamic of its own, often fueled by pressures of the irrational: anger, fear, revenge, and hatred. Thus, the resort to violence rarely remains for long tied to cold, clear political objectives; it can—and has—moved in unexpected directions.

Military professionals operate within an environment that cannot be fully replicated in training. The arena in which military professionals operate is a deadly one. Not only are they attempting, as General George Patton stated, to "make the other poor bastard die for his country," the enemy is attempting to do the same to us. Consequently, war is an arena characterized by extraordinary fear, pain, uncertainty, and suffering.

Enduring Truths

Three enduring truths describe the nature of war. Despite technological advances and the best of plans and intentions, war will never be as straightforward in execution as planned, nor free of unintended consequences. The particular characteristics usually change from conflict to conflict, but the nature of war remains eternal.

- **War is an instrument of policy.** Victory in war is not measured by casualties inflicted, battles won or lost, number of tanks destroyed, or territory occupied, but by the achievement of (or failure to achieve) the strategy and policy objectives of nation states, and often the cultural objectives of all actors (including non- or supra-state entities). More than any other factor, these objectives—one's own and those of the enemy—shape the scope, intensity, and duration of war. To support US national policy objectives, military objectives and operations should be coordinated and orchestrated with nonmilitary and partner nation instruments of power. Prussian philosopher of war Carl von Clausewitz emphasized that war is a continuation of the policies of nations, but not all belligerents in war are organized nation states.
- **War is a complex and chaotic human endeavor.** Irrational and non-rational human impulses and human frailties shape war's nature—it is not deterministic. Uncertainty and unpredictability—what many call the "fog of war"—combine with danger, physical stress, and human fallibility to produce what Clausewitz called "friction," which makes even simple operations unexpectedly and sometimes even insurmountably difficult.
- **War is a clash of opposing wills.** War is collision of two or more living forces. War is not waged against an inanimate or static object, but against a living, calculating, interactively complex, adaptive opponent. The enemy often does not think as we think and often holds different values, motivations, and priorities than ours. Victory results from creating advantages against a thinking adversary bent on creating his own advantages. Allied and enemy resolve—the determination to enforce one's will on one side and to resist on the other—can be the decisive element.

Success in war requires mastery of the art of war as well as the science of war. Warfare is one of the most complex of human activities. Success depends more on intellectual superiority, morale, and determination than it does on numerical and technological superiority. Success thus demands an intricate combination of science (that which can be measured, studied, and controlled) and art (creativity, flexibility, intuition, and the ability to adapt). Sound doctrine, good leadership, effective organization, moral values, and realistic training can lessen the effects of uncertainty, unpredictability, and unreliability that are always present in war.

Traditional and Irregular War

The US' overwhelming dominance in recent conventional wars has made it highly unlikely that most adversaries would choose to fight the US in a traditional, conventional manner. Thus, for relatively weaker powers (including non-state entities) irregular warfare has become an attractive, if not more necessary, option.

- **Traditional warfare** is characterized as "a confrontation between nation-states or coalitions/alliances of nation-states" (JP 1).
- **Irregular warfare** is defined as "a violent struggle among state and non-state actors for legitimacy and influence over the relevant populations. Irregular warfare favors indirect and asymmetric approaches, though it may employ the full range of military and other capabilities in order to erode an adversary's power, influence, and will" (JP 1-02).

Both IW and traditional warfare seek to resolve conflict by compelling change in adversarial behavior. However, they differ significantly in both strategy and conduct. Traditional warfare focuses on dominance over an adversary's ability to sustain its war fighting capability. IW focuses on population-centric approaches that affect actors, behaviors, relationships, and stability in the area or region of interest. Therefore, IW requires a different level of operational thought and threat comprehension.

IW is not a lesser-included form of traditional warfare. Rather, IW encompasses a variety of operations where the characteristics are significantly different from traditional war. There are principally five activities or operations that are undertaken in sequence, in parallel, or in blended form in a coherent campaign to address irregular threats: counterterrorism, unconventional warfare, foreign internal defense, counterinsurgency, and stability operations.

Refer to Annex 3-2, Irregular Warfare, for detailed discussion on IW.

The Role of Culture

The role of culture in establishing the terms of conflict is another vital component that has increased in importance in recent operations. War among Western powers has always been seen as an adjunct to politics and commerce, and often as a dangerous distraction from them. The rewards of war are physical; psychological reinforcement comes predominantly from war's spoils, not from war itself. In general, this view has led Western powers to try to force resolution as quickly and "cheaply" as practicable (in all but comparatively rare civil and religious wars), to seek decisive engagement with the enemy when possible, and to focus warfare upon defeat of the enemy's fielded military forces. This was true even during Industrial Age conflicts, where the total moral and physical power of the nation-state was mobilized for war. This is the cultural legacy that has most heavily influenced the modern use of airpower.

Refer to TAA2: Military Engagement, Security Cooperation & Stability SMARTbook (Foreign Train, Advise, & Assist) for further discussion. Topics include the Range of Military Operations (JP 3-0), Security Cooperation & Security Assistance (Train, Advise, & Assist), Stability Operations (ADRP 3-07), Peace Operations (JP 3-07.3), Counterinsurgency Operations (JP & FM 3-24), Civil-Military Operations (JP 3-57), Multinational Operations (JP 3-16), Interorganizational Cooperation (JP 3-08), and more.

V. Homeland Operations

Ref: Annex 3-0, Operations & Planning (4 Nov 16), pp. 32 to 33.

The Air Force plays a significant role in homeland operations. It employs airpower to assist federal, state, and local governments, as well as other branches of the Department of Defense (DOD) and non-governmental organizations (NGOs) in detecting, helping preempt, responding to, mitigating, and recovering from a full spectrum of threats and incidents, man-made and natural, within the United States and its territories and possessions. Homeland operations consist of two major mission areas: homeland defense and defense support of civil authorities (DSCA), along with the integral subset mission of emergency preparedness. While homeland operations may arguably be considered a subset within the ROMO previously described, Air Force doctrine considers these activities important enough to warrant separate discussion.

See pp. 2-14 to 2-15 for additional discussion.

Homeland Defense

DOD defines homeland defense as "the protection of US territory, sovereignty, domestic population, and critical infrastructure against external threats and aggression." Homeland defense missions include force protection actions; counterintelligence; air, space, and cyberspace warning and control; counter-terrorism; critical infrastructure protection; air, space, cyberspace, and missile defense; and information security operations. Homeland defense also includes protection of military installations and facilities within the United States. In all of these missions, DOD either acts as the designated lead federal agency, or with a high level of autonomy within the national security structure. The most familiar Air Force role here is fulfilling North American Aerospace Defense Command's (NORAD's) air sovereignty mission through defensive counterair.

Defense Support of Civil Authorities (DSCA)

The term DSCA denotes DOD support provided during and in the aftermath of domestic emergencies—such as terrorist attacks or major disasters. DSCA missions include, but are not limited to, preventing or defeating terrorist attacks; response to natural disasters; support to civilian law enforcement agencies; counterdrug operations; border security; and response to civil disturbances or insurrection. It also covers consequence management due to CBRN incidents, including toxic industrial chemicals and materials. In all of these missions, various federal, state, or local environments may be further complicated by the differences in duty status and authority of civilian agencies who are primarily responsible for the management of the particular incident. DOD's involvement is supportive and is normally dependent on a request from the lead agency.

Emergency Preparedness

Emergency preparedness activities are those planning activities undertaken to ensure DOD processes, procedures, and resources are in place to support the President and Secretary of Defense in a national security emergency. This includes continuity of operations, continuity of government functions, and the performance of threat assessments.

Refer to The Homeland Defense & DSCA SMARTbook (Protecting the Homeland / Defense Support to Civil Authority) for complete discussion. Topics and references include homeland defense (JP 3-28), defense support of civil authorities (JP 3-28), Army support of civil authorities (ADRP 3-28), multi-service DSCA TTPs (ATP 3-28.1/MCWP 3-36.2), DSCA liaison officer toolkit (GTA 90-01-020), key legal and policy documents, and specific hazard and planning guidance.

Chap 2

IV. Principles of Joint Operations

Ref: Volume 1, Air Force Basic Doctrine (27 Feb 15), pp. 48 to 64.

The role of the Air Force is to defend the US and protect its interests through airpower, guided by the principles of joint operations and the tenets of airpower. Airmen should understand these fundamental beliefs as they apply to operations across all domains, not just air, space, and cyberspace.

Principles of Joint Operations

In conducting contemporary operations, commanders generally consider 13 broad principles collectively known as "the principles of joint operations." They combine the nine long-standing principles of war with four other principles developed through recent experience in irregular warfare.

I. Principles of War

Throughout the history of conflict, military leaders have noted certain principles that tended to produce military victory. Known as the principles of war, they are those aspects of warfare that are universally true and relevant. As members of the joint team, Airmen should appreciate how these principles apply to all forces, but should fully understand them as they pertain to Air Force forces. Airpower, no matter which Service operates the systems and no matter which type platform is used, provides unique capabilities.

Unity of Command

Unity of command ensures concentration of effort for every objective under one responsible commander. This principle emphasizes that all efforts should be directed and coordinated toward a common objective. Airpower's operational-level perspective calls for unity of command to gain the most effective and efficient application. Coordination may be achieved by cooperation; it is, however, best achieved by vesting a single commander with the authority and the capability to direct all force employment in pursuit of a common objective.

Objective

The principle of objective is to direct military operations toward a defined and attainable objective that contributes to strategic, operational, and tactical aims. In application, this principle refers to unity of effort in purpose, space, and time. In a broad sense, this principle holds that political and military goals should be complementary and clearly articulated.

Offensive

The purpose of an offensive action is to seize, retain, and exploit the initiative. The offensive aim is to act rather than react and to dictate the time, place, purpose, scope, intensity, and pace of operations. The initiative should be seized as soon as possible. The principle of the offensive holds that offensive action, or initiative, provides the means for joint forces to dictate operations. Once seized, the initiative should be retained and fully exploited.

Mass

The purpose of mass is to concentrate the effects of combat power at the most advantageous place and time to achieve decisive results. Concentration of military power is a fundamental consideration in all military operations. At the operational level of war, this principle suggests that superior, concentrated combat power is used to achieve decisive results.

Maneuver

Maneuver places the enemy in a position of disadvantage through the flexible application of combat power in a multidimensional combat space. Airpower's ability to conduct maneuver is not only a product of its speed and range, but also flows from its flexibility and versatility during the planning and execution of operations.

Economy of Force

Economy of force is the judicious employment and distribution of forces. Its purpose is to allocate minimum essential resources to secondary efforts. This principle calls for the rational use of force by selecting the best mix of air, space, and cyberspace capabilities. To ensure overwhelming combat power is available, maximum effort should be devoted to primary objectives

Security

The purpose of security is to never permit the enemy to acquire unexpected advantage. Friendly forces and their operations should be protected from enemy action that could provide the enemy with unexpected advantage. The lethal consequences of enemy attack make the security of friendly forces a paramount concern..

Surprise

Surprise leverages the principle of security by attacking the enemy at a time, place, or in a manner for which they are not prepared. The speed and range of air, space, and cyberspace capabilities, coupled with their flexibility and versatility, allow air forces to achieve surprise more readily than other forces.

Simplicity

Simplicity calls for avoiding unnecessary complexity in organizing, preparing, planning, and conducting military operations. Simplicity ensures that guidance, plans, and orders are as simple and direct as the objective allows. Simple guidance allows subordinate commanders the freedom to operate creatively within their portion of the operational environment, supporting the concept of decentralized execution.

II. Additional Principles of Operations

In addition to the traditionally-held principles of war, an additional set of principles has been developed as a result of experience in contingency operations. These were first cast as "principles of military operations other than war" and later as "the political dimension of smaller-scale contingencies." A distinguishing characteristic of such operations has been the degree to which political objectives influence operations and tactics.

Note that joint doctrine does not contain unity of effort as an additional principle.

Unity of Effort

Often the military is not the sole, or even the lead, agency in contingency operations.

Restraint

Restraint is the disciplined application of military force appropriate to the situation.

Perseverance

The principle of perseverance encompasses the patient, resolute, and persistent pursuit of national goals and objectives, for as long as necessary to achieve them.

Legitimacy

In order to reduce the threat to US forces and to enable them to work toward their objective, the US should be viewed as a legitimate actor in the mission, working towards multi-lateral interests rather than just its own.

Chap 2

V. Tenets of Airpower

Ref: Volume 1, Air Force Basic Doctrine (27 Feb 15), pp. 65 to 75.

The application of airpower is refined by several fundamental guiding truths. These truths are known as tenets. They reflect not only the unique historical and doctrinal evolution of airpower, but also the current appreciation for the nature of airpower. The tenets of airpower complement the principles of joint operations. While the principles of war provide general guidance on the application of military forces, the tenets provide more specific considerations for the employment of airpower.

Tenets of Airpower

- **A** Centralized Control and Decentralized Execution
- **B** Flexibility and Versatility
- **C** Synergistic Effects
- **D** Persistence
- **E** Concentration
- **F** Priority
- **G** Balance

The tenets of airpower are interconnected, overlapping, and often interlocking. Flexibility and versatility necessitate priorities. Priorities determine synergies, levels of concentration, and degrees of persistence. Balance calculations influence all operations. The combinations and permutations of interrelationships between the tenets are nearly endless. However, the oldest tenet of airpower—centralized control and decentralized execution—remains the keystone of success in modern warfare.

As with the principles of joint operations, these tenets require informed judgment in application. They require a skillful blending to tailor them to the ever-changing operational environment. The competing demands of the principles and tenets (for example mass versus economy of force, concentration versus balance, and priority versus objective) require an Airman's expert understanding in order to strike the required balance. In the last analysis, commanders accept the fact that war is incredibly complicated and no two operations are identical. Commanders should apply their professional judgment and experience as they employ airpower in a given situation.

A. Centralized Control and Decentralized Execution

The tenet of centralized control and decentralized execution is critical to effective employment of airpower. Indeed, they are the fundamental organizing principles for airpower, having been proven over decades of experience as the most effective and efficient means of employing it. It enables the principle of mass while maintaining economy of force. Because of airpower's unique potential to directly affect the strategic and operational levels of war, it should be controlled by a single Airman who maintains the broad, strategic perspective necessary to balance and prioritize the use of a powerful, highly desired yet limited force. A single air component commander, focused on the broader aspects of an operation, can best balance or mediate urgent demands for tactical support against longer-term strategic and operational requirements. The ability to concentrate the air effort to fulfill the highest priorities for effects and to quickly shift the effort can only be accomplished through centralized control. On the other hand, the flexibility to take advantage of tactical opportunities and to effectively respond to shifting local circumstances can only be achieved through decentralized execution.

Centralized Control

This tenet is best appreciated as a general philosophy for the command and control (C2) of airpower. The construct of centralized control is an encapsulation of a hard-learned truth: that control of a valuable yet scarce resource (airpower) should be commanded by a single Airman, not parceled out and hardwired to subordinate surface echelons as it was prior to 1943. Tied to this fundamental truth is the recognition that no single Airman is capable of making all decisions, and should thus empower subordinates to respond in accordance with senior leader intent.

Centralized control should be accomplished by an Airman at the functional component commander level who maintains a broad focus on the joint force commander's (JFC's) objectives to direct, integrate, prioritize, plan, coordinate, and assess the use of air, space, and cyberspace assets across the range of military operations. Centralized control may be manifest at different levels within a combatant command depending on how the air component(s) is (are) organized and the nature of the supporting C2 architecture (functional or geographic). Also, due to the dynamics of the operational environment, control over some capabilities may, over time, shift up or down the command chain according to changes in priorities.

Centralized control empowers the air component commander to respond to changes in the operational environment and take advantage of fleeting opportunities, and embodies the tenet of flexibility and versatility. Some would rather this be just "centralized planning and direction." From an Airman's perspective, "planning and directing" do not convey all aspects of control implied in "centralized control," which maximizes the flexibility and effectiveness of airpower. Centralized control is thus pivotal to the determination of continuing advantage. However, it should not become a recipe for micromanagement, stifling the initiative subordinates need to deal with combat's inevitable uncertainties.

Decentralized Execution

Decentralized execution is defined as the "delegation of authority to designated lower-level commanders" and other tactical-level decision makers to achieve effective span of control and to foster disciplined initiative and tactical flexibility. It allows subordinates, all the way down to the tactical level, to exploit situational responsiveness and fleeting opportunities in rapidly changing, fluid situations. The benefits inherent in decentralized execution, however, are maximized only when a commander clearly communicates intent and subordinate commanders frame their actions accordingly.

Centralized control and decentralized execution of airpower provide broad global or theater-wide focus while allowing operational flexibility to meet military objectives. They assure concentration of effort while maintaining economy of force. They exploit airpower's versatility and flexibility to ensure that it remains responsive, survivable, and sustainable.

Execution should be decentralized within a C2 architecture that exploits the ability of front-line decision makers (such as strike package leaders, air battle managers, forward air controllers) to make on-scene decisions during complex, rapidly unfolding operations. Modern communications technology may tempt commanders to take direct control of distant events and override the decisions of forward leaders, even when such control is not operationally warranted. This should be resisted at all costs in all functional components—not just air. Despite impressive gains in data exploitation and automated decision aids, a single person cannot, with confidence, achieve and maintain detailed situational awareness over individual missions when fighting a conflict involving many simultaneous engagements taking place throughout a large area, or over individual missions conducted in locally fluid and complex environments.

There may be some situations where there may be valid reasons for control of specific operations at higher levels, most notably when the JFC (or perhaps even higher authorities) may wish to control strategic effects, even at the sacrifice of tactical efficiency. However, such instances should be rare, as in the short notice prosecution of high-value, time-sensitive targets, or when the operational climate demands tighter control over selected missions due to political sensitivities, such as the potential for collateral damage or mistargeting, or in the case of nuclear employment. In all cases, senior commanders balance overall campaign execution against the pressing need for tactical effectiveness. As long as a subordinate's decision supports the superior commander's intent and meets campaign objectives, subordinates should be allowed to take the initiative during execution.

B. Flexibility and Versatility

Flexibility allows airpower to exploit mass and maneuver simultaneously. Flexibility allows airpower to shift from one campaign objective to another, quickly and decisively; to "go downtown" on one sortie, then hit fielded enemy forces the next; to re-role assets quickly from a preplanned mission to support an unanticipated need for close air support of friendly troops in contact with enemy forces.

Versatility is the ability to employ airpower effectively at the strategic, operational, and tactical levels of war and provide a wide variety of tasks in concert with other joint force elements. Airpower has the potential to achieve this unmatched synergy through asymmetric and parallel operations. Space and cyberspace capabilities are especially able to simultaneously support multiple taskings around the globe and support tasks at all levels of warfare.

C. Synergistic Effects

The proper application of a coordinated force across multiple domains can produce effects that exceed the contributions of forces employed individually. The destruction of a large number of targets through attrition warfare is rarely the key objective in modern war. Instead, the objective is the precise, coordinated application of the various elements of airpower and surface power to bring disproportionate pressure on enemy leaders to comply with our national will (affecting their intent) or to cause functional defeat of the enemy forces (affecting their capability). Airpower's ability to observe adversaries allows joint force commanders to counter enemy movements with unprecedented speed and agility. Airpower is unique in its ability to dictate the tempo and direction of an entire warfighting effort regardless of the scale of the operation.

D. Persistence

Air, space, and cyberspace operations may be conducted continuously against a broad spectrum of targets. Airpower's exceptional speed and range allow its forces to visit and revisit wide ranges of targets nearly at will. Airpower does not have to occupy terrain or remain constantly in proximity to areas of operation to bring force upon targets. Space forces in particular hold the ultimate high ground, and as space systems continue to advance and proliferate, they offer the potential for persistent overhead access; unmanned aircraft systems offer similar possibilities from the atmosphere.

Examples of persistent operations might be maintaining a continuous flow of materiel to peacetime distressed areas; Air Force intelligence, surveillance, and reconnaissance capabilities monitoring adversaries to ensure they cannot conduct actions counter to those agreed upon; assuring that targets are kept continually out of commission; or ensuring that resources and facilities are denied an enemy or provided to an ally during a specified time. The end result would be to deny the opponent an opportunity to seize the initiative and to directly accomplish assigned tasks.

Factors such as enemy resilience, effective defenses, or environmental concerns may prevent commanders from quickly attaining their objectives. However, for many situations, airpower provides the most efficient and effective means to attain national objectives. Commanders must persist in the conduct of operations and resist pressures to divert resources to other efforts unless such diversions are vital to attaining theater goals or to survival of an element of the joint force.

E. Concentration

One of the most constant and important trends throughout military history has been the effort to concentrate overwhelming power at the decisive time and place. The principles of mass and economy of force deal directly with concentrating overwhelming power at the right time and the right place (or places). The versatility of airpower with its lethality, speed, and persistence makes it an attractive option for many tasks. With capabilities as flexible and versatile as airpower, the demand for them often exceeds the available forces and may result in the fragmentation of the integrated airpower effort in attempts to fulfill the many demands of the operation. Depending on the operational situation, such a course of action may court the triple risk of failing to achieve operational-level objectives, delaying or diminishing the attainment of decisive effects, and increasing the attrition rate of air forces—and consequently risking defeat. Airmen should guard against the inadvertent dilution of airpower effects resulting from high demand.

F. Priority

Commanders should establish clear priorities for the use of airpower. Due to its inherent flexibility and versatility, the demands for airpower may likely exceed available resources. If commanders fail to establish priorities, they can become ineffective. Commanders of all components need to effectively prioritize their requirements for coordinated airpower effects to the joint force commander (JFC), and only then can effective priorities for the use of airpower flow from an informed dialogue between the JFC and the air component commander. The air component commander should assess the possible uses of component forces and their strengths and capabilities to support the overall joint campaign. Limited resources require that airpower be applied where it can make the greatest contribution to the most critical current JFC requirements. The application of airpower should be balanced among its ability to conduct operations at all levels of war, often simultaneously.

G. Balance

Balance is an essential guideline for air commanders. Much of the skill of an air component commander is reflected in the dynamic and correct balancing of the principles of joint operations and the tenets of airpower to bring Air Force capabilities together to produce synergistic effects. An air component commander should balance combat opportunity, necessity, effectiveness, efficiency, and the impact on accomplishing assigned objectives against the associated risk to friendly forces.

An Airman is uniquely—and best—suited to determine the proper theater-wide balance between offensive and defensive air operations, and among strategic, operational, and tactical applications. Air, space, and cyberspace assets are normally available only in finite numbers; thus, balance is a crucial determinant for an air component commander.

I. Commanding U.S. Air Force Forces

Ref: Volume 3, Command (22 Nov 16), pp. 43 to 48.

Organization is critically important to effective and efficient operations. Service and joint force organization and command relationships—literally, who owns what, and who can do what with whom, and when—easily create the most friction within any operation. Air Force organization and preferred command arrangements are designed to address unity of command, a key principle of war. Clear lines of authority, with clearly identified commanders at appropriate echelons exercising appropriate control, are essential to achieving unity of effort, reducing confusion, and maintaining priorities.

The key to successful employment of Air Force forces as part of a joint force effort is providing a single Air Force commander with the responsibility and authority to properly organize, train, equip and employ Air Force forces to accomplish assigned functions and tasks. The title of this commander is Commander, Air Force Forces (COMAFFOR).

Operationally, the COMAFFOR should be prepared to employ Air Force forces as directed by the joint force commander (JFC), and if directed be prepared to employ joint air forces as the joint force air component commander (JFACC). In either event, the COMAFFOR should also ensure that Air Force forces are prepared to execute the missions assigned by the JFC. The requirements and responsibilities of the COMAFFOR and JFACC are inextricably linked; both are critical to operational success.

Commander, Air Force Forces (COMAFFOR)

The title of COMAFFOR is reserved exclusively to the single Air Force commander of an Air Force Service component assigned or attached to a JFC at the unified combatant command, subunified combatant command, or joint task force (JTF) level.

If Air Force forces are attached to a JFC, they should be presented as an air expeditionary task force (AETF).

- The AETF becomes the Air Force Service component to the JTF and the AETF commander is the COMAFFOR to the JTF commander. Thus, depending on the scenario, the position of COMAFFOR may exist simultaneously at different levels within a given theater as long as each COMAFFOR is separately assigned or attached to and under the operational control of a different JFC.
- The COMAFFOR provides unity of command. To a JFC, a COMAFFOR provides a single face for all Air Force issues. Within the Air Force Service component, the COMAFFOR is the single commander who conveys commander's intent and is responsible for operating and supporting all Air Force forces assigned or attached to that joint force.
- The COMAFFOR commands forces through two separate branches of the chain of command: the operational branch and the administrative branch.

The COMAFFOR should normally be designated at a command level above the operating forces and should not be dual-hatted as commander of one of the subordinate operating units. This allows the COMAFFOR to focus at the operational level.

I. COMAFFOR Operational Responsibilities

When Air Force forces are assigned or attached to a JFC, the JFC normally receives operational control (OPCON) of these forces. This authority is best exercised through subordinate JFCs and Service component commanders and thus is normally delegated accordingly. If not delegated OPCON, or if the stated command authorities are not clear, the COMAFFOR should request delegation of OPCON. When the COMAFFOR is delegated OPCON of the Air Force component forces, and no joint force air component commander (JFACC) has been designated, the COMAFFOR has the following operational and tactical responsibilities: (Note: if a JFACC is designated, many of these responsibilities belong to that functional component commander.)

- Make recommendations to JFC on proper employment of forces in the Air Force component.
- Accomplish assigned tasks for operational missions.
- Develop and recommend courses of action (COAs) to the JFC.
- Develop a strategy and operation plan that states how the COMAFFOR plans to exploit Air Force capabilities to support the JFC's objectives.
- Develop a joint air operations plan (JAOP) and air operations directive to support the JFC's objectives.
- Establish (or implement, when passed down by the JFC) theater rules of engagement (ROEs) for all assigned and attached forces. For those Service or functional components that operate organic air assets, it should be clearly defined when the air component ROEs also apply to their operations (this would normally be recommended).
- Make air apportionment recommendations to the JFC.
- Plan, coordinate, allocate, and task Service forces and joint forces made available.
- Normally serve as the supported commander for counterair operations, strategic attack, the JFC's overall air interdiction effort, most space control operations, theater airborne reconnaissance and surveillance, and other operations as directed by the JFC. As the supported commander, the COMAFFOR has the authority to designate the target priority, effects, and timing of these operations and attack targets within the entire joint operations area (JOA).
- Normally serve as the supported commander for the following operations as directed by the JFC. As the supported commander, the COMAFFOR has the authority to designate the target priority, effects, and timing of these operations and attack targets across the entire joint operations area (JOA) in accordance with JFC guidance, to include coordinated targets within land and maritime areas of operations (AOs).This includes strategic attack, counterair (to include integrated air and missile defense), counterland, countersea, space control, air mobility, information operations, theater airborne intelligence, surveillance, and reconnaissance (ISR).
- Normally serve as supporting commander, as directed by the JFC, for operations such as close air support (CAS), air interdiction within other components' AOs, and maritime support.
- If so designated, act as airspace control authority (ACA), area air defense commander (AADC), and space coordinating authority (SCA), and electronic warfare control authority, and develop plans and products associated with these responsibilities.
- Coordinate personnel recovery operations, including combat search & rescue (CSAR).
- Direct intratheater air mobility operations and coordinate them with intertheater air mobility operations.
- Coordinate support for special operations requirements with the joint force special operations component commander or the joint special operations task force commander.
- Perform assessments of air component operations at the operational (component) and tactical levels.
- Conduct joint training, including the training, as directed, of components of other Services in joint operations for which the COMAFFOR has or may be assigned primary responsibility, or for which the Air Force component's facilities and capabilities are suitable.

II. COMAFFOR Administrative Responsibilities

Ref: Volume 3, Command (22 Nov 16), pp. 47 to 48.

Commanders of Air Force components have responsibilities and authorities that derive from their roles in fulfilling the Service's administrative control (ADCON) function.

- Within the administrative branch, the COMAFFOR has complete ADCON of all assigned Air Force component forces and specified ADCON of all attached Air Force component forces.
- The specified responsibilities listed below apply to all attached forces, regardless of major command or Air Force component (regular, Guard, or Reserve).
- The COMAFFOR also has some ADCON responsibilities for Air Force elements and personnel assigned to other joint force components (such as liaisons).

As the Service component commander to a JFC, the COMAFFOR has the following responsibilities:

- Organize, train, and sustain assigned and attached Air Force forces for combatant commander (CCDR)-assigned missions.
  - Prescribe the chain of command within the Air Force Service component.
  - Maintain reachback between the Air Force component and other supporting Air Force elements. Delineate responsibilities between forward and rear elements.
  - Provide training in Service-unique doctrine, tactical methods, and techniques.
  - Provide for logistics and mission support functions normal to the command.
- Inform the JFC (and the CCDR, if affected) of planning for changes in logistics support that would significantly affect operational capability or sustainability sufficiently early in the planning process for the JFC to evaluate the proposals prior to final decision or implementation.
- Provide lateral liaisons with Army, Navy, Marines, special operations forces, and coalition partners.
- Maintain internal administration and discipline, including application of the Uniform Code of Military Justice (UCMJ).
- Establish force protection and other local defense requirements.
- Provide Service intelligence matters and oversight of intelligence activities to ensure compliance with laws, executive orders, policies, and directives.

At the CCDR level, the Air Force Service component commander also has the following additional responsibilities:

- Develop program and budget requests that comply with CCDR guidance on warfighting requirements and priorities.
- Inform the CCDR (and any intermediate JFCs) of program and budget decisions that may affect joint operation planning.
- Support the CCDR's theater campaign plans through development of appropriate supporting Service plans.
  - Develop steady-state strategy to support the CCDR's strategy.
  - Contribute to the development of CCDR steady-state campaign plans and security cooperation country plans.
  - Develop campaign support plans in support of CCDR campaign plans.

III. Additional Responsibilities as the Service Component Commander to a Combatant Commander

Ref: Annex 3-30, Command and Control (7 Nov '14), pp. 34 to 35.

When the COMAFFOR is the CCDR's Air Force Service component commander, he/she also has the following additional operational and administrative responsibilities:

- Develop program and budget requests that comply with CCDR guidance on warfighting requirements and priorities.
- Inform the CCDR (and any intermediate JFCs) of program and budget decisions that may affect joint operation planning.
- Support the CCDR's theater campaign plans through development of appropriate supporting Service plans.
  - Develop steady-state strategy to support the CCDR's strategy.
  - Contribute to the development of CCDR steady-state campaign plans and security cooperation country plans.
  - Develop campaign support plans in support of CCDR campaign plans.
  - Develop security cooperation country plans in support of CCDR security cooperation country plans.
  - Recommend and/or implement policy and rules of engagement for the conduct of steady-state operations, including planning, execution, and assessment.
  - Provide commander's intent to inform tactical-level planning, execution, and assessment.
  - Execute and assess steady-state operations.

Chap 2

II. Organizing U.S. Air Force Forces

Ref: Volume 3, Command (22 Nov 16), pp. 49 to 66.

Organization is critically important to effective and efficient operations. Service and joint force organization and command relationships—literally, who owns what, and who can do what with whom, and when—easily create the most friction within any operation.

I. Regional versus Functional Organization

It is important to understand that airpower is flexible in organization and presentation. Because it encompasses a wide range of capabilities and operating environments, it defies a single, general model for organization, planning, and employment.

- Some assets and capabilities provide relatively localized effects and generally are more easily deployable, and thus may organize and operate within a regional model.
- Other assets and capabilities transcend geographic areas of responsibility simultaneously, and thus have global responsibilities. Such forces may be better organized and controlled through a functional model.

However, at the focus of operations within any region, it is possible to place the collective capabilities of airpower in the hands of a single Airman through skillful arrangement of command relationships, focused expeditionary organization, reach-back, and forward deployment of specialized talent.

There will usually be tension between regionally-organized forces and functionally-organized forces. The former seek effectiveness at the point of their operation, while the latter seek effectiveness and efficiency across several regions. At critical times, the requirement for effectiveness may trump efficiency, and additional functional forces may be transferred to the regional command and organized accordingly.

See following page (p. 2-7) for an overview and further discussion. See related discussion on transferring forces and the complete discussion on "Transfer of Functional Forces to a Geographic Command", Annex 3-30, Command and Control. These situations require careful and continuing dialogue between competing senior commanders and their common superior commander.

II. The Air Expeditionary Task Force (AETF)

The air expeditionary task force (AETF) is the organizational structure for Air Force forces in response to operational tasking (i.e., established for a temporary period of time to perform a specified mission). It provides a task-organized, integrated package with the appropriate balance of force, sustainment, control, and force protection.

AETFs may be established as an Air Force Service component to a joint task force (JTF), or as a subordinate task force within a larger Air Force Service component to address specific internal tasks. If an AETF is formed as the former, the AETF commander is also a commander, Air Force forces (COMAFFOR). Otherwise, the AETF commander is not a COMAFFOR, but reports to a COMAFFOR.

A single commander presents a single Air Force face to the joint force commander (JFC) and results in clear lines of authority both ways.

- Internal to the task force, there is only one person clearly in charge; for a JFC, there is only one person to deal with on matters regarding Air Force issues.
- The AETF commander is the senior Air Force warfighter and exercises the appropriate degree of control over the forces assigned, attached, or in support of the AETF.

- Within the joint force, these degrees of control are formally expressed as operational control (OPCON), tactical control (TACON), or support. Within Service lines, the AETF commander exercises administrative control (ADCON).

Appropriate Command and Control Mechanisms

If acting as a COMAFFOR, the AETF commander exercises command in both the operational and administrative branches of the chain of command through an air operations center (AOC), an Air Force forces (AFFOR) staff (sometimes colloquially called an "A-staff"), and appropriate subordinate C2 elements. The AOC and the AFFOR staff are discussed in more detail in Annex 3-30, Command and Control.

Tailored and Fully Supported Forces

The AETF should be tailored to the mission; this includes not only forces, but also the ability to command and control those forces for the missions assigned.

In summary, the AETF is an expeditionary force established for a temporary period of time to perform a specified mission. The AETF provides a tailored package of air, space, and cyberspace capabilities in a structure that preserves Air Force unity of command. An AETF can be tailored in size and composition as appropriate for the mission.

A. AETF Organization

The basic building block of an air expeditionary task force (AETF) is the squadron; however, a squadron normally does not have sufficient resources to operate independently. Thus, the smallest AETF is normally an air expeditionary group (AEG); larger AETFs may be composed of several expeditionary wings.

Within an AETF, the AETF commander organizes forces as necessary into wings, groups, squadrons, flights, detachments, or elements to provide reasonable internal spans of control, command elements at appropriate levels, and to retain unit identity.

See following pages (pp. 2-8 to 2-9) for more complete discussion of internal AETF organization and designation of expeditionary and provisional units, from Annex 3-30, Command and Control.

Expeditionary Elements below Squadron Level

The Air Force may deploy elements below the squadron level for specific, limited functions. These include individuals and specialty teams such as explosive ordnance disposal (EOD) teams, military working dog teams, security forces, liaison teams, etc. They may deploy as part of an AETF or independently of other Air Force units, in remote locations, and may operate directly with other Services.

NOTE: Recent experience has revealed that tracking small, remotely located Air Force elements, especially in the distributed environment encountered in irregular warfare, has posed challenges for the Air Force component headquarters.

- These challenges may range from lack of administrative support to improper employment of small units and individual Airmen in tasks for which they have not been trained.
- The AFFOR staff should take special efforts to maintain effective oversight of such elements in order to fulfill proper ADCON oversight.

Regional versus Functional Organization

Ref: Volume 3, Command (22 Nov 16), pp. 49 to 51.

Regional Organization and Control

All military missions are ultimately under the authority of a joint force commander (JFC) at the appropriate level. If the entire theater is engaged, the combatant commander (CCDR) may be the JFC. If the situation is less than theater-wide, the CCDR may establish a subordinate joint task force (JTF) commanded by a subordinate JFC. In either case, the CCDR should first look to assigned, in-theater forces. If augmentation is required, the JFC should request additional forces through the Secretary of Defense (SecDef). Upon SecDef approval, additional forces transfer into the theater and are attached to the gaining CCDR. The degree of control gained over those forces (i.e., operational control [OPCON] or tactical control [TACON]) should be specified in the deployment orders. The gaining CCDR then normally delegates OPCON of these forces downward to the JTF commander who should, in turn, delegate OPCON to the Service component commanders within the gaining JTF. All Air Force forces should be organized and presented as an air expeditionary task force (AETF).

- Within a joint force, the JFC may organize forces in a mix of Service and functional components. All joint forces contain Service components, because administrative and logistics support are provided through Service components. Therefore, **every joint force containing assigned or attached Air Force forces will have an Air Force Service component in the form of an AETF with a designated commander, Air Force forces (COMAFFOR).**
- The JFC may also establish functional component commands when forces from two or more military Services operate in the same dimension or domain or there is a need to accomplish a distinct aspect of the assigned mission. Functional component commanders, such as the joint force air component commander (JFACC), are established at the discretion of the JFC.
- If functional component commands are established, the Service component commander with the preponderance of forces to be tasked, and with the requisite ability to provide command and control, will normally be designated as that functional component commander. Functional component commanders normally exercise TACON of forces made available for tasking. Through the Air Force component, the Air Force provides a COMAFFOR who is trained, equipped, and prepared to also be the JFACC if so designated by the JFC to whom he/she is assigned or attached.

Functional Organization and Control

Not all Air Force forces employed in an operation may be attached forward to a geographic CCDR. Several aspects of airpower are capable of serving more than one geographic CCDR at a time. Such forces are organized under functional CCDRs to facilitate cross-area of operations (AOR) optimization of those functional forces.

- When such forces are deployed in a geographic CCDR's AOR, they may remain under the OPCON of their respective functional CCDR and operate in support of the geographic CCDR. Within a theater, this support relationship is facilitated through specially designated representatives attached to regional AETFs.
- In some circumstances, after coordination with the owning commander and upon SecDef approval, control of such functional forces may be transferred to a geographic commander and attached with specification of OPCON or TACON.

AETF Organization

Ref: Annex 3-30, Command and Control (7 Nov '14), pp. 58 to 61.

AETFs can be sized and tailored to meet the specific requirements of the mission. The basic building block of an AETF is the squadron; however, a squadron normally does not have sufficient resources to operate independently. Thus, the smallest AETF is normally an air expeditionary group; larger AETFs may be composed of several expeditionary wings. Within an AETF, the AETF commander organizes forces as necessary into wings, groups, squadrons, flights, detachments, or elements to provide reasonable internal spans of control, command elements at appropriate levels, and to retain unit identity.

A. Numbered Expeditionary Air Force (NEAF)

Numbered expeditionary Air Force (NEAF) is the generic title for an AETF made up of multiple expeditionary wings and is the largest sized AETF. NEAFs normally carry an appropriate numerical designation based on NAFs historically associated with the region or command. Subordinate expeditionary units may retain their own numerical designations. Use of the NEAF designation is also intended to provide appropriate unit awards and honors credit for the units and staffs within the NEAF. The NEAF commander is normally a COMAFFOR.

B. Air Expeditionary Task Force-X (AETF-X)

"Air Expeditionary Task Force - X" (AETF-X) is the generic title used when a provisional Air Force command echelon is needed between a NEAF and an air expeditionary wing (AEW). AETF-X is used when a NEAF-level AETF establishes a subordinate provisional command echelon consisting of two or more AEWs. An example of this usage is when the Commander, US Air Forces Central (USAFCENT) established two subordinate AETFs, AETF-Iraq (AETF-I) and AETF-Afghanistan (AETF-A), to provide command over multiple AEWs in their respective JOAs. Depending on why this echelon is established, and its relationship within Service and joint force organizations, the AETF-X commander may or may not be a COMAFFOR.

C. Air Expeditionary Wing (AEW)

AEW is the generic title for a deployed wing or a wing slice within an AETF. An AEW normally is composed of the wing command element and subordinate groups and squadrons. AEWs normally carry the numerical designation of the wing providing the command element. Subordinate expeditionary groups and support squadrons carry the numerical designation of the parent AEW. Subordinate mission squadrons and direct combat support units retain their numeric designation in an expeditionary status. Use of the AEW designation is also intended to provide appropriate unit awards and honors credit for the parent unit. An AEW may be composed of units from different wings, but where possible, the AEW is formed from units of a single wing. AEW commanders report to the COMAFFOR.

D. Air Expeditionary Group (AEG)

Air expeditionary group (AEG) is the generic title for a deployed group assigned to an AEW or a deployed independent group assigned to an AETF. Unlike traditional "home station" groups, which are functionally organized (i.e., operations group, maintenance group, etc.), expeditionary groups that are deployed independent of a wing structure should contain elements of all the functions to conduct semi-autonomous operations. An AEG is composed of a slice of the wing command element and some squadrons. Since Air Force groups are organized without significant staff support, a wing slice is needed to provide the command and control for echelons smaller than the normal wing. An AEG assigned to an AEW carries the numeric designation of the AEW. An independent AEG normally carries the numerical designation of the unit providing the command element and/or the largest portion of the expeditionary organization. Deployed squadrons (assigned or attached) retain their numerical designation and acquire the "expeditionary" designation. Use of the AEG designation is also intended to provide appropriate unit awards and honors credit for the parent unit. An AEG may be composed of units from different wings, but where possible, the AEG is formed from units of a single wing. If deployed as an independent group as part of a larger AETF with other AEGs and/or AEWs, the AEG commander normally reports to the COMAFFOR. If deployed as a group subordinate to an expeditionary wing, the AEG commander reports to the AEW commander. The AEG is normally the smallest independently deployable AETF.

E. Air Expeditionary Squadron (AES)

Air expeditionary squadron (AES) is the generic title for a deployed squadron within an AETF. Squadrons are configured to deploy and employ in support of taskings. However, an individual squadron is not designed to conduct independent operations; it normally requires support from other units to obtain the synergy needed for sustainable, effective operations. As such, an individual squadron or squadron element should not be presented by itself without provision for appropriate support and command elements. If a single operational squadron or squadron element is all that is needed to provide the desired operational effect (for example, an element of C-130s performing humanitarian operations), it should deploy with provision for commensurate support and C2 elements. The structure of this AETF would appear similar to an AEG. In some operations, not all support and C2 elements need to deploy forward with the operational squadron. Some may be positioned "over the horizon," constituting capabilities provided through reachback. A single squadron or squadron element may deploy without full support elements if it is planned to augment a deployed AEW or AEG, and would thus obtain necessary support from the larger units.

F. Expeditionary Elements below Squadron Level

In addition to expeditionary wings, groups, and squadrons, the Air Force may deploy elements below the squadron level for specific, limited functions. These include individuals and specialty teams such as explosive ordnance disposal (EOD) teams, military working dog teams, security forces, liaison teams, etc. They may deploy as part of an AETF or independently of other Air Force units, in remote locations, and may operate directly with other Services. For ADCON purposes, these elements should normally be attached to the commander of a recognizable Air Force entity in the region, either a deployed AETF or the Air Force Service component to the engaged combatant commander. Examples of such deployed elements might be a psychological operations team augmenting a Joint Psychological Operations Task Force (JPOTF), an EOD team augmenting a predominately surface force, or an Air Force element supplementing Army convoy operations. Air Force personnel assigned to a joint staff may also fall in this category.

In many circumstances, elements below squadron level and even individual persons may deploy to provide a specific capability. In such cases, formal establishment and designation of an AETF may not be warranted. However, the Air Force contingent should still be organized as a single entity (perhaps named simply as "Air Force element") and led by the senior Airman in the contingent.

G. Provisional Units

In some instances, expeditionary forces may not form around active numbered units. This may occur, for example, when there are insufficient active numbered units in the AEF rotation to satisfy a very large operation or a single major force provider cannot be identified. In such cases, provisional units may be created using predesignated inactive units. A unit under a single provisional unit designation should also be considered to provide continuity of operations for extended contingency operations in which units are frequently rotated in and out (e.g., Operations NORTHERN and SOUTHERN WATCH, and IRAQI FREEDOM). Upon completion of the operation for which the unit was formed, the unit designation and history are inactivated. Provisional wings, groups, and squadrons are normally generically designated simply as AEWs, AEGs, and AESs.

B. AETF Command and Control Mechanisms

The commander, Air Force forces (COMAFFOR) requires command and control (C2) assets to assist in exercising operational control (OPCON), tactical control (TACON), and administrative control (ADCON). The COMAFFOR normally uses some form of an air operations center (AOC) to exercise control of operations and a Service component staff, commonly called the AFFOR staff, to exercise support operations and administrative control.

The core capabilities of the AOC and AFFOR staff are well established, but they should be tailored in size and function according to the operation.

- Not all operations require a "full-up" AOC with over 1,000 people or a large AFFOR staff. Smaller operations, such as some humanitarian operations, can in fact make do with a small control center that does little more than scheduling and reporting.
- Not all elements of the operations center or AFFOR staff need be forward; some may operate "over the horizon," using reachback to reduce the forward footprint. The goal is to maximize reachback and minimize forward presence.

Air Operations Center (AOC)

In general terms, an AOC is the Air Force component commander's C2 center that provides the capability to plan, direct, and assess the activities of assigned and attached forces.

AOCs do not work in isolation; they require appropriate connectivity to operations centers of higher headquarters (e.g., to the joint force headquarters for the operational branch, and to senior Air Force headquarters for the administrative branch), to lateral headquarters (e.g., other joint force components), to subordinate assigned and attached Air Force units, and to other functional and geographic AOCs as necessary. The overall C2 structure should make maximum use of reachback.

An AOC, along with subordinate C2 elements, should be tailored in size and capability to the mission. An AOC should generally be capable of the following basic tasks:

- Develop the component strategy and requisite planning products.
- Task, execute, and assess day-to-day component operations.
- Plan and execute intelligence, surveillance, and reconnaissance (ISR) tasks appropriate to assigned missions.
- Conduct operational-level assessment.

For an AOC baseline description, refer to Annex 3-30, appendix B.

AFFOR Staff

The AFFOR staff is the mechanism through which the COMAFFOR exercises Service responsibilities and is also responsible for the long-range planning and theater engagement operations that fall outside the AOC's current operational focus.

- An AFFOR staff should be ready to fill one or more roles: that of a theater-wide Air Force Service component, an Air Force warfighting component within a JTF, or the core or "plug" within a JTF headquarters.
- The COMAFFOR should avoid dual- or triple-hatting the AFFOR staff to the maximum extent possible. Dual- or triple-hatting may have detrimental consequences as the staff struggles to focus at the right level of war at the right time. Manning and distribution of workload may limit the staff's ability to cover all involved duties simultaneously and augmentation may be necessary.
- The AFFOR staff's function is to support and assist the COMAFFOR in preparing the Air Force component to carry out the functions and tasks assigned by the joint force commander (JFC).

For a baseline AFFOR staff description, refer to Annex 3-30, appendix C.

III. Reachback/Distributed/Split Operations

Reachback

Reachback is defined as "the process of obtaining products, services, and applications, or forces, or equipment, or material from organizations that are not forward deployed." Reachback may be provided from a supporting/supported relationship or by Service retained forces. This relationship gives the commander, Air Force forces (COMAFFOR) the support necessary to conduct operations while maintaining a smaller deployed footprint.

Distributed Operations

Distributed operations are defined as operations when independent or interdependent forces, some of which may be outside the joint operations area, participate in the operational planning and/or operational decision-making process to accomplish missions and objectives for commanders. While Service-retained forces may provide reachback, forces conducting distributed operations should be assigned or attached to a combatant command.

Split Operations

Split operations is a type of distributed operation conducted by a single command and control (C2) entity separated between two or more geographic locations. A single commander must have oversight of all aspects of a split C2 operation.

Note: The decision to establish distributed or split operations invokes several tradeoffs. For more detail refer to Annex 3-30, Command and Control.

IV. Command Relationship Models

When employing military forces, a combatant commander (CCDR) first turns to those forces already assigned. Assigned forces are delineated in the Secretary of Defense's (SecDef's) "Forces for Unified Commands" memorandum, and the CCDR exercises combatant command (command authority) (COCOM) over them.

Additional forces beyond those assigned to the CCDR may be attached by SecDef action. Under current policies, attached forces may be provided through one of two specific Global Force Management allocation supporting processes: rotational force allocation in support of CCDR annual force needs, and emergent force allocation in support of CCDR emerging or crisis-based requests for capabilities and forces. The deployment order should clearly delineate the degree of command authority to be exercised by the gaining commander. Forces temporarily transferred via SecDef action are normally attached with specification of operational control (OPCON) to the gaining CCDR.

The DEPORD is the primary instrument for transferring forces and establishing supported and supporting relationships between CCDRs. Forces may also be transferred by an execute order which executes an approved operation plan.

For Air Force forces, there are four general models for command relationships. Considerations for these relationships should include the ability of gaining commands to receive the forces and to command and control them appropriately; the characteristics and support requirements of the forces involved, and the operating locations of the forces.

- In-Theater Forces deployed and executing operations within the theater to which they are attached. (Model 1)
- Out-Of-Theater Forces executing missions inside the theater of operations but based outside the theater (i.e., across areas of responsibility [AOR]). (Model 2)
- Functional forces with global missions. (Model 3)
- Transient forces. (Model 4)

V. Integrating Regional and Functional Air Force Forces

Airpower is usually presented through a mix of regional and functional models, with the latter usually supporting the former.

Functional forces usually maintain a separate organization from the supported regional organization, and are integrated in the theater through specially trained liaisons attached to the regional commander, Air Force forces (COMAFFOR).

The most likely functional capabilities to be provided in such a supporting relationship are air mobility operations, space operations, special operations, cyberspace operations, and nuclear operations.

Integrating Air Mobility Operations

Because air mobility forces serve several regions concurrently, their employment should be balanced between regional and intertheater requirements and priorities.

- The air mobility systems performing intratheater and intertheater missions within a given region should operate in close coordination to provide responsive and integrated aerial movement to the supported combatant commander (CCDR).
- Carefully constructed command relationships can allow an interlocking arrangement to manage intratheater and intertheater air mobility operations.
- The Director of Air Mobility Forces (DIRMOBFOR). Within an Air Force component, the DIRMOBFOR is the COMAFFOR's designated coordinating authority for air mobility operations.

See pp. 4-41 to 4-56 for discussion of air mobility operations from Annex 3-17.

Integrating Space Operations

Space presents another form of military operations that, much like air mobility, usually are best presented functionally to a regional commander through a supporting relationship if they are not attached.

- Space command and control brings another level of complexity because many space assets that support military interests come from a variety of organizations, some outside of the Department of Defense (DOD).
- Space Coordinating Authority (SCA). Within a regional operation, the joint force commander (JFC) should designate SCA to facilitate unity of effort with DOD-wide space operations and non-DOD space capabilities.
- The Director of Space Forces (DIRSPACEFOR) serves as the senior space advisor to the COMAFFOR. The DIRSPACEFOR, an Air Force space officer, coordinates, integrates, and staffs activities to tailor space support to the COMAFFOR. In addition, when the COMAFFOR is delegated SCA, the DIRSPACEFOR works the day-to-day SCA activities on behalf of the COMAFFOR. If the COMAFFOR is neither delegated SCA nor designated as the JFACC, the COMAFFOR should establish a space liaison to the JFACC.

See pp. 4-37 to 4-40 for discussion of space operations from Annex 3-14.

Integrating Special Operations

Commander, US Special Operations Command (USSOCOM) exercises combatant command (command authority) (COCOM) of worldwide special operations forces, while the geographic CCDR exercises OPCON of assigned/attached Air Force special operations forces (AFSOF) through the commander of the theater special operations command.

- For conventional missions, the COMAFFOR may receive OPCON or tactical control of Air Force special operations forces (AFSOF) assets when directed

VI. Transfer of Functional Forces to a Geographic Command

Ref: Volume 3, Command (22 Nov 16), pp. 59 to 60. See also Annex 3-30.

In some situations, a geographic commander may request additional functional forces beyond those apportioned or allocated during deliberate or crisis action planning. The decision to transfer functional forces, with specification of operational control (OPCON), to a geographic combatant commander (CCDR) should be balanced against competing needs across multiple areas of responsibility (AORs).

In some cases, the requirement for OPCON over specific forces to accomplish the geographic CCDR's missions may be of higher priority than the competing worldwide mission requirements of the functional CCDR. Therefore, after coordination with the owning functional commander and upon SecDef approval, functional forces may be transferred to the geographic command and organized accordingly. The decision to attach additional functional forces has two parts. First, the decision should consider whether:

- The geographic CCDR will use the forces at or near 100 percent of their capability with little or no residual capability for other global missions.
- The forces will be used regularly and frequently over a period of time, not just for a single mission employment.
- The geographic commander has the ability to effectively command and control the forces.

If the answer to all three questions above is "yes," then the functional forces should be attached to the geographic combatant command. If any of the above questions are answered "no," then the functional forces should remain under the OPCON of the functional CCDR's commander, Air Force forces (COMAFFOR) and be tasked in support. If the decision is to attach forces, the second question is whether the forces should be attached with specification of either OPCON or tactical control (TACON).

Specification of OPCON

OPCON is the more complete—and preferred—choice of control. OPCON "normally provides full authority to organize commands and forces and to employ those forces as the commander in operational control considers necessary to accomplish assigned missions; it does not include authoritative direction for logistics or matters of administration, discipline, internal organization, or unit training."

Specification of TACON

TACON is the more limited choice of control. It is defined as "the authority over forces that is limited to the detailed direction and control of movements or maneuvers within the operational area necessary to accomplish missions or tasks assigned." Joint Publication 1, Doctrine for the Armed Forces of the United States, states "when transfer of forces to a joint force will be temporary, the forces will be attached to the gaining commands and JFCs, normally through the Service component commander, will exercise OPCON over the attached forces." Thus, transfer and attachment with specification of TACON is not the expected norm. While it is possible for the SecDef to attach forces across combatant command lines with the specification of TACON in lieu of OPCON, such action will deviate from joint doctrine established in JP 1 and would result in a more confused chain of command with OPCON and TACON split between two different CCDRs.

Regional COMAFFORs have inherent responsibilities for such issues as local force protection, lodging, and dining. Thus, if a regional COMAFFOR holds OPCON of forces outside the AOR, he or she is not responsible for such issues—that is the responsibility of the COMAFFOR in the region in which they are bedded down. In a parallel fashion, if such out-of-region forces divert into bases in his/her region (for example, for emergencies), that COMAFFOR is now responsible for basic support and protection.

VII. Homeland Organizational Considerations

Ref: Annex 3-30, Command and Control (7 Nov 2014), pp. 72 to 74.

Military operations inside the United States and its Territories fall into two mission areas: homeland defense, for which the Department of Defense (DOD) serves as the lead federal agency and military forces are used to conduct military operations in defense of the homeland; and civil support for which DOD serves in a supporting role to other agencies at the federal, state, tribal, and local levels.

For most homeland scenarios, Air Force forces should be presented as an air expeditionary task force (AETF) under the operational control (OPCON) of a commander, Air Force forces (COMAFFOR), just as in any other theater. Air National Guard (ANG) forces, whether activated and operating in Title 10 status, supporting a Federal mission or operation under Title 32 and attached to a combatant command (CCMD), or remaining under state control in Title 32 or state active duty status, should still be organized and presented within an AETF or equivalent structure.

For homeland operations, 1st Air Force (Air Forces Northern) (1AF [AFNORTH]) is the Air Force component to US Northern Command (USNORTHCOM) and the designated USNORTHCOM joint force air component commander. Within North American Aerospace Defense Command (NORAD), the Commander, 1AF (AFNORTH), is also the commander of the continental US (CONUS) NORAD region.

The command relationships between a joint force commander (JFC) and a COMAFFOR in a homeland context should be as previously described for any other region—although legal and interagency considerations may have significant impact, the homeland is not a special case regarding command and control (C2) or organization of air, space, and cyberspace forces. The COMAFFOR should still be under direct operational control of a designated JFC, should still normally exercise OPCON and administrative control (ADCON) over the Air Force Service component forces, and should still coordinate activities with other components and outside agencies to achieve JFC objectives. Additionally, when the ANG is operating in Title 32 or state active duty status under the authority of a state governor, a similar command relationship exists between the state Adjutant General or joint task force (JTF) commander and the designated ANG air commander.

Additionally, the Secretary of Defense may request State governors to allow their respective ANG personnel or units to support federal operations or missions such as providing intelligence and cyberspace support to combatant commanders (CCDRs) or supporting civil authorities pursuant to Title 32. ANG personnel and units would remain in Title 32 status, but be attached to the Service component of a CCMD, and under the operational authority of the CCDR. The nature, extent, and degree of control exercised by the CCDR and his subordinate commanders, including dual-status commanders, would be set forth in a command arrangements agreement (CAA) agreed upon by the Secretary of Defense and State governors. The CAA would be similar to those negotiated for multinational operations. Administrative authority for ANG personnel and units would remain with the State.

In some civil support operations, a JFC may elect to allocate combat support forces to subordinate functional task force commanders (TFCs) with a specification of OPCON to the TFC. For example, a JFC in a major disaster relief operation might organize forces into separate engineering, transportation, and medical task forces. This organizational scheme—a legacy construct which sidesteps the role of Service components and Service component commanders—divides Air Force assets among other component commanders and fractures Service unity of command. This is not the most operationally effective scheme for achieving unity of command and unity of effort under a single Airman. Ideally, the JFC allows the COMAFFOR to retain OPCON of all assigned and attached Air Force forces. The COMAFFOR then provides direct support to the various functional TFCs with the COMAFFOR as a supporting commander.

In disaster relief operations, particularly in consequence management of a manmade or natural disaster, the Air Force contribution will likely include a Total Force mix of capabilities. ANG forces may be activated under Title 10; support a federal response remaining in Title 32 status, but attached to and under the control of a CCDR; or more normally operate under Title 32 or state active duty status under the authority of their governor.

Each state has a state joint force headquarters (JFHQ-State) that may provide a contingency C2 capability in support of homeland defense, civil support, and other related operations, and may thus function as a bridge between state and federal forces. Additionally, a governor may stand up a JTF-State to provide direction and control of assigned non-federalized National Guard forces and those attached from other states. ANG forces conducting operations in Title 32 or state active duty status should be organized as an AETF or equivalent within their state force structure to provide unity of command, with a single Airman in command of the ANG forces.

State and federal military forces may adopt a parallel command structure.

- A **parallel command** structure exists when state and federal authorities have separate chains of command, and retain control of their deployed forces. Unity of effort and decisions of mutual interest are handled through a coordinated liaison effort of the political and senior military leadership of state and federal forces.
- Federal statutes now provides the capability for a **dual status command** structure, in which a designated commander subordinate to a combatant commander may unify and streamline the command structure by simultaneously serving in Federal and State duty statuses while performing the separate and distinct duties of those statuses over forces in Title 32 as well as forces in Title 10.

See also p. 1-16 for related discussion. For more detailed discussion on homeland operations in general, refer to Annex 3-27, Homeland Operations.

Refer to The Homeland Defense & DSCA SMARTbook (Protecting the Homeland / Defense Support to Civil Authority) for complete discussion. Topics and references include homeland defense (JP 3-28), defense support of civil authorities (JP 3-28), Army support of civil authorities (ADRP 3-28), multi-service DSCA TTPs (ATP 3-28.1/MCWP 3-36.2), DSCA liaison officer toolkit (GTA 90-01-020), key legal and policy documents, and specific hazard and planning guidance.

Refer to CTS1: The Counterterrorism, WMD & Hybrid Threat SMARTbook for further discussion. CTS1 topics and chapters include: the terrorist threat (characteristics, goals & objectives, organization, state-sponsored, international, and domestic), hybrid and future threats, forms of terrorism (tactics, techniques, & procedures), counterterrorism, critical infrastructure, protection planning and preparation, countering WMD, and consequence management (all hazards response).

Refer to CYBER: The Cyberspace Operations SMARTbook (in development). U.S. armed forces operate in an increasingly network-based world. The proliferation of information technologies is changing the way humans interact with each other and their environment, including interactions during military operations. This broad and rapidly changing operational environment requires that today's armed forces must operate in cyberspace and leverage an electromagnetic spectrum that is increasingly competitive, congested, and contested.

by the JFC. However, in most cases, AFSOF will only normally be in a direct support relationship with conventional assets.

- When SOF operate in concert with "conventional" JTFs, they normally take the form of a separate joint special operations task force (JSOTF) within the JTF, commanded by a joint force special operations component commander (JFSOCC).
- The Special Operations Liaison Element (SOLE). The SOLE is a liaison team that represents the JFSOCC to the COMAFFOR. The SOLE synchronizes all SOF air and surface operations with joint air operations via the air tasking process. Additionally, the SOLE deconflicts SOF operations with other component liaisons in the AOC.

See pp. 4-83 to 4-88 for discussion of special operations from Annex 3-05.

Integrating Cyberspace Operations

Global cyberspace capabilities may be presented to a regional commander through a supporting relationship, to supplement regional cyberspace capabilities. US Cyber Command, as a subordinate unified command under USSTRATCOM, is the focal point for providing cyberspace capabilities to other combatant commanders. To support regional operations, USCYBERCOM may also provide cyberspace expertise to regional staffs if necessary.

See pp. 4-89 to 4-94 for discussion of cyberspace operations from Annex 3-12.

VIII. Integrating the Air Reserve Components

The Air Force, under the Total Force construct, has a substantial part of its forces in the Air Reserve Components (ARC), which consists of the Air Force Reserve (AFR) and the Air National Guard (ANG). The ARC provides a strategic reserve and a surge capacity for the Air Force; in some instances, the ARC has unique capabilities not resident within the regular component.

The SecDef may make these forces available during the planning process. While they may seamlessly operate alongside the regular Air Force, they are subject to different levels of activation and different degrees of operational control and administrative control. Also, differences in tour length availability pose continuity challenges for a commander, Air Force forces (COMAFFOR), and planners should carefully consider such issues for any category of activation (whether by volunteerism or mobilization).

Refer to Annex 3-30, Appendix E, The Air Reserve Components, for more discussion on ARC organization and accessing ARC forces.

IX. The Senior/Host Air Force Installation Commander

Recent operations, notably Operations ENDURING FREEDOM and IRAQI FREEDOM, highlighted the nuances in on-base command arrangements and support requirements that result from mixed forces deploying forward, often to bare bases.

An installation commander, regardless of Service, always exercises some authority over and responsibility for forces on his/her base for protection of assigned forces and assets, lodging, dining, and administrative reporting, regardless of the command relations of those forces. These are inherent in his/her responsibilities as an installation commander. The senior Air Force commander on any base where Air Force forces are present has responsibilities for care and provisioning of the Air Force forces on that installation, regardless of organization.

For more detailed discussion, refer to Annex 3-30, The Senior / Host Air Force Installation Commander; refer to AFI 38-101, Air Force Organization, for more specific policy guidance.

III. Air Force Component Within the Joint Force

Ref: Volume 3, Command (22 Nov 16), pp. 67 to

I. Joint Force Organizational Basics

When a crisis requires a military response, the geographic combatant commander (CCDR) will usually form a tailored joint task force (JTF). If Air Force forces are attached to the JTF, they stand up as an air expeditionary task force (AETF) within the JTF. The AETF commander, as the commander, Air Force forces (COMAFFOR), provides the single Air Force face to the JTF commander.

Other Services may also provide forces, and normally stand up as separate Army, Navy, and Marine forces, each with their respective commander (Commander, Army forces [COMARFOR]; Commander, Navy forces [COMNAVFOR]; and Commander, Marine Corps forces [COMMARFOR]).

The designation of joint force air, land, maritime and special operations component commanders (JFACC, joint force land component commander [JFLCC], joint force maritime component commander [JFMCC], and joint force special operations component commander [JFSOCC] respectively) is at the discretion of the joint force commander (JFC). The JFC normally assigns broad missions to the component commanders; with each mission comes a specification of supported commander for that mission. As an example, the JFC may designate the COMAFFOR as the supported commander for strategic attack, air interdiction, and theater airborne intelligence, surveillance, and reconnaissance (among other missions).

The COMAFFOR should establish a close working relationship with the JFC to ensure the best representation of airpower's potential. The commander responsible for a mission should be given the requisite authority to carry out that mission.

II. The Joint Force Air Component Commander (JFACC)

Historically, when Air Force forces have been attached to a joint task force (JTF), the commander, Air Force forces (COMAFFOR) is normally designated as the joint force air component commander (JFACC), not merely due to preponderance of forces but also due to the ability to command and control airpower through an air operations center (AOC), which forms the core of the JFACC's JAOC.

- This is why the COMAFFOR trains to act as the JFACC.
- It is rare that sizeable Air Force forces have been present in a JTF, and the COMAFFOR has not been the JFACC.

If aviation assets from more than one Service are present within a joint force, the joint force commander (JFC) normally designates a JFACC to exploit the full capabilities of joint operations.

- The JFACC should be the Service component commander with the preponderance of forces to be tasked and the ability to plan, task, and control joint air operations (Joint Publication P 3-30, Command and Control for Joint Air Operations).
- If working with allies in a coalition or alliance operation, the JFACC may be designated as the combined force air component commander (CFACC).

Because of the wide scope of joint air operations, the JFACC typically maintains a similar theater-wide or joint operations area (JOA)-wide perspective as the JFC. The JFACC:

- As with any component commander, should not also be dual-hatted as the JFC as the scope of command is usually too broad for any one commander and staff.

Functional component commanders normally exercise tactical control (TACON) of forces made available to them by the JFC. Thus, a COMAFFOR normally exercises operational control (OPCON) of assigned and attached Air Force forces and, acting as a JFACC, normally exercises TACON of forces made available for tasking (i.e., those forces not retained for their own Service's organic operations).

See pp. 3-21 to 3-34 for further discussion of the JFACC.

III. Miscellaneous Joint Notes

Joint Staff Composition

The composition of a truly joint staff should reflect the composition of the subordinate joint forces to ensure that those responsible for employing joint forces have a thorough knowledge of the capabilities and limitations of assigned or attached forces. The presence of liaisons on a single-Service staff does not transform that Service staff into a joint staff. The same general guidelines for joint staffs apply to coalition operations.

IV. Air Force Component Presentation

Ref: Volume 3, Command (22 Nov 16), p, 71.

There are many possible options for presenting forces in support of a joint force commander (JFC). To provide an initial baseline for organizational decisions, there are three general models for presenting an Air Force component in support of a JFC.

Theater-level component

This establishes an AETF at the CCDR level, attached with specification of OPCON and commanded by a theater COMAFFOR/JFACC.

Sub-theater-level component

This establishes an AETF at the subunified command or JTF level, attached with specification of OPCON, with a COMAFFOR (prepared to act as a JFACC) at a level below the CCDR.

Sub-theater-level AETF in support of a JTF

This establishes a dedicated Air Force force in direct support of a subordinate JTF, with OPCON retained by the theater COMAFFOR/JFACC.

The placement of an Air Force component within the CCDR's command structure, as well as the formal command relationships necessary to enable it to interface with other joint forces, requires careful deliberation based on the situation and capabilities available. At times, Air Force forces and capabilities may be best positioned at the theater (i.e., CCDR) level and at other times at the JTF level.

The CCDR decides whether effective accomplishment of the operational mission at the JTF level outweigh competing missions at the CCDR's AOR level and can best be accomplished by attaching Air Force forces with specification of OPCON to a JTF commander. Deliberations should examine the interplay of priority, tempo, intensity, duration, and scope of operations.

Commanding & Organizing

Relationship between Commanders and Staffs

"Commanders command, staffs support." Within a joint force, only those with the title of "commander"—i.e., the joint force commander (JFC), the Service component commanders, and the functional component commanders—may exercise any degree of operational control over forces. Only commanders have the legal and moral authority to place personnel in harm's way. Under no circumstance should staff agencies, including those of the JFC's staff, attempt to command forces. Staff agencies should neither attempt to nor be permitted to directly command or control elements of the subordinate forces. While this guidance is aimed at joint staffs, it also applies to Service staffs.

JFACC Staff

When the commander, Air Force forces (COMAFFOR) is designated the joint force air component commander (JFACC), he/she may need to establish a small joint or combined staff to deal with joint issues beyond the purview of the AFFOR staff.

- Augmentation within each AOC directorate from relevant Service components and coalition partners ensures adequate joint representation on the staff.
- At the discretion of the COMAFFOR, officers from other Services and coalition partners may fill key deputy and principal staff positions.
- For very large and complex operations-as might be encountered with large coalition operations-a COMAFFOR dual-hatted as a JFACC may delegate some aspects of COMAFFOR functions to a subordinate deputy COMAFFOR.

See pp. 3-24 to 3-25 for an overview and discussion of typical JFACC staff and JAOC organization.

V. Multi-Hatting Commanders/Span of Control

Caution should be applied when multi-hatting commanders. Too many "hats" may distract a commander from focusing on the right level of war at the right time, or may simply overwhelm the commander with detail. Of equal importance is the fact that a commander's staff can usually operate effectively only at one level of war at a time. If a commander wears several hats, it is preferable that the associated responsibilities lie at the same level of war.

While it is normally inappropriate for either a Service or a functional component commander to also serve as the joint force commander (JFC), it is entirely appropriate for a joint force air component commander (JFACC) to also serve as the airspace control authority, area air defense commander, and space coordinating authority, since all four functions lie at the operational level and all four functions are supported through the same command node (the JAOC). To alleviate the overload, a multi-hatted commander may delegate some functions (but not the ultimate responsibility) to appropriate deputies.

More challenging are those instances when a commander's hats vertically span several levels of war, as in the case when the JFC (normally acting at the theater-strategic level) is also acting as a functional component commander (operational level), and also as the commander of one of the operating (tactical) units. In such cases, the commander may be inadvertently drawn to the tactical level of detail at the expense of the operational-level fight.

VI. Joint Air Component Coordination Element (JACCE)

Ref: Volume 3, Command (22 Nov 16), pp. 72 to 73.

The commander, Air Force forces (COMAFFOR), when acting as the joint force air component commander may establish one or more joint air component coordination elements (JACCEs) with other component commanders' headquarters to better integrate the air component's operations with their operations, and with the supported joint task force (JTF) headquarters (if the theater COMAFFOR is designated in support to a JTF) to better integrate air component operations within the overall joint force.

The JACCE facilitates integration by exchanging current intelligence, operational data, and support requirements, and by coordinating the integration of COMAFFOR requirements for airspace coordinating measures, fire support coordinating measures, close air support, air mobility, and space requirements. As such, the JACCE is a liaison element, not a command and control (C2) node; thus, the JACCE normally has no authority to direct or employ forces. The JACCE should not replace, replicate, or circumvent normal request mechanisms already in place in the component/JTF staffs, nor supplant normal planning performed by the air operations center and AFFOR staff.

Normally, the JACCE should:

- Ensure the COMAFFOR is aware of each commander's priorities and plans.
- Ensure the COMAFFOR staff coordinates within their surface component/JTF headquarters counterparts to work issues.
- Ensure appropriate commanders are aware of the COMAFFOR's capabilities and limitations (constraints, restraints, and restrictions).
- Ensure appropriate commanders are aware of the COMAFFOR's plan to support the surface commander's scheme of maneuver and JFC's intent and objectives.
- Facilitate COMAFFOR staff processes with the surface/JTF commanders. Provide oversight of other COMAFFOR liaisons to component/JTF headquarters staffs, if directed.
- Ensure information flows properly between the AOC, sister components, and the JFC.

See also p. 3-25 (JACCE).

VII. Control of Other Services' Aviation Capabilities

Ref: Annex 3-30, Command and Control (7 Nov 2014), pp. 38 to 39.

When the commander, Air Force forces (COMAFFOR) is designated as the joint force air component commander (JFACC), he/she may control aviation assets of other Services, in whole or in part, depending on the situation. However, he/she only controls those capabilities "made available for tasking" as directed by the joint force commander (JFC).

Regardless of whether the COMAFFOR, as the JFACC, exercises tactical control of other Services' forces, the COMAFFOR, in the normally expected additional roles of airspace control authority, area air defense commander (AADC), space coordinating authority, and electronic warfare control authority, normally requires inclusion of such forces on the air tasking order (ATO) and airspace control order (ACO). This provides situational awareness of all friendly aviation in the area of responsibility/joint operations area, prevents fratricide, and deconflicts airspace.

Army aviation assets. These assets are normally retained for employment as organic forces within its combined arms paradigm. However, some Army helicopters could be employed in close air support, interdiction, or other missions, in which case they may come under the purview of the COMAFFOR when the COMAFFOR has been tasked to plan and execute the theater interdiction effort. The same can hold true for other systems (such as the Army Tactical Missile System) when employed for interdiction or offensive counterair, depending on tasking and target location. As a minimum, Army aviation elements, including some unmanned systems, should comply with the ACO to deconflict airspace and friendly air defense planning. Placing Army aviation assets on the ATO/ACO reduces the risk of fratricide and provides better overall integration with other joint air component operations. Additionally, Army Patriot surface-to-air missiles and Theater High Altitude Air Defense (THAAD) capabilities should be integrated into the overall theater defensive counterair effort and may operate in a direct support role to the COMAFFOR acting as AADC.

Navy aviation assets. These assets include carrier-based aircraft, land-based naval aircraft, and missiles. They provide a diverse array of power projection capabilities. Such assets, beyond those retained as needed for fleet defense and related naval missions, are usually available for tasking via the air tasking process. Additionally, Navy Aegis air and missile defense capabilities may be integrated into the overall theater defensive counterair effort. As with Army aviation assets, Navy aviation assets, including unmanned systems, should comply with the ACO for airspace deconfliction and air defense planning.

Marine aviation assets. The primary mission of Marine aviation is support of the Marine air-ground task force (MAGTF) ground element. Sorties in excess of organic MAGTF direct support requirements should be provided through the JFC to the COMAFFOR for ATO tasking. (Note: Marine sorties provided for tasking for such theater missions as long range reconnaissance, theater air interdiction or defensive counterair, are not considered "excess" sorties.)

Special operations forces (SOF) aviation assets. The JFC may assign control of SOF aviation forces to either a Service or a functional component commander. When SOF air assets are employed as part of joint SOF operations, the JFC may assign control of those forces to the joint force special operations component commander (JFSOCC), who may in turn designate a joint special operations air component commander responsible for planning and executing joint special air operations.

IV. Multinational & Interagency

Ref: JP 3-16, Multinational Operations (Jul '13) and Annex 3-30 (7 Nov 14), pp. 26 to 28.

Most operations today are not US-only. Many operations involve military forces of allies, and many operations also involve intergovernmental organizations (IGOs), nongovernmental organizations (NGOs), and regional organizations. Managing the myriad interrelationships is necessary, but often challenging. In many instances, direct command over these various entities is not possible, and unity of effort rather than unity of command becomes the goal.

I. Multinational Operations

Multinational operations are operations conducted by forces of two or more nations, and are usually undertaken within the structure of a coalition or alliance.

Alliance

An alliance is the relationship that results from a formal agreement between two or more nations for broad, long-term objectives that further the common interests of the members.

Coalition

A coalition is "an arrangement between two or more nations for common action." Coalitions are formed by different nations with different objectives, usually for a single occasion or for longer cooperation in a narrow sector of common interest.

In a multinational force, the joint commanders become combined commanders; thus, a joint force commander becomes a combined force commander, a joint force air component commander becomes a combined force air component commander, etc. Similarly, an air operations center (AOC) (properly a joint air operations center [JAOC] in joint context) becomes a combined air operations center (CAOC) with representation that, as with a JAOC, reflects the composition of the force.

An important point is that commanders may not have the same defined degree of control over forces (e.g., operational control [OPCON], tactical control [TACON], etc.) as in a US-only force; degrees of control may have to be negotiated. Sometimes, existing non-US controls may be used, as may be encountered in North Atlantic Treaty Organization (NATO) operations by the use of NATO operational command (OPCOM), OPCON, tactical command (TACOM), and TACON; commanders and staff should be aware of the different nuances. Finally, each nation may retain its own chain of command over its forces and separate rules of engagement; thereby further complicating unity of command.

Refer to JFODS5: The Joint Forces Operations & Doctrine SMARTbook (Guide to Joint, Multinational & Interorganizational Operations) for further discussion. Topics and chapters include joint doctrine fundamentals (JP 1), joint operations (JP 3-0), joint planning (JP 5-0), joint logistics (JP 4-0), joint task forces (JP 3-33), information operations (JP 3-13), multinational operations (JP 3-16), interorganizational cooperation (JP 3-08), plus more!

Thus, the challenge in multinational operations is the effective integration and synchronization of available capabilities toward the achievement of common objectives through unity of effort despite disparate (and occasionally incompatible) command and control (C2) structures, capabilities, equipment, and procedures.

Unified Action

Unified action during multinational operations involves the synergistic application of all instruments of national and multinational power; it includes the actions of nonmilitary organizations as well as military forces. This concept is applicable at all levels of command. In a multinational environment, unified action synchronizes, coordinates, and/or integrates multinational operations with the operations of other HN and national government agencies, IGOs (e.g., UN), NGOs, and the private sector in an attempt to achieve unity of effort in the operational area (OA). When working with NATO forces, it can also be referred to as a comprehensive approach. Nations do not relinquish their national interests by participating in multinational operations. This is one of the major characteristics of operating in the multinational environment.

A. Command and Control of U.S. Forces in Multinational Operations

Although nations will often participate in multinational operations, they rarely, if ever, relinquish national command of their forces. As such, forces participating in a multinational operation will always have at least two distinct chains of command: a national chain of command and a multinational chain of command.

National Command

As Commander in Chief, the President always retains and cannot relinquish national command authority over US forces. National command includes the authority and responsibility for organizing, directing, coordinating, controlling, planning employment of, and protecting military forces. The President also has the authority to terminate US participation in multinational operations at any time.

Multinational Command

Command authority for an MNFC is normally negotiated between the participating nations and can vary from nation to nation. In making a decision regarding an appropriate command relationship for a multinational military operation, national leaders should carefully consider such factors as mission, nature of the OE, size of the proposed US force, risks involved, anticipated duration, and rules of engagement (ROE). US commanders will maintain the capability to report to higher US military authorities in addition to foreign commanders. For matters that are potentially outside the mandate of the mission to which the President has agreed or illegal under US or international law, US commanders will normally first attempt resolution with the appropriate foreign commander. If issues remain unresolved, the US commanders refer the matters to higher US authorities.

Multinational Force Commander (MNFC) is a generic term applied to a commander who exercises command authority over a military force composed of elements from two or more nations. The extent of the MNFC's command authority is determined by the participating nations or elements. This authority can vary widely and may be limited by national caveats of those nations participating in the operation. The MNFC's primary duty is to unify the efforts of the MNF toward common objectives. An operation could have numerous MNFCs.

B. Command Structures of Forces in Multinational Operations

Ref: JP 3-16, Multinational Operations (Jul '13), pp. II-4 to II-8.

No single command structure meets the needs of every multinational command but one absolute remains constant; political considerations will heavily influence the ultimate shape of the command structure. Organizational structures include the following:

Integrated Command Structure

Multinational commands organized under an integrated command structure provide unity of effort in a multinational setting. A good example of this command structure is found in the North Atlantic Treaty Organization where a strategic commander is designated from a member nation, but the strategic command staff and the commanders and staffs of subordinate commands are of multinational makeup.

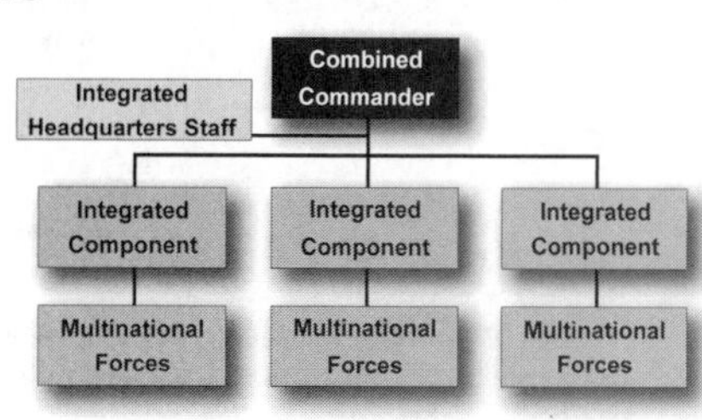

Lead Nation Command Structure

A lead nation structure exists when all member nations place their forces under the control of one nation. The lead nation command can be distinguished by a dominant lead nation command and staff arrangement with subordinate elements retaining strict national integrity. A good example of the lead nation structure is Combined Forces Command-Afghanistan wherein a US-led headquarters provides the overall military C2 over the two main subordinate commands: one predominately US forces and the other predominately Afghan forces.

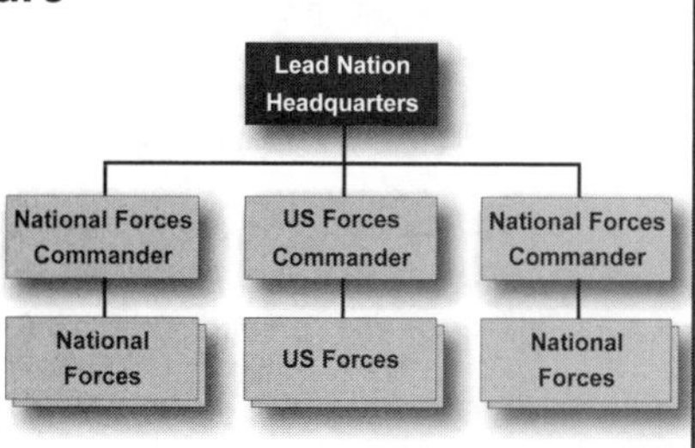

Parallel Command Structures

Under a parallel command structure, no single force commander is designated. The coalition leadership must develop a means for coordination among the participants to attain unity of effort. This can be accomplished through the use of coordination centers. Nonetheless, because of the absence of a single commander, the use of a parallel command structure should be avoided if at all possible.

Commanding & Organizing

II. Interagency Coordination

Interorganizational Cooperation

JP 3-08 describes the joint force commander's (JFC's) coordination with various external organizations that may be involved with, or operate simultaneously with, joint operations. This coordination includes the Armed Forces of the United States; United States Government (USG) departments and agencies; state, territorial, local, and tribal government agencies; foreign military forces and government agencies; international organizations; nongovernmental organizations (NGOs); and the private sector. **Interagency coordination describes the interaction between USG departments and agencies and is a subset of interorganizational cooperation.**

The Department of Defense (DOD) conducts interorganizational cooperation across a range of operations, with each type of operation involving different communities of interest, structures, and authorities. The terms "interagency" and "interorganizational" do not define structures or organizations, but rather describe processes occurring among various separate entities.

JP 3-08, Interorganizational Cooperation (Oct '16).

Interagency coordination is "the coordination that occurs between elements of the Department of Defense [DOD], and engaged US Government agencies and departments for the purpose of achieving an objective" Attaining national objectives requires the efficient and effective use of the diplomatic, informational, economic, and military instruments of national power supported by and coordinated with those of our allies and various IGOs, NGOs, and regional organizations.

> *"A large number of civilian agencies and organizations—many with indispensable practical competencies and significant legal responsibilities—interact with the Armed Forces of the United States and its multinational counterparts. Joint and multinational operations must be strategically integrated and operationally and tactically coordinated with the activities of participating USG agencies, IGOs, NGOs, host nation (HN) agencies, and the private sector to achieve common objectives. Within the context of DOD involvement, interagency coordination is the coordination that occurs between elements of DOD and engaged USG agencies for the purpose of achieving an objective. Interagency coordination forges the vital link between the US military and the other instruments of national power. Similarly, within the context of DOD involvement, interorganizational coordination is the interaction that occurs among elements of the DOD; engaged USG agencies; state, territorial, local, and tribal agencies; foreign military forces and government agencies; IGOs; NGOs; and the private sector. Successful interorganizational coordination enables the USG to build international and domestic support, conserve resources, and conduct coherent operations that more effectively and efficiently achieve common objectives."*

As with multinational operations, C2 is not as straightforward as within a US-only joint force, and unity of effort is the goal.

Refer to JFODS5: The Joint Forces Operations & Doctrine SMARTbook (Guide to Joint, Multinational & Interorganizational Operations) for further discussion. Topics and chapters include joint doctrine fundamentals (JP 1), joint operations (JP 3-0), joint planning (JP 5-0), joint logistics (JP 4-0), joint task forces (JP 3-33), information operations (JP 3-13), multinational operations (JP 3-16), interorganizational cooperation (JP 3-08), plus more!

Command & Control Overview

Ref: Annex 3-30, Command and Control (7 Nov 2014), pp. 11 to 22.

Modern military operations require flexibility in execution to adapt to a wide variety of scenarios; this drives a need to assemble the right mix of forces from the appropriate Services to tailor the operation. This need to assemble the right forces drives a corresponding need for proper organization, clearly defined command relationships, and appropriate command and control mechanisms.

Command and control (C2) and organization are inextricably linked. Forces should be organized around the principle of unity of command. Clear lines of authority, with clearly identified commanders at appropriate echelons exercising appropriate control, are essential to achieving unity of effort, reducing confusion, and maintaining priorities. To this end, commanders should be clearly identified and empowered with appropriate operational and administrative command authorities, and appropriate joint command arrangements should be clearly specified to integrate effects across Service lines. Effective joint and Service organization is "rocket science."

Air Force expeditionary organization and preferred command arrangements are designed to address unity of command. The axiom that "Airmen work for Airmen, and the senior Airman works for the joint force commander (JFC)," not only preserves the principle of unity of command, it also embodies the principle of simplicity. When Air Force forces are assigned or attached to a joint force at any level, the senior ranking Airman qualified for command should be designated as the commander, Air Force forces (COMAFFOR) and the Air Force component should normally be formed as an air expeditionary task force (AETF). As the senior Airman representing the Air Force component, the COMAFFOR provides a single Air Force face to the JFC for all Air Force matters.

Some capabilities may not be organic to the component and may be made available through a supported/supporting command relationship, or be made available through reachback or distributed C2 arrangements.

I. Key Considerations of Command and Control

Commanders should be cognizant of the authorities they are given and their relationships under that authority with superior, subordinate, and lateral force commanders. Command relationships should be clearly defined to avoid confusion in executing operations. The command of airpower requires intricate knowledge of the capabilities and interdependencies of the forces to be employed, and a keen understanding of the joint force commander's (JFC's) intent and the authorities of other component commanders.

A. Unity of Command

Unity of command is one of the principles of war. According to Air Force doctrine Volume 1, Basic Doctrine, "unity of command ensures concentration of effort for every objective under one responsible commander. This principle emphasizes that all efforts should be directed and coordinated toward a common objective."

Unity of command is not intended to promote centralized control without delegation of execution authority to subordinate commanders. Some commanders may fulfill their responsibilities by personally directing units to engage in missions or tasks. However, as the breadth of command expands to include the full spectrum of operations, commanders are normally precluded from exercising such immediate control over all operations in their area of command. Thus, C2 arrangements normally include the assignment of responsibilities and the delegation of authorities between superior and subordinate commanders. A reluctance to delegate decisions to subordinate commanders impedes operations and inhibits the subordinates' initiative. Senior commanders should provide the desired end state, desired effects, rules of engagement (ROE), and required feedback on the progress of the operation and not actually direct tactical operations.

Forces should be organized to assure unity of command in a carefully arranged hierarchy with commanders clearly delineated at appropriate echelons. Within a joint force, the COMAFFOR provides a single Airman in command of assigned and attached Air Force forces.

B. Centralized Control and Decentralized Execution

Centralized control and decentralized execution are key tenets of C2; they provide Airmen the ability to exploit the speed, flexibility, and versatility of airpower. Centralized control is defined as "in joint air operations, placing within one commander the responsibility and authority for planning, directing, and coordinating a military operation or group/category of operations" Decentralized execution is defined as "the delegation of execution authority to subordinate commanders" and other tactical-level decision makers to achieve effective span of control and to foster disciplined initiative and tactical flexibility. Airpower's unique speed, range, and ability to maneuver in three dimensions depend on centralized control and decentralized execution to achieve the desired effects.

Centralized control and decentralized execution are critical to the effective employment of airpower. Indeed, they are the fundamental organizing principles Airmen use for effective C2, having been proven over decades of experience as the most effective means of employing airpower. Because of airpower's potential to directly affect the strategic level of war and operational level of war, it should be controlled by a single Airman at the air component commander level. This Airman should maintain the broad strategic perspective necessary to balance and prioritize use of airpower resources that have been allocated to the theater. A single commander, focused on the broader aspects of an operation, can best mediate competing demands for tactical support against the strategic and operational requirements of the conflict.

Air Force doctrine Volume 1, Basic Doctrine, embodies the Air Force's commitment to the tenet of centralized control and decentralized execution of airpower.

C. Commander's Intent

Two joint C2 concepts that nurture implicit communications are commander's intent and mission-type orders. By expressing intent and direction through mission-type orders, the commander attempts to provide clear objectives and goals to enable subordinates to execute the mission.

Guidance for planning and conducting air component operations is reflected in the commander's intent. Those granted delegated authority must understand the commander's intent, which is disseminated through such products as a JFC's operation plan; a COMAFFOR's air operations plan and air operations directive; air, space, or cyberspace tasking orders produced by appropriate Air Force components; and annexes to such plans and orders that provide specific guidance for specialized functions. Unity of effort over complex operations is made possible through decentralized execution of centralized, overarching plans. Roles and responsibilities throughout the chain of command should be clearly spelled out and understood, not only to ensure proper follow-through of the original mission intent and accountability for mission completion, but also to provide continuity of operations in the event of degraded communications between echelons. Communication between commanders and those to whom authority is delegated is essential throughout all phases of the military operation.

C2 Definitions

Ref: Annex 3-30, Command and Control (7 Nov 2014), pp. 6 to 7.

Command

Command is defined as "the authority that a commander in the Armed Forces lawfully exercises over subordinates by virtue of rank or assignment." The concept of command encompasses certain powers, duties, and unique responsibilities not normally given to leaders in the public or private sector. The art of command must be exercised with care and should be awarded only to those who have demonstrated potential to selflessly lead others. Commanders are given authority and responsibility to accomplish the mission assigned. Although commanders may delegate authority to accomplish the mission, they cannot delegate the responsibility for the attainment of mission objectives.

Control

Control is defined as "authority that may be less than full command exercised by a commander over part of the activities of subordinate or other organizations." Control is the process by which commanders plan, guide, and conduct operations. The control process occurs before and during the operation. Control involves dynamic balances between commanders directing operations and allowing subordinates freedom of action. These processes require strong leaders who conduct assessment and evaluation of follow-up actions. Time and distance factors often limit the direct control of subordinates. Commanders should rely on delegation of authorities and promulgation of commander's intent as methods to control forces. The commander's intent should specify the goals, priorities, acceptable risks, and limits of the operation.

Command and Control (C2)

Command and control (C2) is defined as "the exercise of authority and direction by a properly designated commander over assigned and attached forces in the accomplishment of the mission." C2 is not unique from other military functions. It enables mission accomplishment by collaborative planning and synchronizing integrating forces and operations in time and purpose. Effective C2 enables a commander to use available forces at the right place and time. Fluid horizontal and vertical information flow enables effective C2 throughout the chain of command. This information flow, and its timely fusion, enables optimum decision-making, operationalizing the tenet of centralized control and decentralized execution so essential to effective employment of airpower. A robust and redundant C2 system provides commanders the ability to effectively employ their forces despite the fog and friction of war while simultaneously minimizing the enemy's capability to interfere with the same.

Commander

Neither Air Force nor joint doctrine includes an official definition of the general term "commander." Rather, definitions refer to a specific level or position of commander (e.g., JFC, Service component commander, joint force air component commander). For Airmen, the best official description of a commander is found in Air Force Instruction (AFI) 38-101, Air Force Organization: "an officer who occupies a position of command pursuant to orders of appointment or by assumption of command according to AFI 51-604." AFI 51-604, Appointment to and Assumption of Command, and AFI 38-101 go into the particulars regarding the various levels and types of Air Force units for which a commander may be designated, but neither provides more details about or a definition of an Air Force commander. From the available description, however, one may conclude that an Air Force commander is an Air Force officer in charge of any Air Force unit or organization. Note, however, that an Air Force commander is not the same as a commander, Air Force forces (COMAFFOR): "The title of COMAFFOR is reserved exclusively to the single Air Force commander of an Air Force Service component assigned or attached to a JFC at the unified combatant command, subordinate unified command, or joint task force (JTF) level."

D. Battle Rhythm

Battle rhythm discipline as a concept also enhances control of forces. Effective operations in a theater require the synchronization of strategic, operational, and tactical processes, to ensure mission planning, preparation, execution, and assessment are coordinated. This process is called battle rhythm or operational rhythm. It is essentially a schedule of important events that should be synchronized with the other Service or functional components and combined forces.

Battle rhythm is a deliberate daily cycle of command, staff, and unit activities intended to synchronize and pace current and future operations. Activities at each echelon must incorporate higher headquarters guidance and commander's intent, and subordinate units' requirements for mission planning, preparation, execution, and assessment. If one element of the task force is not following the battle rhythm, it can produce problems in planning and executing operations with other elements of the task force. Every command headquarters has a rhythm regulated by the flow of information and the decision cycle. The keys to capturing and maintaining control over the battle rhythm are simplicity and sensitivity to the superior commander's and the Service components' battle rhythms.

E. Trust

Trust among the commanders and staffs in a joint force expands the senior commander's options and enhances flexibility, agility, and the freedom to take the initiative when conditions warrant. Mutual trust results from honest efforts to learn about and understand the capabilities that each member brings to the joint force: demonstrated competence and planning and training together. Most trust is still built through personal relationships, which are best formed in person rather than over email, telephone, or video-teleconferences.

II. Command Authorities and Relationships

Clear and effective command relationships are central to effective operations and organizations. In order to apply the principles of war and tenets of airpower to any organization, Airmen should fully understand the terms of command and support that underpin today's organizations and operations. A working understanding of command terminology and how forces are gained or placed in support is essential to understanding the relationships among components and the responsibilities inherent in organizations.

The authority vested in a commander should be commensurate with the responsibility assigned. In other words, the commander with responsibility for a particular mission should have the authority necessary to carry out that mission. The four types of operational command relationships are—combatant command (command authority) (COCOM), operational control (OPCON), tactical control (TACON), and support. These authorities flow through joint channels, from the Secretary of Defense to the combatant commanders (CCDRs), to subordinate joint force commanders (JFCs), and to Service and/or functional component commanders. The CCDR attaches various forces to the JFC and specifies the degree of control over each force element in terms of OPCON, TACON, or support. The JFC either retains and exercises these authorities directly, or (more normally) delegates appropriate authorities to the various subordinate component commanders. Thus, a commander, Air Force forces (COMAFFOR) actually exercises only those operational authorities delegated by the JFC. As an example, if the COMAFFOR is delegated OPCON, the COMAFFOR has the authority to move forces as required to accomplish the assigned missions. However, if the JFC retains OPCON, the COMAFFOR only has the authority to move forces as directed by the JFC.

See following pages (pp. 3-6 to 3-8) for an overview and further discussion.

III. Operational and Administrative Branches of the Chain of Command

Ref: Annex 3-30, Command and Control (7 Nov 2014), pp. 11 to 12.

The President and the Secretary of Defense (SecDef) exercise authority and control of the armed forces through two distinct branches of the chain of command and control (C2). One branch runs from the President, through the SecDef to the combatant commanders (CCDRs) for missions and forces assigned to their commands. This is commonly referred to as the "operational" chain of command. The other branch, commonly referred to as the "administrative" chain of command, runs from the President, through the SecDef, to the Secretaries of the military Departments, and as prescribed by the Secretaries, to the commanders of military Service forces.

The Secretaries of the Military Departments exercise administrative control (ADCON) over Service forces through their respective Service chiefs and Service commanders. The Service chiefs, except as otherwise prescribed by law, perform their duties under the authority, direction, and control of the Secretaries of the respective military Departments to whom they are directly responsible.

Air Force Forces within the Chain of Command

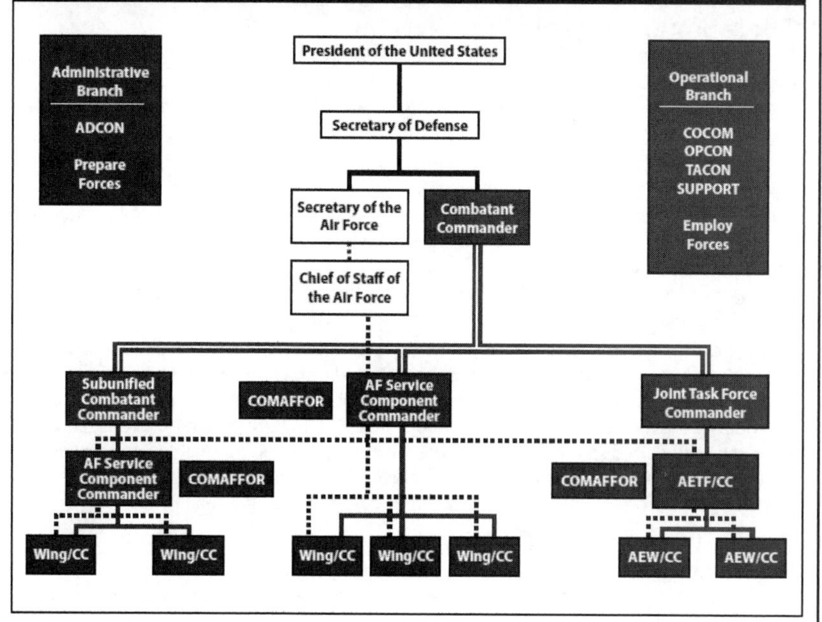

ADCON - administrative control
AETF - air and space expeditionary task force
AEW - air expeditionary wing
CC - commander
COCOM - combatant commander
COMAFFOR- commander, Air Force forces
OPCON - operational control
TACON - tactical control

Ref: Annex 3-30 (7 Nov '14), p. 12, fig: Air Force Forces within the Chain of Command.

Command Authorities and Relationships

Ref: Annex 3-30, Command and Control (7 Nov 2014), app. A.

Clear and effective command relationships are central to effective operations and organizations. A working understanding of command terminology is essential to understanding the relationships among components and the responsibilities inherent in organizations.

Combatant Command (COCOM)

Combatant command (command authority) (COCOM) is defined as "nontransferable command authority, which cannot be delegated, of a combatant commander to perform those functions of command over assigned forces involving organizing and employing commands and forces; assigning tasks; designating objectives; and giving authoritative direction over all aspects of military operations, joint training, and logistics necessary to accomplish the missions assigned to the command."

COCOM is exercised by commanders of combatant commands (CCMDs) as directed by the President or the Secretary of Defense. COCOM should be exercised through the commanders of subordinate organizations such as subordinate joint force commanders and Service and/or functional component commanders. COCOM provides full authority to organize and employ commands and forces as the combatant commander (CCDR) considers necessary to accomplish assigned missions. (Note that the acronym "COCOM" refers only to the command authority, not to an individual or an organization.)

Operational Control (OPCON)

Operational control (OPCON) is defined as "the authority to perform those functions of command over subordinate forces involving organizing and employing commands and forces, assigning tasks, designating objectives, and giving authoritative direction necessary to accomplish the mission."

OPCON is able to be delegated from a lesser authority than combatant command (command authority) (COCOM). OPCON normally provides full authority to organize commands and forces and to employ those forces as the commander in operational control considers necessary to accomplish assigned missions. It does not include authoritative direction for logistics or matters of administration, discipline, internal organization, or unit training. For example, OPCON does not include the authority to change the Service's internal organization of its forces.

Component forces (e.g., the air expeditionary task force and its subordinate mix of expeditionary wings, groups, or squadrons) "should remain organized as designed and in the manner accustomed through training to maximize effectiveness." (Joint Publication 1, Doctrine for the Armed Forces of the United States). OPCON should be exercised through the commanders of subordinate organizations, such as subordinate JFCs and Service and/or functional component commanders. Normally, JFCs exercise OPCON of assigned and attached Air Force forces through the commander, Air Force forces.

Tactical Control (TACON)

Tactical control (TACON) is defined as "the authority over forces that is limited to the detailed direction and control of movements or maneuvers within the operational area necessary to accomplish missions or tasks assigned."

TACON is able to be delegated from a lesser authority than operational control (OPCON) and may be delegated to and exercised by commanders at any echelon at or below the level of combatant command (CCMD). TACON provides sufficient authority for controlling and directing the application of force or tactical use of combat support assets within the assigned mission or task. TACON does not provide organizational authority or authoritative direction for administrative and logistic support.

An example of TACON is when the commander, Air Force forces (COMAFFOR), acting as the joint force air component commander (JFACC), produces an air tasking order that provides detailed instructions for joint air assets made available for tasking. For example, a JFACC functioning as the area air defense commander (AADC) with TACON over Army PATRIOT surface-to-air missile forces would have the authority to specify which asset/battery would be responsible for providing which portion of the air defense coverage for the joint force (exact placement of the assets/battery necessary to achieve the required coverage should normally be left to the Army component commander). The commander exercising TACON is responsible for ensuring communications with the controlled unit.

Support

Support is a command authority that aids, protects, complements, or sustains another force. It is used when neither operational control (OPCON) nor tactical control (TACON) is appropriate. The Secretary of Defense (SecDef) specifies support relationships between combatant commanders (CCDRs); the CCDR may establish support relationships between components assigned or attached to the command.

Over several years of experience, the most common example of this between CCDRs is seen when a functional CCDR (e.g., Commander, USTRANSCOM) is established by the SecDef as a supporting commander and a geographic CCDR (e.g., Commander, USCENTCOM) is established as the supported commander. Within a combatant command, the best example is the last several years of experience within USCENTCOM, in which the commander, Air Force forces (COMAFFOR) (Commander, USAFCENT) is the supporting commander with the joint force commanders in Operations IRAQI FREEDOM (redesignated Operation NEW DAWN) and ENDURING FREEDOM designated by Commander, USCENTCOM as supported commanders.

The supported commander should ensure that the supporting commanders understand the assistance required. The supporting commanders should then provide the assistance needed, subject to a supporting commander's existing capabilities and other assigned tasks. When a supporting commander cannot fulfill the needs of the supported commander, the establishing authority should be notified by either the supported commander or a supporting commander. The establishing authority is responsible for determining a solution.

An establishing directive is normally issued to specify the purpose of the support relationship, the effect desired, and the scope of the action to be taken. It also should include: the forces and resources allocated to the supporting effort; the time, place, level, and duration of the supporting effort; the relative priority of the supporting effort; the authority, if any, of the supporting commander to modify the supporting effort in the event of exceptional opportunity or an emergency; and the degree of authority granted to the supported commander over the supporting effort.

There are four defined categories of support that a CCDR may direct over assigned or attached forces to ensure the appropriate level of support is provided to accomplish mission objectives. These include:

- **General support**. That support which is given to the supported force as a whole rather than to a particular subdivision thereof.
- **Mutual support**. That support which units render each other against an enemy because of their assigned tasks, their position relative to each other and to the enemy, and their inherent capabilities.
- **Direct support**. A mission requiring a force to support another specific force and authorizing it to answer directly to the supported force's request for assistance.
- **Close support**. That action of the supporting force against targets or objectives that are sufficiently near the supported force as to require detailed integration or coordination of the supporting action with the fire, movement, or other actions of the supported force.

A supported relationship does not include authority to position supporting units but does include authority to direct missions or objectives for those units.

Continued on next page

Continued on next page

Continued from previous page

Administrative Control (ADCON)

Administrative control (ADCON) is defined as the "direction or exercise of authority over subordinate or other organizations with respect to administration and support." This includes organization of Service forces, control of resources and equipment, personnel management, unit logistics, individual and unit training, readiness, mobilization, demobilization, discipline, and other matters not included in the operational missions of the subordinate or other organizations.

ADCON is not a warfighting authority like that found in combatant command (command authority), operational control, tactical control, or support relationships. Normally the commander, Air Force forces (COMAFFOR) exercises ADCON over assigned Air Force personnel, and at least those elements of ADCON that are necessary to ensure mission accomplishment over those Air Force personnel attached to the Air Force component command. G-series orders implement Service ADCON authority by detailing those aspects of support that are necessary for the mission, and the relationship the gaining organization possesses over assigned or attached units and personnel. For example, the authority to exercise ADCON could include such elements as building a tent city, ordering supplies and equipment, authorizing training sorties, conducting exercises, working assignment actions for personnel, developing budget requests, protecting personnel, and recommending awards and decorations.

Uniform Code of Military Justice (UCMJ) authority is inherent in command authority, and is distinct from ADCON. However, G-series orders implementing ADCON may incorporate references to UCMJ authority. In specific contingency operations, the G-series order may retain one or more of these authorities in the parent unit. For attached forces, those elements of ADCON that are not specified to be gained by the COMAFFOR to whom the forces are attached, are retained by the parent Service organization to whom the Air Force forces are permanently assigned.

Coordinating Authority

Coordinating authority is defined as "the commander or individual who has the authority to require consultation between the specific functions or activities involving forces of two or more Services, joint force components, or forces of the same Service or agencies, but does not have the authority to compel agreement." In the event that essential agreement cannot be obtained, the matter shall be referred to the appointing authority. Coordinating authority is a consultation relationship, not an authority through which command may be exercised.

Coordinating authority may be exercised by commanders or individuals at any echelon at or below the level of combatant command. Coordinating authority may be granted and modified through a memorandum of agreement to provide unity of effort for operations involving Reserve component and regular component forces engaged in interagency activities. The common task to be coordinated should be specified in the establishing directive without disturbing the normal organizational relationships in other matters.

Coordinating authority is more applicable to planning and similar activities than to operations. Coordinating authority is not in any way tied to force assignment. Assignment of coordinating authority is based on the missions and capabilities of the commands or organizations involved.

Continued from previous page

Direct Liaison Authorized (DIRLAUTH)

Direct liaison authorized (DIRLAUTH) is defined as "that authority granted by a commander (any level) to a subordinate to directly consult or coordinate an action with a command or agency within or outside of the granting command."

DIRLAUTH is more applicable to planning than operations and always carries with it the requirement of keeping the commander granting DIRLAUTH informed. DIRLAUTH is a coordination relationship, not an authority through which command may be exercised. DIRLAUTH is most appropriately used to streamline communications and operations between tactical elements without relinquishing command by the higher authority.

Chap 3

I. Air Operations Center (AOC)

Ref: Annex 3-30, Command and Control (7 Nov 2014), app. B and AFI 13-1AOCV3, Operational Procedures—Air Operations Center (2 Nov 11), w/Chg 1.

The AOC provides operational-level C2 of air component forces as the focal point for planning, executing, and assessing air component operations. The AOC can be tailored and scaled to a specific or changing mission and to the associated task force the COMAFFOR presents to the JFC. Thus, for smaller scale operations, the Air Force may not necessarily provide all of the elements described in the following sections if the situation does not warrant them.

I. AOC Primary Functions

The primary functions of the AOC are to:

- Develop air component operations strategy and planning documents that integrate air, space, and cyberspace operations to meet COMAFFOR objectives and guidance.
- Task, execute, and assess day-to-day air component operations; provide rapid reaction, positive control, and coordinate and deconflict weapons employment as well as integrate the total air component effort.
- Receive, assemble, analyze, filter, and disseminate all-source intelligence and weather information to support air component operations planning, execution, and assessment.
- Integrate space capabilities and coordinate space activities for the COMAFFOR when the COMAFFOR is designated as space coordinating authority.
- Issue airspace control procedures and coordinate airspace control activities for the airspace control authority (ACA) when the COMAFFOR is designated the ACA.
- Provide overall direction of air defense, including theater missile defense (TMD), for the area air defense commander (AADC) when the COMAFFOR is designated the AADC.
- Plan, task, and execute the theater air- and space-borne intelligence, surveillance, and reconnaissance (ISR) mission.
- Conduct component-level assessment to determine mission and overall air component operations effectiveness as required by the JFC to support the theater assessment effort.
- Plan and task air mobility operations according to the theater priorities.

The Air Operations Center (AOC) provides operational-level C2 of air, space, and cyberspace operations. It is the focal point for planning, directing, and assessing air, space, and cyberspace operations to meet JFACC operational objectives and guidance. The regional scope of Geographic AOCs and disparate, global scope of Functional AOCs, require AOCs to be tailored to efficiently and effectively plan and execute their steady-state missions. Although the USAF provides the core manpower for the AOC, other service components provide personnel in support of exercises and contingency operations. The AOC coordinates closely with superior and subordinate C2 nodes, as well as the headquarters of other functional and service component commands to integrate the numerous aspects of air, space, and cyberspace operations and accomplish its mission.

II. AOC Organization & Functional Teams

Ref: AFI 13-1AOCV3, Operational Procedures—Air Operations Center (w/Chg 1, 18 May 12), pp. 12 to 16 and Annex 3-30, Command and Control (7 Nov 2014), app. B

The baseline AOC organization includes an AOC commander, five divisions (strategy, combat plans, combat operations, ISR, and air mobility), and multiple support/specialty teams. Each integrates numerous disciplines in a cross-functional team approach to planning and execution. Liaisons from other Service and functional components may be present to represent the full range of joint air, space, and cyberspace capabilities made available to the COMAFFOR. The following provides a summary of the major elements of an AOC.

Note: The AOC is an AF unit. The AOC Commander is responsible for the day-to-day readiness of the AOC. When the AOC is employed in contingency operations, the AOC should be prepared to transition, with appropriate joint augmentation, to a JAOC and the AOC commander should be prepared to serve as the JAOC Commander.

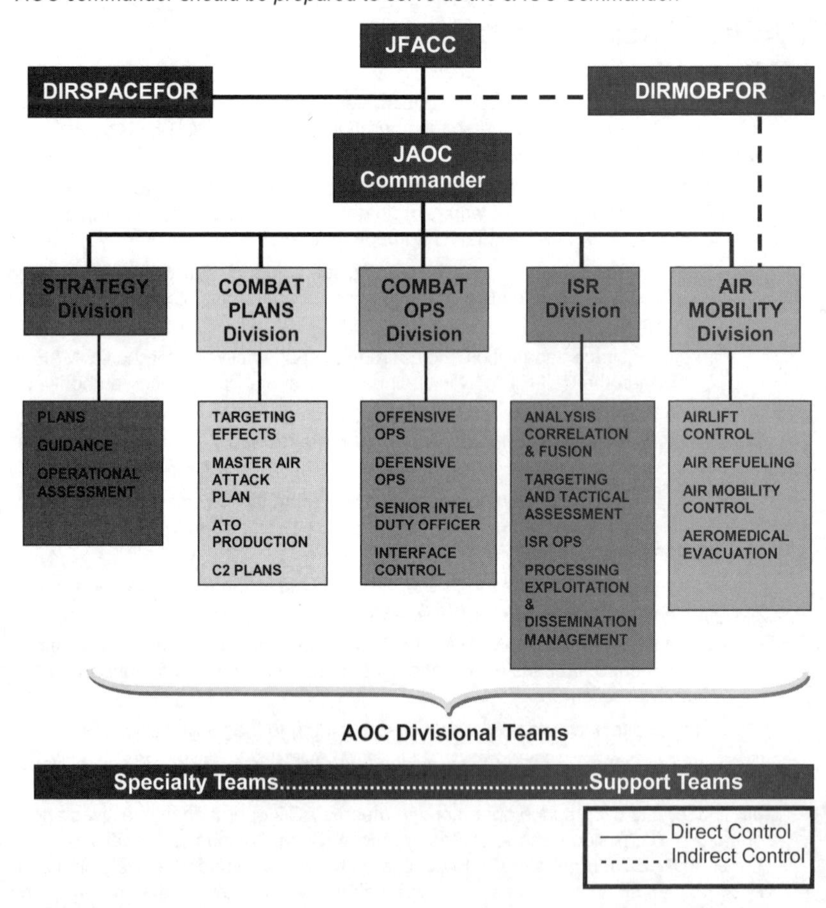

Ref: AFI 13-1AOCV3 (w/Chg 1, 18 May 12), fig. 2.2. JAOC Organization and Functional Teams.

See pp. 3-14 to 3-15 for a listing and discussion of joint liaisons in the AOC.

1. The AOC Commander

The AOC commander is charged with effectively managing air component operations and establishing the AOC battle rhythm. The AOC commander develops and directs processes to plan, coordinate, allocate, task, execute, and assess air component operations in the area of operations/joint operations area based on JFC and COMAFFOR guidance. The AOC commander commands the AOC weapons system (but not air expeditionary task force forces) and should be prepared to direct a joint AOC (JAOC) when the COMAFFOR is designated as the joint force air component commander.

2. The Strategy Division

The strategy division concentrates on long-range planning of air component operations to achieve theater objectives by developing, refining, disseminating, and assessing progress toward achieving the COMAFFOR component strategy. The strategy division is normally task organized into four functionally oriented core teams: the strategy plans team, the strategy guidance team, the operational assessment team, and the information operations team. Key products include the joint air operations plan, the air operations directive (AOD), and other COMAFFOR guidance.

3. The Combat Plans Division

The combat plans division applies operational art to develop detailed execution plans for air component operations. The combat plans division is normally task organized into four functionally oriented core teams: the targeting effects team; the master air attack plan (MAAP) team; the air tasking order (ATO) production team; and the C2 planning team. The division's key products are an area air defense plan, airspace control plan, and a daily ATO, airspace control order (ACO), special instructions, and joint integrated prioritized target list.

4. The Combat Operations Division

The combat operations division monitors and executes current operations. The combat operations division is also the focal point for monitoring the execution of joint and combined operations, such as time-sensitive targeting, TMD, joint suppression of enemy air defense supported by theater forces, and joint air attack team. The combat operations division is normally task-organized into four functionally oriented core teams: offensive operations, defensive operations, senior intelligence duty officer team, and interface control team. The division's main products are daily ATO/ACO changes, the airspace control plan, and air defense plan.

5. The ISR Division

The ISR division, in conjunction with the other AOC divisions, plans and executes airborne ISR operations and provides combat ISR support to air component planning, execution, and assessment activities. The ISR division has four core teams: the analysis, correlation and fusion team; the targets/tactical assessment team; the ISR operations team; and the processing, exploitation, and dissemination management team. Major products include: the reconnaissance, surveillance, and target acquisition annex to the ATO (or the ISR collection plan); updated intelligence preparation of the operational environment; air component target nomination list; and intelligence summaries.

6. The Air Mobility Division

The air mobility division (AMD) plans, coordinates, tasks, and executes the theater air mobility mission. Unlike the other AOC divisions that work solely for the AOC commander, the AMD coordinates with the director, air mobility forces (DIRMOBFOR) but remains responsive to the tempo and timing of the AOC commander's operation. The DIRMOBFOR is responsible for integrating the total air mobility effort for the COMAFFOR and, in this capacity, coordinates with the AMD on behalf of the COMAFFOR to execute the air mobility mission. The AMD coordinates with the theater deployment distribution operations center and the 618 AOC. The AMD is comprised of four core teams: the airlift control team, the air refueling control team, the air mobility control team, and the aeromedical evacuation control team. Major products include airlift apportionment plans and air refueling inputs to the MAAP, ATO, ACO, and special instructions.

III. JFACC Responsibilities (through JAOC)

Ref: AFI 13-1AOCV3, Operational Procedures—Air Operations Center (w/Chg 1, 18 May 12), pp. 6 to 8.

When conducting combined and joint air, space, and cyberspace operations, the JFC normally designates a JFACC to exercise tactical control (TACON) over air capabilities and forces in accordance with JP 3-30, Command and Control for Joint Air Operations. The JFACC responsibilities, executed through the JAOC, include, but are not limited to:

- Development of the Joint Air Operations Plan (JAOP).
- Apportionment recommendation for the joint air effort, in consultation with other component commanders.
- Providing centralized direction for the allocation and tasking of capabilities and forces made available.
- Providing control, oversight, and guidance during the execution of joint air, space, and cyberspace operations.
- Coordinating and integrating joint air operations with the operations of other component commanders and forces.
- Assessing results of joint air operations.
- Functioning as supported and supporting commander as directed.

JFACC Additional Responsibilities (when supported by a JAOC)

Normally, the JFC assigns additional responsibilities associated with air operations to the JFACC. Additionally, the JFACC may provide specialized support to other components. When delegated these responsibilities, the JFACC becomes the supported commander for these theater functions.

Area Air Defense Commander (AADC)

Since airspace operations, and defense of that airspace and everything below it, are inherently linked and are an integral part of joint air operations, the JFC normally designates the JFACC as the AADC. The AADC is assigned overall responsibility for air and missile defense of the Joint Operations Area (JOA). As the AADC, the JFACC employs the JAOC to coordinate with other components and develop the Area Air Defense Plan (AADP) for JFC approval. Once approved, the JAOC plans, coordinates, and manages air defense and missile operations.

Airspace Control Authority (ACA)

The JFC normally designates the JFACC as the ACA. The ACA has overall responsibility for controlling the airspace in the JOA and operation of the Airspace Control System in the Airspace Coordination Area. The ACA coordinates, develops, and issues the Airspace Control Plan (ACP) and Airspace Control Orders (ACO), which provide guidance and procedures for use and control of airspace activities.

Collection Management Authority (CMA)

CMA is the authority to establish, prioritize and validate theater collection requirements, establish sensor tasking guidance and develop theater collection plans. IAW JP 2.01, Joint and National Intelligence Support to Military Operations, Collection operation management (COM) is the authoritative direction, scheduling, and control of specific collection operations and associated processing, exploitation, and dissemination resources. COM is the inherent responsibility of the JFACC when delegated the responsibility for theater airborne intelligence, surveillance, and reconnaissance (ISR) operations. The JFACC will also likely exercise Collection Requirements Management (CRM) of assigned units and receive collection requirements from other CRM authorities. IAW JP 2.01, Joint and National Intelligence Support to Military Operations, CRM is the authoritative development and control of collection processing, exploitation, and/or reporting requirements that normally result in either the direct tasking of assets over which the collection manager has authority, or the generation of tasking requests to CMA at higher, lower, or lateral echelon to accomplish the collection mission.

Space Coordinating Authority (SCA)

The SCA is responsible for coordinating and integrating space capabilities and is responsible for joint space operations planning. In a joint force, the SCA, normally supported by assigned/attached embedded space personnel, serves as the focal point for gathering space requirements in support of the JFC's campaign. To ensure prompt and timely support, the supported Geographic Combatant Commander (GCC) and Commander, USSTRATCOM may authorize direct liaison between the SCA and applicable components of US Strategic Command. These requirements include requests for space services and capabilities to achieve space effects. The SCA develops a recommended prioritized list of space requirements for the JTF based on JFC objectives.

Jamming Control Authority (JCA)

IAW JP 3-13.1, Electronic Warfare, the JCA is the commander designated to assume overall responsibility for the operation of Electronic Attack (EA) assets in the Area of Responsibility (AOR)/JOA. This responsibility assumes the JCA can access the Joint Communications-Electronics Operating Instructions (JCEOI) and Joint Restricted Frequency List (JRFL), can analyze immediate jamming requests for frequency interference issues (to include harmonic interference), and can ensure positive C2 of jamming assets in order to start/stop jamming activity.

Supported Commander for Personnel Recovery (PR)

IAW JP 3-50, Personnel Recovery, the supported commander for PR is the joint force component commander designated by the JFC with the overall authority to plan, coordinate, and conduct joint PR operations and activities within the JFC's operational area. If the JFC designates the JFACC as the supported commander for PR, the Joint Personnel Recovery Center (JPRC) should be integrated into the JFACC's JAOC. The JFACC must also retain a Personnel Recovery Coordination Cell (PRCC) capability to plan and conduct combat search and rescue (CSAR) missions in support of JFACC operations.

Joint Air Component Coordination Element (JACCE)

The JFACC may establish and deploy a JACCE to each land, maritime, and special operations component commanders' headquarters to better integrate air, space, and cyberspace operations with surface/subsurface operations. The JFACC may establish and deploy a JACCE to the JFC headquarters to better integrate air, space, and cyberspace operations within the overall joint or combined force. The JACCE will be sourced from the C-NAF AOC or AFFOR staff, augmentation units or AEF Center/AFPC-identified personnel as well as SMEs from each of the sister services as needed. If possible, the C-NAF commander should establish a working relationship with JACCE personnel before deployment and execution. The JACCE director acts as the air component commander's personal liaison and primary representative to the other commanders in the operation. The JACCE team facilitates interaction and communication between the respective staffs. The JACCE performs a liaison function and is responsible for understanding (and participating in, if possible) the JFC's/ JFACC's initial planning and for understanding the other commanders' plans. The JACCE team works with their respective counterparts in the AOC and AFFOR staff (JACCE reachback) to provide the other HQs commander information on the best way to employ air and space power. This is a two-way relationship in that the JACCE not only provides information flow to the JFACC but must also help ensure JFACC information is flowing to and understood by the JFC, JFLCC, JFMCC, and/or JFSOCC as applicable. JACCEs will participate in training events and exercises with appropriate headquarters organizations to maintain mission readiness.

IV. Joint Liaisons in the AOC

Ref: AFI 13-1AOCV3, Operational Procedures—Air Operations Center (w/Chg 1, 18 May 12), pp. 81 to 83.

The specialty/support functions provide the AOC with diverse capabilities to help orchestrate theater air, space, and cyberspace operations power. Many of these capabilities are provided to the AOC from agencies external to the AOC organization. It is crucial to the success of the AOC that these capabilities are integrated into the air, space, cyberspace, and IO planning and execution process to ensure the best use of available assets. The AFFOR staff coordinates regularly with the AOC and often provides specialized expertise. Specialty/support functions are listed in the following paragraphs.

Component Liaisons

Component liaisons work for their respective component commanders and with the JFACC and staff. Each component normally provides liaison elements (e.g., BCD, SOLE, NALE, MARLE, etc.) that work within the AOC. These liaison elements consist of experienced warfare specialists who provide component planning and tasking expertise and coordination capabilities. They help integrate and coordinate their component's participation in joint air, space, and cyberspace operations. The USAF component may require other liaison augmentation to support AOC functions such as Coast Guard, space forces, DIA, NSA, CIA, USAF Intelligence, Surveillance, and Reconnaissance Agency (AFISRA), National Reconnaissance Office (NRO), and FAA in various operational and support areas.

Refer to JP 3-30, Command and Control for Joint Operations, for additional discussion.

Battlefield Coordination Detachment (BCD)

The BCD supports the integration of air, space, and cyberspace operations with ground maneuver. BCD personnel are integrated into AOC divisions to support planning, operations, air defense, intelligence, airlift/logistics, airspace control, and communications. In particular, the BCD coordinates ground force priorities, requests, and items of interest. One of the BCD's most important functions is to coordinate boundary line and fire support coordination line (FSCL) changes and timing. The BCD brings ground order of battle (GOB) (friendly and enemy) situational awareness and expertise into the AOC and will normally brief the ground situation/intelligence update. The BCD may also provide current ground situation inputs to AOC teams for incorporation into daily briefings and intelligence summaries.

Air and Missile Defense Commander (AAMDC)

The AAMDC is normally under the OPCON of the ARFOR commander or joint forces land component commander (JFLCC). When directed by the JFC, AAMDC assets may be placed in direct support of the JFACC/AADC as appropriate. The roles of the commanding general of the AAMDC are Senior Army ADA commander, theater Army air and missile defense coordinator (TAAMDCOORD), and deputy area air defense commander (DAADC). Coordination and liaison functions between all three are essential to effective air and missile defense operations within a given theater. The AAMDC and AOC intelligence personnel build a collaborative TAMD IPB, which serves as the basis for JTAMD strategies and plans. The AAMDC (attack operations section in coordination with the intelligence section) submits TM target nominations directly to the AOC for inclusion as JFACC nominated targets. The AAMDC also sends a robust LNO team (active defense, intelligence, and attack operations personnel) to support the JFACC, AADC, and DAADC requirements and may deploy the AAMDC TOC (Main) to the JFACC, AADC location. As the senior Army air defense element at the AADC's location, the AAMDC LNO team serves as the primary interface at the AOC for all land-based active air defense.

Naval and Amphibious Liaison Element (NALE)

NALE personnel from the maritime components support the JAOC in integrating naval air, naval fires, and amphibious operations into theater air operations and monitor and interpret the maritime battle situation for the AOC.

Marine Liaison Element (MARLE)

The MARLE represents the Commander, Marine Corps Forces (COMMARFOR) and his associated Aviation Combat Element Commander. The MARLEs will support the JFACC in integrating Marine Air-Ground Task Force (MAGTF) fires, maneuver, and Marine air into the theater campaign and supporting JAOP. This team will be well versed in the MAGTF Commander's guidance, intentions, schemes of maneuver, and direct support aviation plan.

Special Operations Liaison Element (SOLE)

The Joint Forces Special Operations Component Commander (JFSOCC) provides a SOLE to the JFACC to coordinate and integrate SOF activities in the entire battlespace. This joint SOLE is comprised of representatives from SOF aviation, intelligence, airspace, logistics, Air Force STTs, Army Special Forces, Navy Sea-Air Land Teams (SEAL) and Marine Special Operations Forces, as required. Depending upon command structure agreements, the SOLE may or may not represent coalition or allied SOF. SOLE personnel coordinate, integrate and synchronize with various AOC functional areas to ensure that all SOF targets, SOF teams, and SOF air tasks and/or missions are deconflicted, properly integrated, and coordinated during planning and execution phases. The prevention of fratricide is a critical product of the SOLE's efforts.

Specific SOLE functions include, but are not limited to, inputs into the JFACC strategy development; inputs into the ATO development; inputs into the ACO development; real-time mission support coordination with the Joint Special Operations Air Component Commander (JSOACC) with special emphasis on airspace deconfliction; operational and intelligence inputs into the targeting process; and close coordination with the RCC/JPRC.

As the JFSOCC and the JFACC share a common environment throughout the entire battlespace, it is imperative that SOF aviation and surface forces are integrated into joint air, space, and cyberspace operations planning and execution to prevent fratricide, duplication of effort, and conflict. Active SOLE participation in the development of air, space, and cyberspace operations strategy and the supporting plans to the theater campaign plan ensures that SOF efforts will, in fact, be a force multiplier for the theater campaign plan.

SOF normally pursues SOF-unique objectives, which prepare, shape or enhance broader JFC objectives, they may be tasked to operate in support of conventional objectives or require conventional support of their objectives.

Additionally, SOLE has the following responsibilities: provide inputs and guidance to the IO team; act as the focal point for raising JFACC concerns or MISO objective/tasking to the JFC for consideration, planning, and execution; provide support to the IO team MISO effort to synchronize and deconflict MISO into the air, space, and cyberspace operations campaign (e.g., leaflet drops, message broadcasts, and aircraft mission are included into the ATO, etc.).

Coalition/Allied Liaison Officers

LNOs representing coalition/allied surface forces may improve AOC situational awareness regarding the disposition of friendly forces, especially when those forces do not have a mature TACS. They are also essential for unity of effort for coalition air defense operations and airspace deconfliction. When teamed with linguists, they can help overcome language barriers with remote allied/coalition forces. In force projection scenarios into an immature theater, the AOC Commander must anticipate the need for LNOs and actively seek them out via the JFC staff, in-country military group, staff country team, or direct contact with coalition forces, if necessary.

Actions at the joint force level establish the requirements for the Theater Air Ground System (TAGS), to include the CCDR's guidance, perspective, and strategy for the AOR (the JFC's JOA strategy [if the JFC is not the CCDR], command organization, and relationships; the campaign plan; assignment of objectives; and apportionment of forces). Personnel assigned to, or working with, the TAGS must understand the decision processes and problems associated with the operational and tactical levels of command. Armed with this knowledge, commanders and staffs will better understand TAGS functions and how to work within the system to receive or give support. The AOC is the senior element of the Theater Air Control System (TACS) which, along with the Army Air to Ground Systems (AAGS), Marine Air Command and Control System (MACCS), Navy Tactical Air Control system (NTACS), and the Special Operations Air Ground System (SOAGS) comprise the TAGS. The TACS is composed of airborne and ground-based C2 elements. Airborne elements of the theater air control system (AETACS) are the Airborne Warning and Control System (AWACS) and the Joint Surveillance Target Attack Radar System (JSTARS). The ground elements are the AOC, Control and Reporting Center (CRC), Air Support Operations Center (ASOC), and Tactical Air Control Party (TACP). To effectively integrate the TACS elements, the AOC develops and establishes theater-wide C2 guidance of regular and irregular warfare (IW), providing overarching direction to all the TACS elements.

Based on the tenet of centralized planning and control, and decentralized execution, the AOC enables the JFACC to exercise operational-level C2 of air and space forces. When multinational operations are involved, the JFACC becomes a combined force air and space component commander (CFACC).

V. Specialty/Support Functions

The specialty/support functions provide the AOC with diverse capabilities to help orchestrate theater air, space, and cyberspace operations power. Many of these capabilities are provided to the AOC from agencies external to the AOC organization. It is crucial to the success of the AOC that these capabilities are integrated into the air, space, cyberspace, and IO planning and execution process to ensure the best use of available assets. The AFFOR staff coordinates regularly with the AOC and often provides specialized expertise.

- Joint Liaisons in the AOC. *See previous pages (pp. 3-14 to 3-15).*
- Combat Reports Cell
- Common Tactical Picture (CTP) Management Cell
- Airspace Management Team
- Space Operations Specialty Team (SOST)
- Information Operations Team (IOT)
- Judge Advocate (JA)
- Wx Specialty Team (WST)
- Combat Support Team (CST)
- Knowledge Operations (KO) Team
- Personnel Rescue Coordination Cell (PRCC)
- Special Technical Operations (STO) Team
- Regional Air Movements Control Center (RAMCC)
- Combat Information Cell (CIC)
- Director of Space Forces (DIRSPACEFOR)
- AOC Communications Team (ACT)
- ACT Plans and Programs Function
- Cyber Operations Liaison Element (COLE)

II. (AFFOR) Air Force Forces Staff

Ref: Annex 3-30, Command and Control (7 Nov 2014), app. C and AFI 13-103, AFFOR Staff Operations, Readiness and Structures (19 Aug '14).

I. Air Force Forces (AFFOR) Staff

An Air Force forces (AFFOR) staff (sometimes also called an A-Staff) supports the commander, Air Force forces (COMAFFOR) at the combatant command, subordinate unified command, or joint task force level.

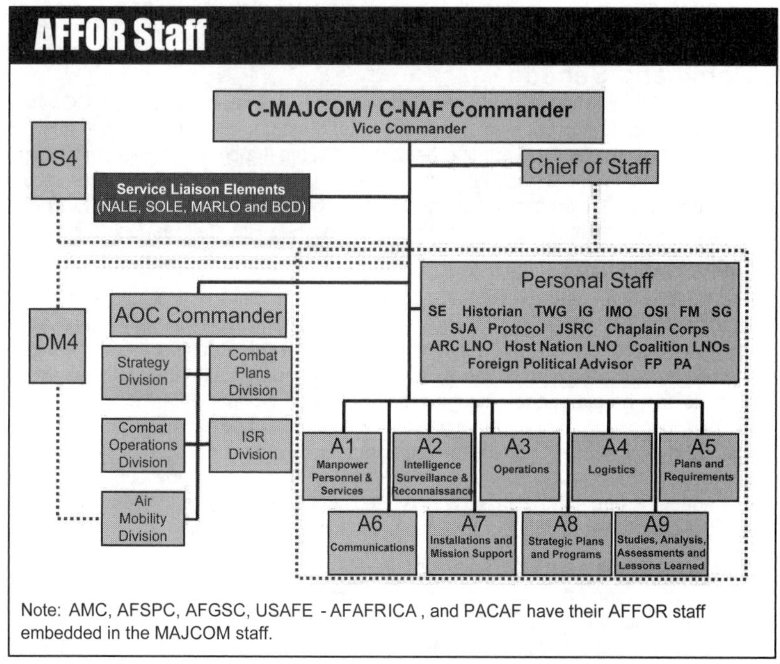

Ref: AFI 13-103, AFFOR Staff Operations, Readiness and Structures (19 Aug '14), fig. 1. C-MAJCOM/C-NAF Headquarters Organization.

The AFFOR staff is the vehicle through which the COMAFFOR fulfills operational and administrative responsibilities for assigned and attached forces across the range of military operations, from steady-state operations in the engagement phase through major operations and campaigns. In the steady state, the AFFOR staff performs administrative responsibilities (organize, train, and equip), and also plans, executes, and assesses operations in support of the CCDR's theater campaign strategies and plans. The AFFOR staff is also responsible for the operational planning that occurs outside the air tasking cycle (e.g., deliberate planning). The AFFOR staff consists of functionally oriented directorates, a command section, a personal staff, and any required liaisons. The AFFOR staff issues mission type orders on behalf of the COMAFFOR to direct subordinate units to execute actions outside of the scope of the air tasking order (ATO). Examples of such orders may include setting a baseline force protection condition, directing the move of a unit to another operating base, and overseeing the execution of steady-state or security cooperation operations.

Command & Control

II. AFFOR Organization & Staff Directorates

Ref: Annex 3-30, Command and Control (7 Nov 2014), app. C.

The differing mission requirements of any given operation may dictate different task emphasis and staff arrangements. Very large or complex operations, for example, may require all staff directorates. In some cases, senior component liaison elements may not be needed; in other cases, some of the required support may be obtained through reachback. For very small or limited operations, a full AFFOR staff may not be required. As a rule of thumb, the size and span of the AFFOR staff should normally be held to the smallest number of divisions necessary to handle the demands of the operation; in some cases, the COMAFFOR may combine some leadership positions (e.g., A-3/5; A-4/7). Other operations may employ an AFFOR staff split into forward and rear elements, using reachback to maintain unity of effort. In each case, based upon regional requirements, the COMAFFOR determines the size, shape, and location of the AFFOR staff and air operations center (AOC) to best support the operation.

Command Section

The command section is normally composed of the commander (i.e., the COMAFFOR), vice commander, chief of staff, command chief master sergeant, executive assistant, and appropriate administrative support personnel. Within the command section, the chief of staff coordinates and directs the daily activities of the AFFOR staff; approves actions, orders, and plans, as authorized by the COMAFFOR; and ensures COMAFFOR decisions and concepts are implemented by directing and assigning staff responsibilities.

Personal Staff

The COMAFFOR has several staff activities that normally function outside the AFFOR staff directorates. These activities fulfill specific responsibilities usually related to providing close, personal advice or services to the commander, or assist the commander and the component staff with technical, administrative, or tactical matters. These activities may include the commander's legal advisor, political advisor (POLAD), public affairs advisor, inspector general, protocol advisor, historian, chaplain, counterintelligence and special investigations, financial management, force protection, air mobility operations (DIRMOBFOR), space operations (DIRSPACEFOR), medical, knowledge operations management, and safety. Based on the needs of the operation and the requirements of the AFFOR staff, some of these activities may be located within the AFFOR staff directorates.

Senior Component Liaisons

The senior liaison officer (LNO) from each component represents his or her respective commander to the COMAFFOR. Subordinate LNOs from each component may perform duties throughout the staff as required, providing weapon system expertise. LNOs should be knowledgeable of the capabilities and limitations of their units and Service.

Manpower and Personnel (A–1)

The director of manpower, personnel, and services is the principal staff assistant to the COMAFFOR for total force accountability, personnel policy and procedures, the establishment and documentation of manpower requirements, organizational structures, mortuary affairs, food and force beddown operations, the coordination of exchange services, and the provision of quality of life programs to enable and sustain forces assigned and attached to the COMAFFOR.

Intelligence (A–2)

The director of intelligence, surveillance, and reconnaissance (ISR) is the principal staff assistant to the COMAFFOR for policy and guidance for all Air Force ISR operational architectures, personnel, systems, and training. The A–2 provides intelligence support to forces within the assigned area of operations. The A–2 does not normally direct ISR collection assets when an ISR division is resident in the AOC; this is normally directed by ISR division chief.

Operations (A–3)

The director of operations serves as the principal staff assistant to the COMAFFOR in the direction and control of all assigned and attached Air Force forces. When operational control (OPCON) of Air Force units is formally transferred to the COMAFFOR, the A-3 ensures they are capable of performing tasked missions. This includes monitoring unit deployments and bed-down locations, combat readiness, mission rehearsals, force protection, and training activities. The A-3 is the focal point for executing component operations outside the purview of the AOC.

Logistics (A–4)

The director of logistics is the principal staff assistant to the COMAFFOR for logistics and sustainment support of assigned and attached Air Force forces. This includes oversight, integration, and operational level planning for and management of logistics capabilities for deploying units and the AOC, and similar support to other US government agencies, nongovernment organizations (NGOs), and private voluntary organizations as appropriate. Most of the challenges confronting this division will likely be Air Force component-unique.

Plans and Requirements (A–5)

The director of plans and requirements serves as the principal staff assistant to the COMAFFOR for all consolidated planning functions. In coordination with the A–4, the A–5 conducts comprehensive force-level movement and execution planning throughout the campaign. This involves preparation and subsequent refinement of the force flow, beddown, and redeployment in the time-phased force and deployment data. The A–5 is the focal point for planning not under the purview of the AOC, to include the COMAFFOR campaign support plan and security cooperation country plans. This planning is normally preceded by the development of a COMAFFOR strategy. The A-5 is also the focal point for the operational assessment of such plans. In addition, the A-5 leads in the development of the organizational structure and command relationships for the Air Force component within the framework of the joint operation. The A–5 normally publishes the Air Force component operations order to support the JFC's campaign.

Communications (A–6)

The director of communications is the principal staff assistant to the COMAFFOR for communications-electronics and information capabilities. This includes establishing the theater communications and automated systems architecture to support operational and command requirements.

Installations and Mission Support (A–7)

The director of installations and mission support is the COMAFFOR's primary advisor for installations; mission support; force protection; civil engineering; explosive ordnance disposal; firefighting; emergency management; chemical, biological, radiological, and nuclear passive defense and response; contracting; and all cross-functional expeditionary combat support. Additionally, the A–7 works in coordination with the A–4 and A-1 on formulation of beddown plans and coordination and supervision of force beddown.

Strategic Plans and Programs (A–8)

The director of strategic plans and programs provides the COMAFFOR comprehensive advice on all aspects of strategic planning and programming. The A-8 also conducts program assessment and provides coordinated resource inputs to the supporting MAJCOM's Program Objective Memorandum processes.

Studies, Analyses, Assessments, and Lessons Learned (A–9)

The director of studies, analyses, assessments, and lessons learned, collects, documents, reports, and disseminates critical information necessary to analyze, assess, and document Air Force aspects of campaigns and contingencies, and to document lessons identified. (Note: A–9 functions do not include campaign operational assessment, a task performed within the AOC). This information provides the primary source documents for both contemporary and future Air Force planning and analysis. Moreover, they serve as an official permanent record of component mission accomplishment.

The AFFOR staff comprises assigned and attached personnel tailored to the specific needs of the COMAFFOR. A command section, personal staff, A-staff and a variety of cross-functional teams may be used to support the COMAFFOR. The AFFOR staff coordinates across the C-MAJCOM/C-NAF, AOC/OC when assigned, and with the CCMD staff as required, to help fulfill the COMAFFOR's full range of responsibilities and to synchronize overall Service component staff efforts within the COMAFFOR's battle rhythm. Depending upon level/type of mission, configure each AFFOR staff appropriately and integrate various disciplines in a cross-functional team approach for planning and execution.

III. AFFOR Staff Responsibilities

Air Force component tasks and responsibilities may include command section support, public affairs office, protocol office, legal section, strategic level theater security cooperation planners, and base operating support liaisons (as required to ensure the component receives the required support). The AFFOR staff must establish processes to ensure integration with the CCMD staff, other service component staffs, the C-MAJCOM, the C-NAF, partner nations, governmental and non-governmental agencies for all phases of military operations. Command relationships should be highlighted. Perhaps the greatest single challenge for the AFCHQs is the transition to contingency/combat ops. Establishing a sound transition plan to include command structures prior to a crisis scenario is essential to continuity of operations. As an example, the formation/reception of new units (coalition, joint, service) within a theater and the ability to effectively command and control those units is an essential element of planning. The AFFOR staff's primary function is to support the COMAFFOR at the operational level as the Air Force service component to a JFC.

The AFFOR staff, in coordination with the AOC/OC, supports USAF requirements and integrates air component capabilities into the CCDR's joint force planning. The AFFOR staff implements and establishes COMAFFOR policies and procedures (supplemental to CCDR policies and procedures) within the theater of operations. The AFFOR staff also plans, organizes, conducts and assesses steady-state campaign activities in support of the CCDR's campaign plan and conducts service, joint and multinational exercises. Refer to members designated to perform these functions as ASO or ASP.

Refer to Air Force Instruction 13-103, AFFOR Staff Operations, Readiness and Structures for more information and specific guidance.

IV. JFACC STAFF

When the commander, Air Force forces (COMAFFOR) is designated the joint force air component commander (JFACC), he/she may need to establish a small joint or combined staff to deal with joint issues beyond the purview of the AFFOR staff.

Additionally, some AFFOR staff personnel may be present in the air operations center (AOC) to provide access to Air Force component information; normally, such AFFOR staff personnel should not be dual-hatted within the AOC. Augmentation within each AOC directorate from relevant Service components and coalition partners ensures adequate joint representation on the staff. At the discretion of the COMAFFOR, officers from other Services and coalition partners may fill key deputy and principal staff positions. Finally, for very large and complex operations—as might be encountered with large coalition operations—a COMAFFOR dual-hatted as a JFACC may delegate some aspects of COMAFFOR functions to a subordinate deputy COMAFFOR to ensure that they receive the proper attention.

See following section (pp. 3-21 to 3-34) for an overview and further discussion of the JFACC.

III. (JFACC) Joint Force Air Component Commander

Ref: JP 3-30, Command & Control for Joint Air Operations (Feb '14), chaps. 1 to 2.

JFCs organize forces to accomplish the mission based on their vision and a concept of operations (CONOPS) developed in coordination with their component commanders and supporting organizations. JFCs provide direction and guidance to subordinate commanders and establish command relationships to enable effective spans of control, responsiveness, tactical flexibility, and protection. The JFC's air component should be organized for coordinated action (through unity of command) using the air capabilities of the joint force. Centralized control and decentralized execution are key considerations when organizing for joint air operations. While JFCs have full authority, within establishing directives, to assign missions, redirect efforts, and direct coordination among subordinate commanders, they should allow Service tactical and operational groupings to generally function as they were designed. The intent is to meet the needs of the JFC while maintaining the tactical and operational integrity of the Service organizations.

A JFC has three basic organizational options affecting C2 of joint air operations. In each case a key task includes organizing the staff, C2 system, and subordinate forces that will plan, execute, and assess joint air operations.

- A JFC may designate a JFACC
- A JFC may designate a Service component commander
- A JFC may retain C2

When designated, the JFACC is the commander within a combatant command, subordinate unified command, or joint task force (JTF) responsible for tasking joint air forces, planning and coordinating joint air operations, or accomplishing such operational missions as may be assigned. The JFACC is given the authority necessary to accomplish missions and tasks assigned by the establishing commander.

Joint air operations are performed by forces made available for joint air tasking. Joint air operations do not include those air operations that a component conducts as an integral and organic part of its own operations. Though missions vary widely within the operational environment and across the range of military operations, the framework and process for the conduct of joint air operations must be consistent.

Joint air operations are normally conducted using centralized control and decentralized execution to achieve effective control and foster initiative, responsiveness, and flexibility. In joint air operations centralized control is giving one commander the responsibility and authority for planning, directing, and coordinating a military operation or group/category of operations. **Centralized control** facilitates the integration of forces for the joint air effort and maintains the ability to focus the impact of joint air forces as needed throughout the operational area. **Decentralized execution** is the delegation of execution authority to subordinate commanders. This makes it possible to generate the required tempo of operations and to cope with the uncertainty, disorder, and fluidity of combat.

Mission Command

Mission command is the conduct of military operations through decentralized execution based upon mission-type orders and is a key component of the C2 function. Its intent is for subordinates to clearly understand the commander's intent and to foster flexibility and initiative at the tactical level to best accomplish the mission. While philosophically consistent with historical C2 of air operations, modern joint air operations and their unique aspects of speed, range, and flexibility demand a balanced approach to C2. This approach is best codified in centralized control and decentralized execution.

The Joint Force Commander (JFC)

The JFC has the authority to organize assigned/attached forces to best accomplish the assigned mission based on the CONOPS. The JFC establishes subordinate commands, assigns responsibilities, establishes or delegates appropriate command relationships, and establishes coordinating instructions for subordinate commanders. When organizing joint forces, simplicity and clarity are critical.

C2 Options

When contemplating C2 options for joint air operations within the operational area, the JFC can choose to exercise C2 through a functional component commander by designating a JFACC, one of the Service component commanders, or the joint force staff. Many factors will weigh on the JFC's selection – most notably the type and availability of forces/capabilities to accomplish the assigned mission. Additional factors may include:

- Span of control is the JFC's ability to effectively manage the actions of subordinates. Span of control is based on the number of subordinates, number of activities, range of weapon systems, force capabilities, and the size and complexity of the operational area.
- When joint air operations are the only operations or the duration and scope of air operations are of a very limited nature, the JFC may elect to plan, direct, and control joint air operations through the joint forces staff.
- Expertise in effective and efficient employment of joint air assets to accomplish the JFC's mission is available. If the JFC elects to conduct joint air operations through the joint staff, the staff must be properly manned and adequately equipped with both the personnel expertise and the C2 equipment and processes necessary to direct and control the joint air effort.
- Complexity and Scope of Joint Air Operations. If the scope or complexity of the operations is significant, the JFC should consider designating a JFACC. This will allow the JFC time to focus on the overall campaign vice spending it on directing air operations.

Theater-Level Considerations

The geographic combatant commander (GCC) will weigh the operational circumstances and decide if available air forces/capabilities can be most effectively employed at the GCC level, the subordinate JFC level, or some combination thereof.

I. Joint Force Air Component Commander (JFACC)

The JFC normally designates a JFACC to establish unity of command and unity of effort for joint air operations. The JFC will normally assign JFACC responsibilities to the component commander having the preponderance of forces to be tasked and the ability to effectively plan, task, and control joint air operations. However the JFC will always consider the mission, nature, and duration of the operation, force capabilities, and the C2 capabilities in selecting a commander.

A. Authority

The JFC delegates the JFACC the authority necessary to accomplish assigned missions and tasks. The JFACC will normally exercise tactical control (TACON) over forces made available for tasking. Service component commanders will normally retain operational control (OPCON) over their assigned and attached Service forces. Since the JFC will normally designate one of the Service component commanders as the JFACC, the dual-designated Service component commander/JFACC will exercise OPCON over their own Service forces as the Service component commander and TACON over other Services' forces made available for tasking. The JFC may also establish support relationships between the JFACC and other components to facilitate operations. The JFACC conducts joint air operations in accordance with the JFC's intent and CONOPS.

B. JFACC Responsibilities

Ref: JP 3-30, Command & Control for Joint Air Operations (Feb '14), pp. II-2 to II-4.

The responsibilities of the JFACC are assigned by the JFC. These include, but are not limited to:

- Develop a joint air operations plan (JAOP) to best support the JFC's CONOPS or OPLAN
- Recommend to the JFC air apportionment priorities that should be devoted to the various air operations for a given period of time, after considering objective, priority, or other criteria and consulting with other component commanders
- Allocate and task the joint air capabilities and forces made available by the Service components based on the JFC's air apportionment decision.
- Provide the JFACC's guidance in the air operations directive (AOD) for the use of joint air capabilities for a specified period that is used throughout the planning stages of the joint air tasking cycle and the execution of the ATO. The AOD may include the JFC's apportionment decision, the JFACC's intent, objectives, weight of effort, and other detailed planning guidance that includes priority of joint air support to JFC and other component operations
- Provide oversight and guidance during execution of joint air operations, to include making timely adjustments to taskings of available joint air forces. The JFACC coordinates with the JFC and affected component commanders, as appropriate, or when the situation requires changes to planned joint air operations.
- Assess the results of joint air operations and forward assessments to the JFC to support the overall assessment effort.
- Perform the duties of the airspace control authority (ACA), if designated.
- Perform the duties of the area air defense commander (AADC), if designated.
- Perform the duties of the space coordinating authority (SCA), if designated. The SCA is responsible for planning, integrating, and coordinating space operations support in the operational area and has primary responsibility for joint space operations planning, to include ascertaining space requirements within the joint force. If the individual designated to be the JFACC is also designated to be the SCA, he/she will normally designate a senior space officer who facilitates coordination, integration, and staffing activities for space operations on a daily basis.
- Perform the duties of the PR coordinator, as required.
- In concert with the above responsibilities, perform tasks within various mission areas to include, but not limited to:
  - Defensive counterair (DCA) and offensive counterair (OCA)
  - CAS
  - Airborne ISR and incident awareness and assessment
  - Air mobility operations
  - Strategic attack
  - Air interdiction

C. Typical JFACC Staff & Joint Air Operations Center Organization

Ref: JP 3-30, Command & Control for Joint Air Operations (Feb '14), pp. II-11 to II-13.

The JAOC is structured to operate as a fully integrated command center and should be staffed by members of all participating components, to include key staff positions, to fulfill the JFACC's responsibilities. A JAOC provides the capability to plan, coordinate, allocate, task, execute, monitor, and assess the activities of assigned or attached forces. Through the JAOC, the JFACC monitors execution of joint air operations and directs changes as the situation dictates.

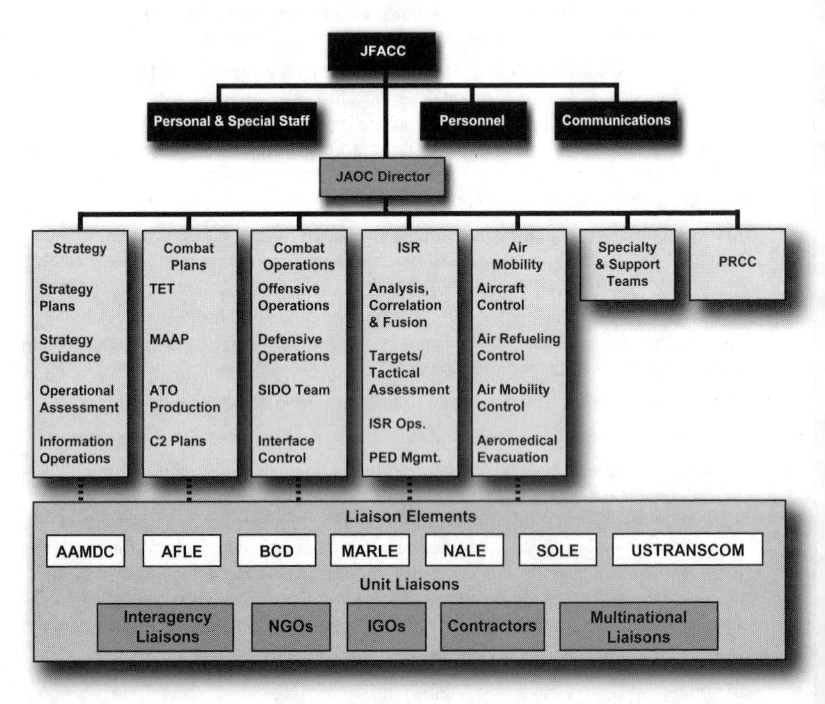

JAOC organizations may differ. Elements that should be common to all JAOCs are the strategy division (SD), combat plans division (CPD), ISR division, air mobility division (AMD), and combat operations division (COD). Divisions, cells, or teams within the JAOC should be established as needed.

Refer to JP 3-30, Appendix E, "Joint Air Operations Center Divisions and Descriptions," for further discussion.

The JAOC director is responsible to the JFACC for integrating the planning, coordinating, allocating, tasking, executing, and assessing tasks for all joint air operations, and coordinates with the director of mobility forces (DIRMOBFOR) to meet airlift and tanker priorities with support of United States Transportation Command (USTRANSCOM) mobility forces. Planning future joint air operations and assessing the effectiveness of past operations is usually the responsibility of the SD, while the CPD is usually devoted to near-term planning and drafting of the daily ATO. Execution of the daily ATO is carried out by the COD and closely follows the action of current joint operations, shifting air missions from their scheduled times or targets, and making other adjustments as the situation requires.

The AMD is normally responsible for integrating intertheater and intratheater airlift, aerial refueling, and aeromedical evacuation (AE) into air plans and tasking orders, and for coordinating with the JFC movement requirements and control authority and Air Mobility Command's Tanker Airlift Control Center. The ISRD provides the JFACC with predictive and actionable intelligence, targeting support, and collection management expertise to support the air tasking cycle.

Liaisons

Each of the JAOC's major activities relies on expertise from liaisons (e.g., battlefield coordination detachment [BCD], AAMDC liaison team, NALE, Air Force liaison element [AFLE], SOLE, Marine liaison element [MARLE]) to coordinate requests or requirements and maintain a current and relevant picture of the other component operations.

See also pp. 3-14 to 3-15. For more on liaisons, refer to JP 3-30, appendix F.

Functional Area and Mission Experts

Functional area experts (such as force protection, intelligence, meteorological and oceanographic, logistics, space operations, legal, airspace, plans, and communications personnel) provide the critical expertise in support, plans, execution, and assessment functions. Mission experts in air-to-air, air-to-ground, ground-to-air, information operations (IO), intelligence, air refueling, PR, and other areas provide the technical warfighting expertise required to plan for joint air operations and employ capabilities/forces made available by the components. Functional and mission experts from all components will provide manning throughout the JAOC and at all levels of command, and may be organized in special teams.

Preparation

The nucleus of the JFACC's staff should be formally educated and trained to perform fundamental operational activities in joint air operations and be representative of the joint force. Staff augmentation with manning as identified above ensures joint representation throughout the JAOC. The JFACC, in coordination with other component commanders, will determine specific manning requirements based on the size and scope of the operation, force list, and personnel availability.

Intelligence

Finally, the role of intelligence is extremely important and is an integral part of the daily functions of the JAOC. Intelligence personnel monitor and assess adversary capabilities and intentions, especially WMD threats, and provide assistance in target, weapon, fuse, and platform selection, including UA recommendations, and WMD response. In coordination with the SD's operational assessment team, they also conduct an assessment of the effectiveness of combat operations and provide an up-to-date picture of the adversary, expected adversary operations, and the status and priority of assigned targets to assist in execution day changes.

Joint Air Component Coordination Elements (JACCEs)

The JFACC may establish one or more joint air component coordination elements (JACCEs) with other commanders' headquarters to better integrate joint air operations with their operations. When established, the JACCE is a component-level liaison that serves as the direct representative of the JFACC. The JACCE does not perform any C2 functions and the JACCE director does not have command authority over any air forces. The JACCE is established by the JFACC to better integrate with other component's senior deployed headquarters. JACCE may also be assigned to the supported JTF headquarters (if the theater JFACC is designated in support to a JTF) to better integrate air component operations within the overall joint force. When established, the JACCE acts as the JFACC's primary representative to the respective commander and facilitates interaction among the respective staff to communicate, advise, coordinate, and support effective interplay. JACCE expertise should include plans, operations, ISR, space, airspace management, air mobility, and administrative and communications support.

See also p. 2-21. Refer to JP 3-30, appendix G for further discussion.

D. Tasking Component Forces

The JFC directs and controls the joint air effort by providing broad guidance, prioritized objectives, targeting priorities, procedures, and assigning appropriate authorities to subordinate commanders. Each component commander may be tasked to support other components and/or to provide support to the joint force as a whole. The JFACC, in consultation with other component commanders, is responsible for the air apportionment recommendation to the JFC, who makes the air apportionment decision.

Forces are tasked by the JFACC based on the JFC's approval of the JFACC's air apportionment recommendation (e.g., CAS, interdiction). In the case of a theater JFACC, the GCC will decide what air capabilities/forces are provided to each subordinate JFC. The air apportionment decision referenced here is made by each subordinate JFC.

Component forces must comply with the ROE, ACP, ACO, AADP, and SPINS. The ATO includes all known and prospective aircraft (alert, medical evacuation PR, etc.) that may fly during that ATO day, even if not apportioned to the JFACC. The overarching intent is to identify all potential flights to airspace managers and users in the JOA. Some smaller (Group 1) UASs may not be included on the ATO based on use and mission requirements. The inclusion of air assets in the ATO does not imply any change in command relationships or tasking authority over them, nor does it restrict component commanders' flexibility to respond to the dynamics of the operational environment.

E. Options for Establishing a Joint Force Air Component Command (JFACC)

Options available to a GCC during contingency operations to establish a joint force air component command include: designating a sub-theater (e.g., subunified or JTF) JFACC; a theater JFACC operating in support of subunified or JTF-level commands; or a combination of a theater and sub-theater JFACCs.

1. Designated JFACC for Sub-Theater Commands

When a GCC establishes a subordinate JTF to conduct operations and air forces are attached, with specification of OPCON to the subordinate JFC, the JFC has the option of designating a JFACC. This option will place dedicated air assets and independent C2 capability under the OPCON of the JFC for whom they are performing the mission. It provides unity of command over the forces employed within the assigned JOA and greater direct control and predictability as to which air assets are available. When there is only one JTF in a given theater, this is normally the preferred option.

2. Theater JFACC

When a GCC establishes multiple JTFs within the area of responsibility (AOR), the GCC normally will retain C2 of joint air forces at the GCC level. Joint air forces will be controlled to support the multiple JTF commanders according to the JTF commanders' objectives and the GCC's AOR-wide priorities. In this situation, joint air forces are controlled at the theater level, under the direction of the "theater JFACC," subordinate to the GCC. The theater-level JFACC provides flexibility in managing limited air assets to meet the requirements of the GCC and multiple JTFs. When there are multiple JTFs in one GCC's AOR, this is normally the preferred option.

- The theater JFACC will be the supporting commander to the GCC's subordinate JTF commanders' joint air operations within their respective JOAs. Per JP 1, Doctrine for the Armed Forces of the United States, an establishing directive should be promulgated to clearly delineate support command relationships. Unless limited by the establishing directive, the supported JTF commanders will have the authority to exercise general direction of the supporting effort. (General direction includes the designation and prioritization of targets or objectives, timing and duration of the supporting action, and other instructions necessary for coordination and efficiency.)

- The theater JFACC, as the supporting commander, determines the forces, tactics, methods, procedures, and communications to be employed in providing this support. The JFACC will advise and coordinate with the supported JTF commanders on matters concerning the employment and limitations (e.g., logistics) of such support, assist in planning for the integration of such support into the supported JTF commanders' efforts as a whole, and ensure that support requirements are appropriately communicated within the JFACC's organization. When the JFACC cannot fulfill the needs of the supported JTF commander, the GCC will be notified by either the supported JTF commander or JFACC. The GCC is responsible for determining a solution. For their operations, these JTF commanders—as JFCs—will exercise approval authority for products normally generated for "JFC approval" (including products generated by the theater JFACC for their JOA).
- The theater JFACC may deploy one or more JACCEs to the JTF headquarters and other component headquarters as needed to ensure they receive the appropriate level of joint air support. The JACCE will provide on-hand air expertise to the JTF commanders and the direct link back to the theater JFACC and the JAOC.

3. Combination Theater and Sub-Theater Level JFACCs

There may be a theater and sub-theater level JFACC. While in some cases, this may be the most operationally desirable option, it is also the most demanding on available C2 resources (manpower and equipment).

Between these three options presented there can be other potential organizational variations. While it is impossible to assemble a complete list of all potential C2 arrangements, two additional options that commanders may consider follow:

- Multiple JFACCs Sharing a Theater JAOC.
- Theater JFACC or JTF's JFACC to operate concurrently with a JSOACC assigned to a commander, joint special operations task force (CDRJSOTF).

The options discussed above contain combatant commander (CCDR)/subordinate relationships. Approval authority is inherent in command; therefore, it is imperative that subordinate JFCs exercise approval authority over those processes affecting operations within their JOAs regardless as to whether the products are developed by resources allocated to the command or by other headquarters. This includes, but is not limited to: air apportionment decision, targeting products, joint air estimates, JAOP, AOD, ATO, ACP, ACO, and AADP.

F. Joint Force Staff Option

In operations of limited scope, duration, or complexity, or in which air operations are a relatively small aspect of the overall joint force, the JFC may plan, direct, and control joint air operations with the assistance of the JFC staff. In this situation, the JFC would retain command authority and responsibility and would normally request augmentation from appropriate components to perform the C2 air function and assist in planning and coordinating joint air operations. In the joint force staff option all previously discussed JFACC responsibilities will be accomplished by the joint force staff as directed by the JFC.

The JFC staff operates out of the joint operations center (JOC). Under the JFC staff option, the JOC also functions as the C2 node for joint air operations. The composition of a joint staff should reflect the composition of the subordinate joint forces to ensure those responsible for employing forces have a thorough knowledge of their capabilities and limitations. The presence of liaisons on a single-Service staff does not transform that Service staff into a joint staff. The joint staff should be composed of appropriate members in key positions of responsibility from each Service or functional component having significant forces assigned to the command. The same general guidelines for joint staffs apply to multinational operations. Key staff positions ought to be a representative mix of US and multinational officers with shared responsibilities and trust.

G. JFACC Basing and Transition

Procedures for joint air operations are designed to exploit the flexibility of air power to achieve joint force objectives while providing support to component operations. Joint air operations scenarios may vary, and each scenario requires extensive planning when transition of JFACC responsibilities is necessary.

Land-based JFACC

In large-scale air operations, land-based JFACCs and JAOCs are normally desired because of the enhanced logistics and communications provided by additional equipment and workspaces that may not be available on sea-based facilities.

Sea-based JFACC

The JFACC and JAOC may be sea-based when any one of the following conditions:

- Maritime forces provide the preponderance of air assets and have the organizational construct, operating experience, and management functions capability to effectively plan, task, and control joint air operations
- Land-based facilities or sufficient infrastructure does not exist
- A secure land-based area is not available and ground support forces are forced to withdraw

JFACC Transition

Effective joint air operations planning must contain provisions to transition JFACC responsibilities between components of the joint force and/or JFC's staff. The JFACC transition should be identified in the JAOP.

H. JFACC Communications System

The JFACC is responsible for identifying and validating joint air requirements that affect the JFC's mission and allow accomplishment of the JFC's directives. The ability to exchange information via reliable secure communications with the JFC, joint force staff, and other component commanders is key to the successful integration of the joint air effort.

II. Airspace Control Authority (ACA)

The ACA is a commander designated by the JFC to assume overall responsibility for the operation of the airspace control system (ACS) in the airspace control area. Developed by the ACA and approved by the JFC, the ACP establishes general guidance for the control of airspace and procedures for the ACS for the joint force operational area. The ACO implements specific control procedures for established time periods. It defines and establishes airspace for military operations as coordinated by the ACA, and notifies all agencies of the effective time of activation and the structure of the airspace. The ACO is normally published either as part of the ATO or as a separate document, and provides the details of the approved requests for airspace coordinating measures (ACMs). All air missions are subject to the ACO and the ACP. The ACO and ACP provide direction to integrate, coordinate, and deconflict the use of airspace within the operational area. (Note: This does not imply any level of command authority over any air assets.) Methods of airspace control vary by military operation and level of conflict from positive control of all air assets in an airspace control area to procedural control of all such assets, or any effective combination.

See p. 3-36 for discussion of airspace control authority (ACA) from Annex 3-52.

Airspace Control Considerations

Airspace control is provided to reduce the risk of friendly fire, enhance air defense operations, and permit greater flexibility of operations. The JFC will determine the degree of airspace control required in the joint operations area (JOA). Depending on the mission, ROE, and weapons engagement zones, the degree of control of air assets may need to be rigorous, close, and restrictive, especially in an operational environment that can transition quickly from combat to noncombat and back again.

Airspace control may require a combination of positive control, procedural control, and real-time joint battle management to control the operational activity of the joint force including strict constraints on the forces, weapons, and tactics employed. Additionally, airspace control planning should include contingency operations to account for adversary interference. Such interference may inhibit positive control. In such a scenario, procedural control measures should be used. The JFC may set a coordinating altitude for designated airspace in the JOA. No matter what methods the JFC chooses, they need to be continually evaluated for effectiveness and efficiency as the environment and mission change.

As a matter of controlling joint air operations, the JFC may require all air missions, including fixed-wing, rotary-wing, tiltrotor, manned and unmanned (except small hand-held systems) of all components, to appear on the appropriate ATO and/or flight plan.

See pp. 3-35 to 3-52 for detailed discussion of airspace control from Annex 3-52.

Airspace Control Procedure Objectives

- Enhance effectiveness in accomplishing the joint force commander's objectives
- Prevent mutual interference
- Facilitate air defense identification
- Safety accommodate and expedite the flow of all air traffic in the operational area
- Prevent friendly fire
- Facilitate dynamic targeting

Ref: JP 3-30 (Feb '14), fig. II-1. Airspace Control Procedures Objectives.

ACA Responsibilities

The ACA achieves airspace control through positive or procedural methods. This includes centralized direction of the ACP, with the authority of the ACOs, supplemented by ACMs, and coupled with an ACS. The ACA should coordinate with joint force components' liaisons prior to commencement of operations. The ACA must integrate and coordinate the airspace requirements of all the components. The ACA does not have the authority to approve or disapprove combat operations. That authority is only vested in operational commanders. The ACA assumes responsibility for the ACS in the designated operational area. Subject to the authority and approval of the JFC, the broad responsibilities of the ACA include:

- Coordinate and integrate the use of the airspace control area.
- Develop broad policies and procedures for airspace control and for the coordination required among all users of airspace within the airspace control area.
- Establish an ACS that provides for integration of host and other affected nations' constraints and requirements.
- Coordinate and deconflict airspace control area user requirements.
- Promulgate ACS policies and procedures via the JFC-approved ACP.

A key responsibility of the ACA is to provide the flexibility needed within the ACS to meet contingency situations that necessitate rapid employment of forces as well as dynamic changes made by component staffs. The ACO is published either as part of the ATO or as a separate document.

See pp. 3-35 to 3-52 for detailed discussion of airspace control from Annex 3-52.

III. Area Air Defense Commander (AADC)

The AADC is responsible for DCA operations, which include the integrated air defense system for the JOA. DCA and OCA operations combine as the counterair mission, which is designed to attain and maintain the degree of air superiority desired by the JFC. In coordination with the component commanders, the AADC develops, integrates, and distributes a JFC-approved joint AADP. The AADP should be integrated with the ACP to ensure airspace control areas/sectors are synchronized with air defense regions/sectors. Typically, for forces made available for DCA, the AADC retains TACON of air sorties, while surface-based air and missile defense forces (e.g., Patriot missile systems) may be provided in support from another component commander. As such, the US Army Air and Missile Defense Command (AAMDC) should be collocated with the joint air operations center (JAOC), if established, and conduct collaborative counterair intelligence preparation of the battlespace (IPB), planning, and execution control. In distributed operations, the AAMDC may not be in the JAOC but is still functionally tied to it. The Navy component commander (NCC) (or JFMCC, if designated) exercises OPCON of maritime multi-mission and missile defense ships. When designated, these air and missile defense capabilities are in direct support of the AADC for C2 and execution of air defense.

See p. 3-44 for further discussion.

Area Air Defense Considerations

DCA operations are integrated with other air operations within the operational area through the AADP (see sample AADP in JP 3-01, Countering Air and Missile Threats). The AADC normally is responsible for developing an integrated air defense system by integrating the capabilities of different components with a robust C2 architecture. Because of their time-sensitive nature, DCA operations require streamlined coordination and decision-making processes, facilitated by the AADP. The AADP is the integration of active air defense design, passive defense measures, and the C2 system to provide a comprehensive approach to defending against the threat. It should address command relationships, the adversary and friendly situations, the AADC's intent, CONOPS, and logistics and C2 requirements, as well as detailed weapons control and engagement procedures. Weapons control procedures and airspace control procedures for all air defense weapon systems and forces must be established. These procedures must facilitate DCA operations while minimizing the risk of friendly fire. Planners must understand they routinely will be required to modify the AADP due to the dynamic nature of joint counterair operations. Ideally, as the JFC's operation/campaign progresses and the AADP is refined, the combination of DCA and OCA operations should diminish the enemy's ability to conduct air and missile attacks, reducing the requirement for DCA operations and the threat to the JFC's freedom of action.

AADC Responsibilities

AADC responsibilities include planning, integration, synchronization, and coordination of DCA operations with other tactical operations throughout the JOA. This may be facilitated by the JFC's designation of regional and sector air defense commanders. Additional AADC responsibilities include:

- Develop, integrate, and distribute a JFC-approved AADP in coordination with Service and functional components.
- Develop and execute a detailed plan to disseminate timely air and missile

warning and cueing information to components, forces, multinational partners, and civil authorities, as appropriate, in coordination with the intelligence directorate of a joint staff (J-2), the operations directorate of a joint staff (J-3), and the communications system directorate of a joint staff.

- Develop and implement identification and engagement procedures that are appropriate to the air and missile threats.
- Establish timely and accurate track reporting procedures among participating units to provide a consistent common operational picture.
- Establish air defense sectors or regions, as appropriate, to enhance decentralized execution of DCA.
- Establish a framework to prevent friendly fire.
- Coordinate the protection of those assets listed on the defended asset list (DAL).

Implementation

Implementation of the AADP takes place through the AOD, weekly SPINS, and the SPINS annex of the ATO (i.e., daily SPINS).

IV. Joint Air Operations C2 System

The C2 system for joint air operations will vary depending on the operational area and specific missions. Given the flexibility of modern C2 capabilities, geographic considerations have less of an impact on organizational structure today than in the past. The entire C2 system may be spread across the operational area or concentrated in a specific location, either in close proximity to the fight or far from it. Ultimately, there is no standard template for C2 design.

Normally, the joint air operation C2 system will be built around the C2 system of the Service component commander designated as the JFACC. Each of the Service commanders has an organic system designed for C2 of their air operations. Whether it is the Air Force's theater air control system (TACS), the Army air-ground system (AAGS), the Navy's composite warfare commander (CWC)/Navy tactical air control system (NTACS), Marine air command and control system (MACCS), or the special operations air-ground system (SOAGS) that serves as the nucleus for C2 of joint air operations, the remainder will be integrated to best support the JFC's CONOPS.

Theater Air-Ground Control Systems (TACS)

When all elements of the TACS, AAGS, CWC/NTACS, MACCS with fire support coordination center hierarchy, and SOAGS integrate, the entire system is labeled the TAGS. Technology has improved the JFACC's ability to command and control joint air power. The speed of modern warfare, as well as the precision of today's weapons, dictates close coordination in the operational area among the JFC's components. The JFACC ensures that the C2 architecture supports joint air operations by primarily relying upon assigned subordinate C2 elements (e.g., Air Force forces TACS). Other TAGS elements may be made available for tasking to enhance C2 of joint air operations if required. The other components' elements facilitate C2 of their component's operations as an integrated part of the TAGS, and do not receive directions/tasking direct from the JFACC or JAOC unless made available to support/augment the JFACC's ability to command and control joint air operations or accomplish other JFC delegated authorities. TAGS elements roles, responsibilities, and authorities should be clearly spelled out in theater-wide documents such as the AADP, ACP, SPINS, etc., particularly when tasks exceed their component commander's normal scope of operations.

See following pages (pp. 3-22 to 3-33) for an overview and further discussion of Theater Air Control Systems (TACS).

Theater Air-Ground Control Systems (TACS)

Ref: Annex 3-30, Command and Control (7 Nov 2014), app. D.

The theater air control system (TACS) is the Air Force's mechanism for commanding and controlling theater airpower. It consists of airborne and ground elements to conduct tailored command and control (C2) of airpower operations throughout the range of military operations, including counterair and counterland operations, airspace control, and coordination of space mission support not resident within theater.

When the TACS is combined with other components' C2 elements, such as the Army air-ground system, the Navy tactical air control system, and the Marine Corps air command and control system, they become the theater air-ground system (TAGS), and collectively support the JFACC.

For a description of each Service's TAGS element, refer to AFTTP 3-2.17, TAGS.

The TACS is divided into ground and airborne elements as described below.

Key Air Force and Army Components of the Theater Air Control System: Army Air-Ground System

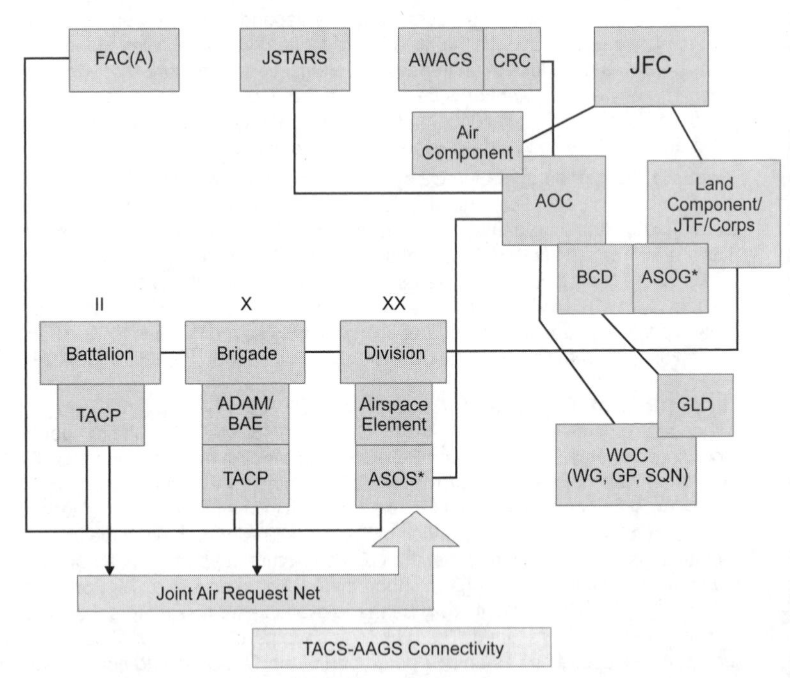

*Exact make up and capabilities of the ASOG/ASOS tailored to match the mission assigned to the corps/division. The ASOC is normally collocated with the senior Army tactical echelon.

NOTE:
Coordination is effected between all organizations for effective/efficient operations.

Ref: JP 3-30 (Feb '14), fig. II-3. Key Air Force and Army Components of the Theater Air Control System: Army Air-Ground System.

Ground TACS Elements

Ground TACS elements include the CRCs, the ASOC, and TACPs.

- The **control and reporting center (CRC)** is subordinate to the AOC and conducts air surveillance and supports strategic attack, counterair, counterland, air refueling operations, and other airpower functions/missions as directed. Responsibility as the region/sector air defense commander may be decentralized to the CRC, which acts as the primary integration point for air defense fighters and air defense artillery (ADA) fire control in its assigned area. It also enhances the joint forces' situational awareness by disseminating the air picture over data-links. The CRC may deploy mobile radars and associated communications equipment to expand radar coverage and communications range within its assigned operating area.
- The **air support operations center (ASOC)**, which reports to the AOC, receives, coordinates, and processes air support requests from subordinate TACPs, which are transmitted through the joint air request net (JARN). ASOCs distribute allocated sorties to satisfy requests for air support and integrate those missions with the supported units' fires and maneuver. An ASOC is normally tasked to support an Army unit but can also support units from other organizations (e.g., special operations, coalition forces). It may also augment other missions requiring C2 of air assets (e.g., humanitarian efforts). The AOC is the senior element within the theater air control system (TACS). The TACS includes the AOC plus subordinate ground and airborne elements, and is directly involved in the command and control of most air missions. Collectively, the TACS has the capability to plan, direct, integrate, and control all air, space, and cyberspace forces assigned, attached, or made available for tasking; monitor the actions of both friendly and enemy forces; plan, direct, coordinate, and control air defense and airspace control; and coordinate for required space and cyberspace support.
- **Tactical air control party (TACPs)** are aligned with Army maneuver elements, battalion through division level. They are primarily responsible for decentralized execution of close air support (CAS) operations. TACPs request, coordinate, and control CAS missions as required. *For more information on TACPs and ASOCs, refer to Annex 3-03, Counterland.* *See also pp. 4-17 to 4-32 for discussion of CAS.*

Airborne TACS Elements

Airborne elements of the TACS include AWACS, JSTARS, and the (FAC [A]).

- **AWACS** is subordinate to the AOC and conducts air and maritime surveillance and supports strategic attack, counterair, counterland, countersea, air refueling operations, and other airpower functions/missions as directed. Responsibility as the region/sector air defense commander may be decentralized to AWACS, which acts as the primary integration point for air defense fighters and ADA fire control in its assigned area. It also enhances the joint forces' situational awareness by disseminating the air and maritime picture over data-links.
- **JSTARS** conducts ground and maritime surveillance and supports strategic attack, counterair, counterland, countersea, and other airpower functions/missions as directed. It primarily provides dedicated support to ground commanders and attack support functions to friendly offensive and defensive air elements and may be employed as an airborne extension to the ASOC. It also enhances the joint forces' situational awareness by disseminating the ground and maritime picture over datalinks.
- The **forward air controller(airborne) [FAC(A)]** is an airborne extension of the TACP and has the authority to direct aircraft delivering ordnance to a specific target cleared by the ground commander. The FAC(A) provides additional flexibility in the operational environment by enabling rapid coordination and execution of air operations. It also enhances the TACS' situational awareness by disseminating information on the flow of aircraft on target.

V. C2 of Joint Air Operations for DSCA and Homeland Defense

Ref: JP 3-30, Command & Control for Joint Air Operations (Feb '14), pp. II-24 to II-26.

Commander, United States Northern Command (USNORTHCOM) and Commander, United States Pacific Command (USPACOM) share the primary mission for US HD and defense support of civil authorities (DSCA) within their assigned AORs. For US-only HD and DSCA air operations within continental United States, the 1st Air Force Commander is designated the JFACC. To facilitate operations in Alaska, USNORTH-COM has established JTF-Alaska. The Commander, Pacific Air Forces is the JFACC for the USPACOM AOR. USNORTHCOM is assisted by North American Aerospace Defense Command (NORAD), the bi-national command (US and Canada) that conducts aerospace warning, control, and maritime warning in defense of North America. NORAD is divided into three regions: Continental NORAD Region, Alaskan NORAD Region, and Canadian NORAD Region. Each region has a commander triple-hatted as JFACC, airspace coordination authority, and AADC for executing DCA missions.

C2 of joint air operations during routine HD and most DSCA non crisis operations is conducted under peacetime rules. As with the international community, when no combat operations have been declared, a civil organization is usually the ACA. For the US, this agency is the FAA and for Canada it is NAV CANADA. These agencies are charged by During DSCA operations, at the request of the lead federal agency and with approval of the Secretary of Defense, the appropriate JFACC, working in support of the FAA, assists in developing procedural control rules for safe use of the airspace by the multitude of participating agencies. Joint forces will coordinate all DSCA air operations with the JFACC to facilitate unity of effort, even while other air activities may be occurring elsewhere in the nation.

The Department of Homeland Security (DHS) has overall responsibility for US homeland security (HS), and the DOD may be asked to provide support to DHS for HS operations. DHS has the largest civilian government air force in the world and has peacetime oversight of the United States Coast Guard (USCG). In HS operations it is possible for the JFACC to have the preponderance of military air assets, but not the preponderance of USG air assets to coordinate. During HS operations with a threat emanating from United States Southern Command's (USSOUTHCOM's) AOR, DOD may direct USSOUTHCOM to lead DOD support to DHS, with the USAF Southern Command commander designated as the JFACC.

Although numerous interagency partners are responsible for DSCA mission accomplishment, most air operations fall under the FAA or are impacted by FAA authority. For instance, aerial search and rescue may be accomplished by state (e.g., Army National Guard, Air National Guard, or state police helicopters), federal (e.g., USAF or USCG), or civil (e.g., Civil Air Patrol) agencies, but all still operate within the procedures established by the FAA.

DSCA support is often provided by National Guard forces, either federalized under Title 10, United States Code (USC), Title 32, USC, or state active duty status. Each state, territory, and the District of Columbia has the latitude to develop aviation C2 procedures pursuant to the appropriate governor's directives. Often, these C2 architectures will incorporate many joint concepts. To enhance unity of effort during emergencies, USNORTHCOM's JAOC has pre-coordinated airspace C2 procedures with both the FAA and through the National Guard Bureau as the channel of communications with the numerous state adjutants general. During DSCA operations, the state governors typically retain C2 of civil and National Guard air assets, and execute operations through National Guard joint force headquarters-state or the state's emergency operations center.

IV. Airspace Control

Ref: Annex 3-52, Airspace Control (23 Aug '17).

The complexity of today's airspace environment grows with each advance in technology. Clearly defined airspace control concepts, forces, and capabilities help identify how best to use them for commanders at the strategic, operational, and tactical levels of military operations. The growth of military integrated air defense systems and the advent of cruise missiles and unmanned aircraft systems (UAS) continue to complicate theater airspace control requirements. Increasing multinational operations with partner and allied nations will add complications to airspace control in order to attain interoperability for more complex chains of command, communications, sensor and weapons interfaces, and planning. In addition to military users, current and future operations can expect a multitude of other air-intense operations either near or within a joint operations area (JOA). In such operations, civilian users, nongovernmental organizations, and relief agencies may require the use of combat zone airspace to conduct operations. Complicating matters, indirect fire systems (e.g., artillery), are recognized airspace users and today range higher, farther, and with greater volume of fire than ever before. These increased user demands require an integrated airspace control system to enable flight safety and prevent friendly fire incidents and unintended engagements against civil and neutral aircraft while enabling mission accomplishment and minimizing risk.

Airspace Control

Airspace control is defined as "capabilities and procedures used to increase operational effectiveness by promoting the safe, efficient, and flexible use of airspace."

Joint Publication [JP] 3-52, Joint Airspace Control.

Properly employed, airspace control maximizes the effectiveness of combat operations without unduly restricting the capabilities of any Service or functional component. Never static, airspace control operations may begin prior to combat operations, continue after, and may transition through varying degrees of civil and military authority. The airspace control procedures within the JOA are approved by the joint force commander (JFC) and are derived entirely from the JFC's authority. Airspace control does not infringe on the authority vested in commanders to approve, disapprove, or deny combat operations.

Airspace control is extremely dynamic and situational, but to optimize airspace use, that control should accommodate users with varied technical capabilities. In addition to expected threat levels, the available surveillance, navigation, and communication technical capabilities of both the airspace users and controlling agencies often determine the nature and use of coordination measures (CM). Airspace coordinating measures (ACM) is one category of a CM. Generally, limited technical capabilities result in increased airspace coordinating measure requirements with an inversely related normally result in decreased airspace coordinating measure requirements and an associated increase in airspace efficiency. Areas with the greatest air traffic congestion and risk of mid-air collisions often correspond to heavily accessed points on the ground (e.g., navigation aids, airports, drop zones, targets, and ground firing systems). Adherence to the JFC's guidance on ACMs should prevent airspace planners from exceeding the JFC's risk tolerance. This integration of ACMs into operations deconflicts airspace usage while decreasing potential friendly fire incidents.

To better organize operational airspace three characterizations exist:

Permissive Combat Airspace

A low risk exists for US and coalition aircraft operations within the airspace of interest. Operations can expect little to no use of adversary electronic warfare, communications jamming, anti-aircraft systems, or aircraft. Air superiority or air supremacy has been achieved.

Contested Combat Airspace

A medium risk exists to US and coalition aircraft within the airspace of interest. Expect the enemy to employ fighters, anti-aircraft systems, and electronic jamming. US and coalition aircraft can achieve localized air superiority for operations within portions of the airspace. Enemy air defense assets are neither fully integrated nor attrited.

Denied-Access Combat Airspace

A high risk exists for many, but not all, US and coalition aircraft from integrated air defense systems, radars, anti-aircraft systems, electronic warfare, and fighter aircraft. The airspace is characterized by pervasive enemy activity. Expect operations to result in high losses or denial of sustained operations until a measure of air superiority can be achieved.

II. Airspace Control Authority (ACA)

Normally, the joint force commander (JFC) designates a joint force air component commander (JFACC) as the commander for joint air operations. The JFACC role is normally filled by the commander, Air Force Forces (COMAFFOR). The JFC may also concurrently designate an airspace control authority (ACA) and area air defense commander (AADC). The ACA is "the commander designated to assume overall responsibility for the operation of the airspace control system (ACS) in the airspace control area." The AADC, on the other hand, is the commander assigned overall responsibility for air defense with the preponderance of air defense capability and the command, control, and communications capability to plan and execute integrated air defense operations (Joint Publication [JP] 3-52, Joint Airspace Control). Because these related authorities are so integral to air operations, the COMAFFOR, as the JFACC, is normally assigned ACA and AADC responsibilities.

As the ACA, the JFACC is responsible for planning, coordinating, and developing airspace control procedures and operating the ACS. The ACA does not have the authority to approve or disapprove combat operations. Airspace control procedures within the JOA are approved by the JFC and are derived entirely from JFC authority. If the ACA and an affected component commander are unable to obtain agreement on an airspace issue, the issue should be referred to the JFC for resolution.

In most operations, the COMAFFOR is designated as the JFACC, ACA, and AADC, largely due to the Air Force's ability to concurrently command and control (C2) these activities. In those joint operations where separate commanders are designated, close coordination is essential for unity of effort; prevention of fratricide and unintended engagements against civil and neutral aircraft; and joint air operations deconfliction (JP 3-52). Because such separate arrangements are rare, the remainder of this publication assumes the COMAFFOR has been designated as the JFACC, ACA, and AADC. This, in fact, is the preferred Air Force construct for which Airmen are trained.

See pp. 3-28 to 3-30 for related discussion of the ACA.

III. Airspace Control System (ACS)

The airspace control system (ACS) is an arrangement of those organizations, personnel, policies, procedures, and facilities required to perform airspace control functions. A system of systems, the ACS enables multiple component air-ground systems to support the joint force commander's (JFC) planning and execution of air-ground operations. The ACS combines each component's command and control (C2) and airspace control system supporting the JFC. Into this arrangement, the Air Force brings its theater air control system (TACS) with deployable air traffic control and landing system elements. The Air Force TACS, along with the Army, Navy, Marine, and Special Operations air ground systems combine to form the military's portion of the ACS. In many operations, wide-ranging interagency and nongovernmental organization (NGO) operations may be involved and challenge unity of command. A coordinated and integrated combat ACS is essential to the conduct of successful operations because any action taken by one airspace user may impact other users. An Airspace Control Authority (ACA)-established ACS supports JFC objectives and facilitates unity of effort.

Airspace control should be executed through a responsive ACS capable of real time control that includes surface and airborne assets, as necessary (e.g., control and reporting center [CRC] and airborne warning and control system). The ACS requires timely exchange of information through reliable, secure, and interoperable communications networks. Elements of the ACS may have dual roles as defensive counterair assets. For example, a CRC can be a regional or sector air defense commander responsible for air and missile defense in addition to their airspace control duties.

The ACA normally delegates airspace control authority to elements of the ACS. Each component normally provides airspace control elements to an ACS. Their associated air traffic control (ATC) functions provide International Civil Aviation Organization-approved traffic and separation standards as required. All of these separate agencies are ultimately governed by the host nation's rules and regulations. However, as operations transition between peace time and combat operations, peacetime airspace rules and organizations change. The nature of those changes will vary from theater to theater.

See following pages (pp. 3-38 to 3-39) for an overview and discussion of airspace control system fundamentals.

IV. Airmen's Perspective

Airmen think and operate on theater and global dimensions. Comprehensive awareness at these levels is fundamental to an Airman's way of thinking. A remotely piloted aircraft (RPA) flown over Iraq during Operation IRAQI FREEDOM, operationally controlled through a combined air operations center while providing direct support to the joint force commander (JFC) and multiple ground units is an example of the Airman's perspective applied operationally. Airmen share the JFC's theater-wide focus. While exploiting airpower's speed, range, and flexibility, Airmen provide capabilities from outside an area of responsibility (globally in some cases). They then provide control for those capabilities where and when they are required in a given operation. This has direct implications for airspace control because airspace control plans should be developed, integrated, and possibly implemented across adjacent regions while supporting several operations simultaneously. Airspace has many users and uses which should be carefully integrated, coordinated, and deconflicted to ensure safe and effective operations; this demonstrates the need for some form of centralized control. This is the key reason airspace control authority is normally vested in a single commander. The need for effective integration is greatest in major combat operations, where manned and unmanned fixed-, tilt- and rotary-winged combat aircraft, military airlift, missiles, artillery, and commercial airspace users all vie for the same airspace.

Airspace Control System Fundamentals

Ref: Annex 3-52, Airspace Control (23 Aug '17), pp. 6 to 8.

A common ACS facilitates accurate and timely coordination of airspace operations among friendly forces. Common equipment, a common understanding of Service and joint doctrine, and familiarity with procedures through joint exercises and training can enhance airspace control operations within the joint operations area (JOA).2 Standardized airspace procedures rely upon an effective mix of identification and control measures. Identification requirements for airspace control should be integrated with those for air defense. Airspace control, air defense, ATC, and supporting command and control (C2) procedures, equipment, and terminology should be compatible, mutually supporting, and integrated to ensure commonality of procedures for airspace users and control agencies. Airspace control agencies should work out procedural agreements and establish required communication links to ensure effective interagency coordination.

Effective airspace control means securing the systems enabling that control. The systems comprising our ACS include, but are not limited to, sensors, communications, data processing, and common operating databases. Information assurance programs such as communications security, physical security, emissions security, and defensive cyberspace operations are methods to protect airspace control systems and information. Due to the US military's dependence on, and the general vulnerability of, electronic information and its supporting systems, information assurance is essential to airspace control. Additionally, when developing communication policies and procedures, it is imperative operations security practices are applied.

Airspace Control Procedures

Airspace control is a mix of procedural and positive control. Airspace control procedures provide maximum flexibility through an effective mix of positive and procedural control measures. The capabilities of the organization executing control over a given section of airspace will normally drive the composition's mix. The control structure should encourage close coordination among joint force components allowing a rapid concentration of combat power. An ACS should be adaptable to changing requirements and priorities as operations progress through various operational phases.

Procedural Control

Procedural control is a method of airspace control relying on a combination of previously agreed to and promulgated orders and procedures. It establishes the basic common criteria and concepts for airspace control. This form of control relies on common published procedures, designated airspace, and promulgated instructions by an authorized control agency to deconflict and activate ATC areas, airspace coordinating measures, fire support coordination measures, and air defense control measures. Controlling agencies activate airspace with a defined time and volume through standard airspace coordinating measures or weapons control statuses. These procedures deconflict both aircraft and airspace use from other airspace users. When appropriate communications exist, an authorized airspace control agency can provide procedural control instructions in real time to increase operational flexibility for airspace users. This method is considered effective for low density airspace saturations and in areas lacking positive control coverage but is not normally as efficient as positive control. Procedural control measures should be uncomplicated, readily accessible to all forces, and disseminated through the airspace control order and special instructions of the air tasking order. Use of these single-source documents is essential for integrating rotary-wing, fixed wing, fires, and unmanned aircraft operations.

Positive Control

Positive control is a method of airspace control relying on the positive identification, tracking and direction of aircraft within a given airspace. It is normally conducted by electronic means by an agency having the authority and responsibility therein (Joint Publication 3-52, Joint Airspace Control). This form of control relies on surveillance, accurate identification, and effective communications between a designated airspace control agency and the airspace user. It is normally conducted by ATC agencies equipped with radar; identification friend or foe interrogators and receivers; beacons; track processing computers; digital data links; and communications equipment. Positive airspace control requires the means to locate and identify airspace users in real time, and the ability to maintain continuous communications with them to pass required control instructions. This positive control method still requires predetermined, standing transition procedures to procedural control should positive control systems become degraded or unavailable. Those procedures should also account for the differences between civil and military communications and surveillance systems.

Cost versus Risk

When discussing procedural and positive control, there is a continuum of efficiency, level of effort, resources required, and risk to be addressed (see figure below).

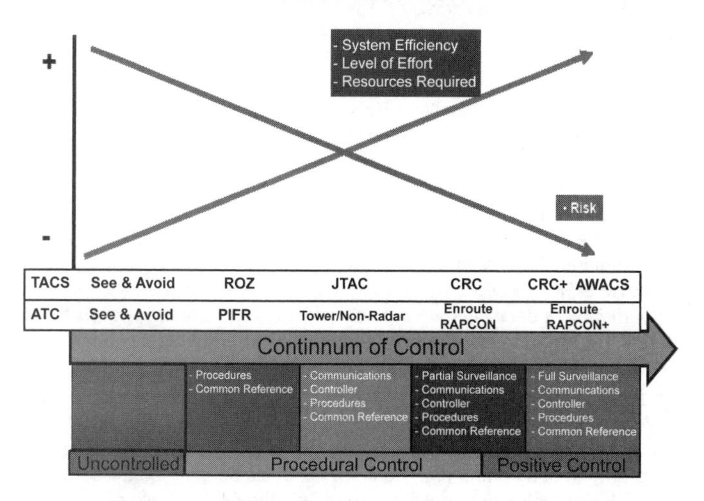

Ref: Annex 3-52 (23 Aug '17), fig: Notional Airspace Continuum of Control.

The minimum requirements for surveillance, identification, and communications equipment can vary by theater and operation, but are likely to be driven by a combination of military and civil aviation regulations and the level of risk the JFC is willing to accept. Assuming a constant air traffic volume, uncontrolled airspace exerts a small drain on resources, but carries increased risk. For that same airspace, standing airspace procedures, such as a restricted operations zone, not only incrementally increase control and resources required, but also reduce risk. Full military or civilian positive control provides the greatest risk mitigation, but exerts a significant drain on resources. Ideally, the entire airspace control area would be under positive control with radar and communication coverage. However, limited resources or other factors, such as terrain, may make this goal unrealistic.

Airpower's Contributions

Airpower has added a vertical flank to the modern battlefield creating a maneuver space throughout a given theater to be taken and exploited. Few missions (land, sea, or air) can be accomplished without at least localized air superiority. Fundamental to discussing airspace is the understanding that air superiority is implicit in establishing even the most limited forms of airspace control. If enemy aircraft target friendly aircraft or ground troops, deconfliction measures between friendly airspace users may be severely challenged until those threats have been neutralized. Airpower operations such as close air support, interdiction, and other supporting efforts are likewise compromised without first establishing at least localized air superiority.

Airspace control deconfliction through the use of intelligence, surveillance, and reconnaissance platforms permits freedom to find, fix, track, and target high value targets while coordinating with fixed-wing and rotary-wing aircraft to complete each operation.

V. Command and Organization

A. Command and Control of Joint Air Operations

Air Force and joint doctrine supports the command and control of joint air operations through three options: a functional component commander, (e.g., a joint force air component commander [JFACC]), a Service component commander, or a staff option. Additionally, both Air Force and joint doctrine support command of airpower through a theater-level commander, Air Force forces (COMAFFOR) who is normally also the JFACC, a joint task force (JTF)-level COMAFFOR-JFACC, or a mix of the two. The latter requires careful consultation between the respective joint force commanders (JFCs) and their COMAFFOR-JFACCs. The geographic combatant commander should provide guidance for the interaction between theater-level components and subordinate JTFs. This should include clarity of supported and supporting command relationships between the JTFs and theater COMAFFOR-JFACC, together with clear priorities of effort, support, and apportionment. The theater COMAFFOR-JFACC should then allocate effort across the area of responsibility (AOR) using combatant commander's (CCDR's) guidance and priorities. The CCDR sets the conditions for success by clearly stating and emphasizing the supported command status of subordinate JTFs and the supporting command role of a theater-level COMAFFOR-JFACC. The CCDR is the ultimate arbiter for prioritization and apportionment decisions among subordinate JTF commanders and the theater COMAFFOR-JFACC to provide sufficient guidance for the theater COMAFFOR-JFACC's subsequent allocation decisions.

B. Air Operations Center (AOC)

The air operations center (AOC) is organic to the Air Force and is the senior Air Force command and control (C2) node for C2 of air, space, and cyberspace forces. When employed for joint or coalition operations, the AOC is known as a joint AOC or combined AOC for coalition operations. In the case of a single JFACC, airpower's inherent flexibility can be leveraged through a single commander and subsequently controlled by a single JAOC, even if C2 is exercised over multiple joint operations areas (JOAs) across a single AOR. However, key personal relationships become harder to maintain under this organization due to heavy reliance on virtual vice physical presence and a greater geographical distance between the JAOC and the joint theater headquarters.

See pp. 3-9 to 3-16 for detailed discussion of the air operations center (AOC).

VI. Cross-Domain Integration

Ref: Annex 3-52, Airspace Control (23 Aug '17), pp. 12 to 13.

"Airpower produces synergistic effects." The Air Force leverages asset capabilities from multiple domains to create effects. As the focal point for air operations, the Air Force's air operations center (AOC) integrates air, space, and cyberspace capabilities into airspace operations. The airspace control plan (ACP), produced within the AOC, is a product of that integration. Satellite communication capabilities resident within the AOC enable the communication necessary to develop and disseminate the ACP to the joint force.

Airspace Control

Airspace control relies heavily on space-based capabilities. Space-based systems provide the positioning, navigation, and timing for airspace users. Coordination measures can become marginalized if airspace users cannot establish their position. The controlling agencies and their systems should be capable of locating, identifying, tracking, and communicating with airspace users within their areas of control. As users travel across mountains, oceans and deserts these capabilities diminish. In these cases, the benefits space brings to airspace control increase in proportion to the relative isolation of the airspace.

The capabilities inherent within the cyberspace domain enable the AOC's coordination process. The integration of the cyberspace domain into airspace control is intuitive. The networked systems and constantly updated information on display are made possible because of this integration. Airspace control's impact upon the cyberspace domain is not as obvious. Airspace control creates effects within cyberspace because the assets using the airspace are free to execute their missions. For example, the kinetic destruction of a cell tower can create some of the same effects as a malicious virus. Both can limit a network's capability to transmit information between users. Likewise, the purposeful jamming of specific frequencies can temporarily disable the triggering system on explosives intended to harm US forces.

Airspace control also plays a large part within the maritime domain. The Navy controls the airspace around its battle groups. Aircraft carriers, while offering platforms to project national power, also simultaneously control and defend the airspace around the battle group. While defending the battle group, airborne assets allow supported assets the freedom to conduct operations. The battle group's success depends on being able to control who enters and exits their airspace, where those users interact, and when those interactions take place.

The effects of airspace control on the land domain are obvious. Whether planning surveillance orbits, establishing air-refueling tracks, executing kill box procedures, or protecting remote special operations teams, airspace control greatly affects land domain operations. OIF and OEF highlight the synergistic effect of airspace deconfliction with respect to unmanned aircraft systems (UAS) and other airspace users in the land domain. Identification of UAS orbits as a priority and allocating airspace control organizations and resources against that problem allowed controlling agencies to deconflict multiple airspace users.

Airspace Management

Airspace management supports airspace control through the coordination, integration, and regulation of airspace users by airspace control elements within airspace of defined dimensions. Airspace management and control procedures enhance effective airspace operations in support of joint force commander's (JFC) objectives. All joint air and space force components have legitimate mission requirements for airspace that should be integrated, coordinated, and deconflicted within the airspace control system. Airspace control is required to prevent friendly fire incidents and avoid excessive collateral damage, to include unintended engagements against civil and neutral aircraft, enhance air defense operations, facilitate fire support, and maximize the effectiveness of operations conducted from and through the air to accomplish the JFC's mission objectives.

C. Roles and Responsibilities of the Commander, Staff, and Subordinate Organization(s)

Airmen, in conjunction with joint and coalition partners, are responsible for planning and integrating airspace control systems in accordance with JFC guidance. Airspace control systems should maximize the combat effectiveness of all forces, while reducing the risk of friendly fire incidents and excessive collateral damage, to include unintended engagements against civil and neutral aircraft. To do this, Air Force commanders train their personnel to employ risk management principles and to be knowledgeable of all component systems and procedures. The primary purpose of command relationships is to establish a chain of command so all involved understand who is in charge, who is supported, and who is supporting.

Joint Force Commander (JFC)

The JFC is responsible for airspace control within the JOA. Although components may use portions of airspace to accomplish the mission, they do so only with the approval of the JFC and in accordance with the JFC's policies and procedures. The control procedures and authorities for the airspace within the JOA are codified in the JFC's airspace control plan (ACP) and executed by the airspace control authority (ACA). The ACP in combination with the airspace control order (ACO) expresses how airspace will be used to support mission accomplishment. The air operations directive (AOD) then establishes the priorities among airspace users and missions.

Commander, Air Force Forces (COMAFFOR)

The COMAFFOR provides the Air Force theater air control system and airspace control expertise and resources to the JFC. Each airspace control system can be tailored to support centralized control and decentralized execution of air forces throughout the range of military operations. The Air Force provides the COMAFFOR with the resources necessary to assume the roles of ACA and area air defense commander (AADC), usually as the JFACC. Unifying the roles of ACA and AADC ensures unity of effort in all aspects of theater airspace operations. If the JFACC is designated from another component, the COMAFFOR ensures Air Force forces are employed in accordance with the JFACC's guidance and tasking. If the JFC decides not to organize functionally, the COMAFFOR should expect to fulfill the roles of ACA and AADC.

Joint Force Air Component Commander (JFACC)

The JFC normally designates the COMAFFOR as the JFACC and the JFACC is normally designated as the AADC and ACA, since air defense and airspace control are an integral part of joint air operations. By design, the AOC is a natural command and control node to integrate these operations. Although the separation of the AADC and the ACA function is not routine, the JFC may designate a separate AADC and ACA (e.g., when a single commander is not capable of performing both roles).

Airspace Control Authority (ACA)

The COMAFFOR, as JFACC (or other component commander) designated as ACA, assumes overall responsibility for the operation of the ACS in the JOA and should be the commander with the preponderance of airspace management and control capability, including the ability to plan, promulgate, execute, and assess integrated airspace control operations. The ACA, on behalf of the JFC, develops broad policies and procedures for airspace control and for the coordination required among units within the operational area. When approved by the JFC, these policies and procedures are promulgated via the JFC's ACP.

See facing page for further discussion of the ACA.

Airspace Control Authority (ACA)

Ref: Annex 3-52, Airspace Control (23 Aug '17)., pp. 16 to 17.

The COMAFFOR, as JFACC (or other component commander) designated as ACA, assumes overall responsibility for the operation of the ACS in the JOA and should be the commander with the preponderance of airspace management and control capability, including the ability to plan, promulgate, execute, and assess integrated airspace control operations. The ACA, on behalf of the JFC, develops broad policies and procedures for airspace control and for the coordination required among units within the operational area. When approved by the JFC, these policies and procedures are promulgated via the JFC's ACP.

A key responsibility of the ACA is to provide an effective and adaptive airspace control system to meet contingency situations and facilitate the rapid employment of forces in support of the JFC's mission. Matters on which the ACA is unable to obtain agreement are referred to the JFC for resolution. Key ACA responsibilities include, but are not limited to:

- Identifying and coordinating airspace access required for the JFC's mission.
- Providing effective and timely integration of the airspace control system with that of the host nation, coordinating and deconflicting airspace user requirements to include conduct of operations in support of normal air commerce operators as governed by host nation and International Civil Aviation Organization guidance.
- Developing the ACP in accordance with JFC guidance.
- Developing ACOs in accordance with the ACP.
- Disseminating the ACP and ACO in a timely manner to all associated joint and coalition units.
- Integrating Service and joint airspace management and control personnel and systems to support required JFC mission.
- Establishing liaison with interagency, host nation, regional, or international airspace agencies as required for the deconfliction of civil and military airspace use.
- Delegating control of a portion of airspace to a commander to accomplish a specified mission or to facilitate decentralized execution (e.g., an amphibious objective area or air defense sector).
- When delegating airspace, retaining overall responsibility for specified airspace and recalling that airspace when required for higher JFC priorities or when the delegated commander can no longer exercise C2 over the delegated airspace.

Airspace Control Authority Products

To execute airspace control effectively, the ACA provides guidance on airspace use through the ACP and ACO. The ACA also provides airspace usage inputs to the JFACC's ATOs, air operations directives (AODs), and Special Instructions. The ACP uses the JFACC's operational guidance provided in the AOD to establish airspace control procedures throughout the JOA. The AOD conveys JFC and JFACC guidance concerning acceptable levels of risk with respect to airspace control. The ACP also establishes the airspace control system, the control nodes, and airspace procedures. The ACO executes the ACP and may contain airspace priorities discussed in the AOD. The AOD, like the ACP, is directive in nature.

The ACP and area air defense plan should complement each other to provide effective airspace control. The ACP should consider procedures and interfaces with the international or regional air traffic systems necessary to effectively support air operations, augmenting forces, and JFC objectives.

Refer to JP 3-52, Joint Airspace Control, or Appendix A, for topics to consider when developing an ACP.

Area Air Defense Commander (AADC)

The JFC designates the AADC, who is normally the COMAFFOR, acting as the JFACC. The AADC is responsible for defensive counterair (DCA) operations, which includes both air and missile defense. The AADC should identify those volumes of airspace and control measures that support and enhance DCA operations, identify required airspace management systems, establish procedures for systems to operate within the airspace, and ensure they are incorporated into the airspace control system. The AADC may also designate regional air defense commanders and sector air defense commanders to allow for ease of command and control of airspace based on the size and scope of the mission or operation. The successful conduct of air defense operations requires the integrated operation of all available air, land, and maritime-based defense systems. The AADC develops the AADP with JFC approval and ensures it is promulgated. The AADP and ACP should be complementary.

Other Component Commanders

In support of JFC airspace management guidance, component commanders may be required to:

- Provide airspace control in areas designated in the ACP.
- Forward requests for airspace coordinating measures to the ACA in accordance with the ACP.
- Develop component-specific airspace control instructions, plans, and procedures in accordance with guidance in the ACP; coordinate these plans and procedures with the ACA to ensure consistency with JFC-approved airspace control guidance.
- Provide facilities and personnel for airspace control functions in assigned operational areas and identify those facilities and personnel to the ACA for inclusion in the ACP.
- When required, provide component airspace liaison personnel to the JFACC or senior air control facility.

VII. Planning Considerations

A. Policy and Other Guidance

In general, airspace policy ascribes two responsibilities to the Air Force and the Department of Defense (DOD) airspace users: protect the public from flight operations and return unused airspace back to the pubic when no longer required. Specifically, Air Force Instruction (AFI) 13-201, Airspace Management, directs the Air Force to "protect the public to the maximum extent practicable from the hazards and effects associated with flight operations." Internationally, DOD Instruction 4540.01, Use of International Airspace by US Military Aircraft and for Missile/Projectile Firings, directs US military aircraft operating in international airspace to observe International Civil Aviation Organization flight procedures when practical and compatible with the mission. Furthermore, DOD Directive (DODD) 5030.19, DOD Responsibilities on Federal Aviation, states that it is DOD policy that airspace designated for military use will be released to the Federal Aviation Administration (FAA) or to other navigation service providers, as appropriate, when the airspace is not needed for military requirements. AFI 13-201 extends those same policies to host nations when using their airspace.

Specific to homeland operations, the FAA is granted statutory authority to regulate the national airspace system. However, DOD may direct and implement emergency security control of air traffic in certain specified circumstances in accordance with 32 CFR Part 245 Plan for the Emergency Security Control of Air Traffic (ESCAT). Additionally, during wartime, the President may transfer FAA responsibilities to DOD in accordance with 49 United States Code (USC) 40107 and Executive Order 11161, as amended.

B. Basic Planning Considerations

Ref: Annex 3-52, Airspace Control (23 Aug '17), pp. 36 to 39.

Airspace control provides joint and coalition forces air domain advantages to create effects across multiple domains. The unmatched speed, range, and flexibility of airpower enables the joint force to create asymmetric and synergistic effects while providing the concentration and priority called for by the JFC via the ATO. Consequently, potential airspace control system modifications should be considered during all planning phases. Airspace control considerations should be integrated into contingency and crisis action planning (CAP) to ensure joint/combined force effectiveness. The ACP should be consistent with specific operation plans (OPLANs) and operation orders developed by the JFC.

Contingency Planning

The operation plan (OPLAN) serves as the foundational employment concept for an operational area. Airspace planners should consider that all operations will not smoothly transition between operational phases. Depending on the nature of the conflict, national political objectives, and joint force commander (JFC) intent, operations may cease prior to the beginning of engagement, cooperation, and deterrence operations. Transferring airspace control authority from civilian to military control, adapting the airspace control system to the JFC's needs during each phase, and eventually returning it to civil authority are complex tasks requiring joint military, diplomatic, and interagency efforts. Since a crisis may occur unexpectedly, airspace control and management activities should be a part of contingency and CAP from the beginning. For instance, moving C2 and airspace control equipment (e.g., control and reporting centers [CRCs] or air traffic control [ATC] facilities) is a time-phased force and deployment data consideration. Since much of this equipment is subject to deployment airlift (or other lift) constraints, a coherent plan from the beginning is required to ensure critical airspace control capability is available at the appropriate time and phase of the operation.

Crisis Action Planning

Unlike contingency planning, CAP is based on emerging events and is conducted in time-sensitive situations. Plans are based on existing circumstances at the time planning occurs. Contingency planning supports CAP by anticipating potential crises and facilitating development of joint operation plans to facilitate the rapid development and selection of a course of action (COA). This is especially crucial for certain airspace control operations that may need substantial coordination in advance with host nation or regional and international airspace or aviation agencies. Required airspace control actions should be fully integrated into the development of all COAs. During COA development, planners should identify tasks for airspace access and airspace control systems to support operational objectives. In addition, planners should examine the role and contributions of airspace control functions through all phases of an operation.

Joint Operation Planning

The joint operation planning process (JOPP) is an orderly analytical process that consists of a logical set of steps to analyze a mission; develop, analyze, and compare alternative COAs against criteria of success and each other; select the best course of action; and produce a joint operation plan or order. A major element of the JOPP is campaign planning, which is the process whereby combatant commanders and subordinate JFCs translate national or theater strategy into operational concepts through the development of an OPLAN for a campaign. Campaign planning may begin during contingency planning when the actual threat, national guidance, and available resources become evident, but is normally not completed until after the President or Secretary of Defense selects the course of action during CAP.

VIII. Execution Considerations

As a guiding principle for all operations, a host nation retains airspace control authority and the joint forces primarily use existing international or host nation aeronautical information publications for airspace procedures or guidelines. Airspace, navigation services and radio frequencies are the sovereign right and responsibility of the host nation. General considerations when addressing airspace functions across the range of military operations include:

- Command and control (C2)/Air traffic control (ATC) and airspace planners should be involved from the outset in planning and executing C2, air traffic control, and airspace management. This ensures airspace requirements are coordinated and approved by the proper agencies.
- Planning should consider the establishment of an ATC cell to liaise with the current host nation infrastructure. Establishing relationships with key host nation and neighboring nations' air traffic control is critical. Establishing an aircraft diplomatic clearance process (e.g., for US Embassy personnel) should be accomplished as early as possible during the planning process. Key issues to resolve during this planning include:

  - Identifying key personnel and their contact information.

  - Identifying existing agreements (e.g., aeronautical information publications and site surveys).

  - Identifying rules, regulations, and existing international, multilateral, or bilateral agreements or arrangements governing proposed operations (e.g., International Civil Aviation Organization, Federal Aviation Administration, regional organization, or host nations). For planning purposes, this type of information may be located in the Foreign Clearance Guide (Authorized by Department of Defense (DOD) Directive 4500.54E, DOD Foreign Clearance Program).

  - Identifying special operating rules or waivers needed for certain types of aircraft or operations that will need to operate within host nation airspace (e.g., rules for unmanned aircraft systems [UAS]).

  - Establishing requirements to integrate liaison officers, equipment, processes, and functions.

Airspace Implementation

The airspace control plan (ACP) provides specific planning guidance and procedures for the airspace control system throughout the joint operations area (JOA). The ACP may be distributed as a separate document or as an annex to the operations plan. The airspace control order (ACO), which implements the ACP, is normally disseminated as a separate document. The ACO provides the details of coordination measures for the next air tasking cycle and includes fire support coordination measures (FSCM), air defense areas, and air traffic areas along with other airspace information. Changes to the ACO are published on an as-needed basis.

Airspace Deconfliction Procedures

Airspace deconfliction at the operational level normally occurs within the air operations center (AOC). The AOC's combat plans division usually resolves airspace conflicts during the theater air tasking cycle pre-air tasking order (ATO) or ACO publication while the combat operations division handles post ATO/ACO publication and real-time airspace control order changes. Deconfliction at the tactical level is executed by elements of the airspace control system capable of providing airspace control functions (e.g., control and reporting center [CRC], airborne warning and control system (AWACS), joint surveillance target attack radar system, air support operations center (ASOC) and joint air ground integration centers, tactical air control party, ATC) and achieved by directing time, position, altitude, and other deconfliction methods to airspace users.

Integration with Air Defense

The air defense functions of weapons control, surveillance, and identification are inherent in the theater air control system, from the oversight and direction provided by the AOC, down through the execution capability of the AWACS and the CRC. The area air defense plan should provide detailed engagement procedures consistent with the ACP; the area air defense plan should incorporate air defense capabilities from all functional components and airspace control system elements.

Integrated Air and Missile Defense (IAMD)

IAMD (a subset of the counterair construct) is the integration of capabilities and overlapping operations to defend the homeland and US national interests, protect the joint force, and enable freedom of action by negating an adversary's ability to achieve adverse effects from their air and missile capabilities. IAMD activities include direct actions such as ballistic missile defense, counter rockets, artillery and mortars, offensive counterair (OCA) attack operations and air and cruise missile defense, as well as foundational support functions such as intelligence, networking, C2 and logistics, and passive defense measures.

Airspace Coordinating Measures (ACMs)

Airspace coordinating measures are employed to facilitate efficient use of airspace to accomplish air operations and fires and simultaneously provide safeguards for friendly forces. ACMs are approved by the ACA and promulgated via the ACO. ACMs support the most efficient use of airspace in support of JFC objectives. Use of ACMs should include an awareness of risks associated with engagement of targets. ACMs have specific usages that further help refine use and assist with effective planning, integration and execution. ACMs are restrictive to fires and flying objects through the ACM airspace, however, with coordination with the appropriate C2 agency, fires and flying objects may transit the ACM. The ACP should list other coordination measures (CM) categories besides ACMs. Examples of CM categories are fire support coordination measures (FSCMs), air reference measures, air defense measures, maneuver control measures, maritime defense measures and air traffic control measures.

Fire Support Coordination Measures (FSCMs)

FSCMs are measures employed by commanders to facilitate the rapid engagement of targets and simultaneously provide safeguards for friendly forces (JP 3-0, Joint Operations). FSCMs are usually activated for a limited time and refer to areas where indirect fires may be active, restrictive, or prohibited. FSCMs define a boundary area on the ground and relate to airspace because of ordnance flight paths. The requirement to deconflict airspace in support of ground fire missions requires the determination of the firing locations, the impact location, and the airspace impacted by the projectile during flight. Those projectile parameters are deconflicted with other airspace users. Service liaisons and airspace control agencies work closely to ensure that appropriate airspace coordinating measures and FSCMs deconflict surface operations and other airspace operations.

Providing an Air Picture for the Joint Force

The JFACC is normally expected to incorporate data from various air, ground, and space sensors into a recognized air picture, to enable planning and decision making for air operations in the JOA. Feeds from data links are managed by the joint interface control officer (JICO) at the AOC and combined with other sources into a common operating picture. This fused picture is shared for mission planning and execution at all appropriate levels of command.

Integration of Air Defense and Airspace Control in the AOC

Airspace control and air defense functions are integrated in both the combat plans and combat operations divisions of the AOC. In the combat plans division, the C2 plans officers integrate air defense considerations such as minimum-risk route; identification friend or foe/selective identification feature procedures, and missile, fighter,

Operational Considerations

Ref: Annex 3-52, Airspace Control (23 Aug '17), pp. 40 to 62.

Steady-State Operations

While normal and routine, operations designed around engagement, cooperation and deterrence discourage potential adversaries and assure or solidify relationships with friends and allies. Various joint, multinational, and interagency airspace activities are executed with the intent to enhance international legitimacy and gain cooperation in support of defined military and national strategic objectives. They are designed to assure success by shaping perceptions and influencing the behavior of both adversaries and allies, developing allied and friendly military capabilities for self-defense and coalition operations, improving information exchange and intelligence sharing, and providing US forces with peacetime and contingency airspace access.

Additionally, the host nation may retain overall airspace control or the ACA may transfer airspace control to the host nation giving the JFC and ACA a less direct voice in the daily conduct of airspace control for continuing JFC operations.

Refer to Annex 3-52, pp. 47 to 49 for further discussion.

Command & Control

Normal and Routine Military Airspace Considerations

In addition to ensuring the continuation of routine DOD flight operations, joint force airspace planners should establish effective relationships with key AO airspace authorities, develop specific ACPs in preparation for future operations, and build airspace planning expertise. Regular DOD or joint force interaction with host nation authorities and participation in regional airspace conferences establishes relationships with the host nation for quick resolution of issues and effective coordination of airspace requirements.

Development of ACPs should include airspace control considerations from peace to combat operations and through all follow-on phases of the operations plan. Additionally, the ACP should integrate known international or host nation air traffic airspace and air defense capabilities. Primary planning considerations include identification of airspace required for joint force operations and the proposed coordination process for obtaining that airspace. Joint operation planning should consider procedures to transfer airspace control authority from the host nation to the ACA. This would include: rerouting of airways, ACA responsibilities for continuity of civil aviation operations, and host nation notification of ACA areas of control through notices to Airmen (NOTAMs) or aeronautical information publication entries.

Developing joint force airspace control expertise for the design of airspace control systems and procedures is also crucial during normal, steady-state operations. Airspace managers should receive formal training prior to arriving at an Air Force component headquarters or AOC, preferably at the AOC formal training unit. Additionally, theater-specific training on airspace control ensures full mission qualification. Exercises provide key opportunities for airspace control planners to practice joint C2 procedures and familiarize themselves with the basic operation plan. Bilateral or regional exercises with host nations are effective in improving cooperation with and understanding of host nation capabilities for improved planning accuracy and interoperability.

Refer to Annex 3-52, pp. 49 to 50 for further discussion.

Deterrence

Normally a demonstration of joint force capabilities and resolve, deterrence seeks to avert undesirable adversary action. Largely characterized by preparatory actions, deterrence operations specifically facilitate the execution of consecutive operations or theater campaigns. Airspace control contributes to these operations by supporting the combatant commander's deterrence strategy. Specific airspace actions may include developing the finalized ACP and airspace database for ACO publication; obtaining initial overflight and

irspace permission; and assignment of joint force airspace liaison personnel to Depart-
nent of State, US embassies, multinational, or host nation organizations to coordinate
irspace requirements for subsequent phases of the operation.

Refer to Annex 3-52, pp. 50 to 51 for further discussion.

Homeland Operations

Natural or man-made disasters and special events can temporarily overwhelm local, state, and non-military federal responders. The DOD has a long history of supporting civil authorities in the wake of catastrophic events. When directed by the President or the Secretary of Defense (SecDef), US Northern Command (USNORTHCOM) and Service components respond to the requests of civil authorities to save lives, prevent human suffering, and mitigate great property damage. The Joint Strategic Capabilities Plan (JSCP) directs the Commander, USNORTHCOM to prepare a plan to support the employment of DOD forces to provide defense support of civil authorities (DSCA) in accordance with the National Response Framework, applicable federal law, DOD directives, and other policy guidance. The plan should include those hazards defined by the national planning scenarios not addressed by other JSCP tasked plans. DSCA is a subset of DOD civil support that is performed within the parameters of the National Response Framework.

Refer to Annex 3-52, pp. 51 to 53 for further discussion.

Major Operations and Campaigns

Traditional Warfare

Transitioning to traditional combat operations from a position of deterrence may be accomplished on the joint force commander's (JFC's) initiative or in response to an enemy attack. During combat operations, peacetime airspace rules and organizations change and the nature of these changes will vary from theater to theater. The airspace control plan (ACP) should contain instructions to transition from peacetime to combat in simple, clear steps. The ACP should include the airspace control concepts for transition to combat operations and robust procedural control methods for potential degraded operations. Airspace planners should be integrated into development of the master air attack plan to ensure required airspace is designed for combat operations and the transition from peacetime to combat airspace control is seamless.

Refer to Annex 3-52, pp. 54 to 62 for further discussion.

Irregular Warfare Operations

Recognizing that aspects of irregular warfare can occur both before and after traditional combat operations, this document addresses airspace considerations across the range of military operations during all phases of operations to include Phase 0 (shape), which includes steady-state and ongoing operations; Phase 1 (deter); Phase 2, (seize the initiative); Phase 3 (dominate); and the post traditional warfare phases, Phase 4 (stability) and Phase 5 (enable civil authority).

As combatants conclude major combat operations and transition to a legitimate post-conflict government, military operations may continue with the goal of reducing the threat (military or political) to a level manageable by the host nation's authorities. During this operational phase, the joint force may be required to perform local governance until legitimate local entities are functioning. The joint force air component commander (JFACC) could be required to perform roles traditionally associated with a host nation aviation authority and may include the development of aeronautical information (e.g., instrument procedures, publications, notices to Airmen), civil flight planning procedures, certification of procedures, aviation safety investigation, training of host nation or contract personnel, or operation of airspace infrastructure systems. The regional air movement control center can play the critical lead role during these phases as the volume of non-military traffic increases.

Refer to Annex 3-52, pp. 63 to 67 for further discussion.

Command & Control

and joint engagement zones. In the combat operations division, the airspace manager is responsible for the execution of airspace control while the senior air defense duty officer is responsible for the execution of air defense operations.

Integration and Synchronization with Surface Operations

Airspace control procedures increase in complexity and detail when air forces operate in proximity to, or in conjunction with, surface forces. To prevent both air-to-surface and surface-to-air friendly fire incidents, integrated joint operations are necessary. Liaison elements are vital when integrating air and surface elements in close proximity. Each surface component's area of operations (AO) may be defined with specific boundaries. These boundaries are normally defined by maneuver control measures including fire support coordination line, forward line of own troops, fire support coordination measures, airspace coordinating measures, or multiples of these during nonlinear operations.

Electromagnetic Spectrum (EMS) Use

Airspace operations rely heavily on equipment using the EMS – GPS, Radio Navigation, ATC Radar, Weather Radar, Voice, etc. Use of spectrum-dependent technologies is constrained by the finite nature of the EMS. The rapid growth of sophisticated weapons systems, as well as intelligence, operations, and communications systems, greatly increases demand for EMS access. Lack of proper, pre-planned EMS coordination and consideration of electromagnetic environmental effects (E3) will have an adverse effect upon the safe, efficient, and flexible use of airspace. EMS operations are further constrained by international law, which protects sovereign rights to utilize and regulate the EMS within national borders.

Integration of Expeditionary Airfields

As an operation flows through its various phases, expeditionary airfields normally open and close as forces reposition. These airfield changes should be integrated and synchronized with ongoing airspace control and regional air movement coordination center procedures. An AFFOR airfield operations cell normally stands up to facilitate the opening of new airfields. Their key actions include installation of required airfield systems, sourcing of personnel, and the development and inspection of flight procedures. The timely establishment of all-weather instrument procedures is crucial for base logistics and operations.

Communication and Information

Although the airspace C2 infrastructure has not changed much over time, the communications network has improved significantly, enhancing the reliability, security, and timeliness of information flow. Information that might previously have stopped at the AWACS or CRC is now sent to the AOC enhancing situational awareness. In addition, broad-bandwidth communications using satellite and internet protocol communications have substantially increased both the ACS C2 coverage and 'reachback' capabilities.

Communications Planning

Planning is an essential element of effective airspace C2. Detailed radar and radio signal analysis ensures that surveillance and communications systems provide appropriate coverage within the airspace using a combination of fixed and mobile systems. Detailed analysis of joint network and joint infrastructure requirements is crucial in order to enable system integration across component and allied operational capabilities. These needs are normally met by installing a combination of organic and commercial communications systems prioritized to meet the commander's mission. The goal is to maximize the use of military capabilities and expand use of commercial systems to increase capacity and reliability and to generate greater freedom of action. Communications planners should perform an operations security (OPSEC) vulnerability analysis to determine procedures that will protect sensitive unclassified information from exploitation.

Coordination with the Host Nation, Regional Authorities and ICAO

Ref: Annex 3-52, Airspace Control (23 Aug '17), pp 48 to 49.

When the host nation retains airspace control authority, joint forces primarily use existing international, host nation, or DOD aeronautical information publications for airspace procedures or guidelines. Airspace and navigation services are the sovereign right and responsibility of the host nation. Joint forces operating within the airspace of any host nation use these airspace services with the sovereign consent of that nation, under the provisions of respective national aeronautical information publications or other appropriate agreements.

Although combat operations may not be in execution, the JFC should consider appointing an ACA, (normally the commander, Air Force forces [COMAFFOR]) for airspace management, air traffic control, and navigation aids issues within the combatant commander's area of responsibility (AOR) or the joint operations area (JOA). The commander, Navy forces is normally assigned responsibility for airspace procedures applicable to fleet air operations over international waters within the operational area and only advises the JFC's lead agent as appropriate. As lead agent, the COMAFFOR is delegated the authority to develop joint force airspace requirements in coordination with the other Service components and represent those joint force airspace requirements to the DOD, interagency, international, or host nation authorities as appropriate. Additionally the lead executive agent normally serves as the focal point to:

- Provide assistance to the JFC, components, Services, and supporting commands on airspace, air traffic, and navigation aid matters.
- Develop appropriate coordinating measures in support of JFC contingency planning to include airspace requirements for UAS.
- Ensure current and future airspace and navigation aid availability for components and supporting commands through joint mission essential task listing inputs.
- Coordinate host nation navigation aids inspections with Headquarters Air Force Flight Standards Agency, the Federal Aviation Administration (FAA)/ICAO aviation system standards, and the DOD program management office for flight inspection.
- Ensure navigation aids are included on the DOD essential foreign-owned navigation aids list if deemed an enduring requirement.
- Develop and establish procedures for airspace actions or issues that cannot be resolved by component commands consistent with applicable DOD, JFC, component, international, and host nation guidance.
- Ensure altitude reservations are coordinated for all DOD aircraft transiting or operating within the operational area.
- Develop friendly host nation airspace capabilities through the joint force theater engagement plan, training, and exercises.
- Submit changes to DOD aeronautical/flight information publications to the National Geospatial-Intelligence Agency on a timely basis.

Airspace Communications Systems

Airspace control agencies primarily communicate with airspace users via voice communications. Principal transmission should be through secure and jam-resistant radio equipment. The TAGS communication capabilities include line of sight, beyond line of sight and satellite systems, but planners should also ensure that radio relays are considered to enhance over-the-horizon radio communications. Networking technologies may also increase the capabilities of C2 nets to disseminate information with unprecedented speed and accuracy.

IX. Limitations of Airspace Control

Despite the exponential growth in airspace control capabilities across the Department of Defense (DOD), airspace control is not without its flaws and limitations. Because of the complexity of most modern airspace control systems, multiple factors can affect the overall effectiveness of any particular system. Factors limiting the airspace control system's effectiveness are the environment, communications, and politics.

Environmental characteristics can have a large impact on airspace control. Afghanistan's mountains presented severe radar tracking and communications problems to ground-based US and coalition forces conducting airspace control for Operation ENDURING FREEDOM. Radars and communications are only effective when objects are within line of sight. If aircraft are over the horizon or behind terrain, radar tracking and communications are not possible. Elevated airspace control systems (i.e., the Airborne Warning and Control System, and the Joint Surveillance Target Attack Radar System) are not affected by terrain obstacles to the same degree as ground-based systems. Major storms (sand, snow, and rain) can also obscure radar and radio signals if the particulate is numerous and dense enough. Intense solar activity, because of its effects on electronics, can also negatively impact airspace control.

Communications at all levels can affect the airspace control systems' performance. Airspace control system personnel and equipment have their own limitations. A lack of proper training or outdated or poorly maintained systems limits their effectiveness. Alternatively, the airspace control system will never reach its potential if all airspace users are not required to use and "feed the system" with information. At the user level, all participants should have access to the same information and should adhere to the same equipment requirements before entering a given section of airspace. Operationally, disconnects between internal elements of the theater airspace control system and external component systems can impose unexpected barriers to airspace control system performance. Equipment compatibility between Service components and host nation systems can present multiple issues.

Politics can also play a role in limiting the airspace control system. Policies as simple as those limiting the number of personnel granted access to a location may deny a needed element of the airspace control system. Even if that element is able to establish itself elsewhere, the optimum placement has already been affected. On a different note, political decisions to upgrade or purchase equipment can affect an airspace control system's effectiveness by impacting an element's maintenance rate, stability, or compatibility with other, newer systems.

The benefits and risk mitigation provided by airspace control rely upon the active participation of airspace users and controlling agencies. The technologies allowing forces to identify, spatially locate, and then communicate with airspace users require resources; ultimately all rely on human beings to operate. To properly execute an airspace control plan, trained personnel and resources should first be made available. In addition, the airspace control systems' placement, design, and setup impact an airspace control system's efficiency and effectiveness. Planners should establish airspace control systems that account for each of the limitations listed above and still operate effectively.

Airpower (Overview)

Ref: Annex 3-0, Operations & Planning (4 Nov 16), pp. 2 to 39.

Editor's Note: For the purposes of this publication (AFOPS2), the material from Annex 3-0 Operations and Planning is presented in two separate chapters, with chapter four (this chapter) focusing on airpower and chapter five focusing on strategy, effects-based approach to operations, and the common operations framework (operational design, planning, execution, and assessment).

Air Force Doctrine Annex 3-0 is the Air Force's foundational doctrine publication on strategy and operational design, planning, employment, and assessment of airpower. It presents the Air Force's most extensive explanation of the effects-based approach to operations (EBAO) and contains the Air Force's doctrinal discussion of operational design and some practical considerations for designing operations to coerce or influence adversaries. It presents doctrine on cross-domain integration and steady-state operations–emerging, but validated concepts that are integral to and fully complement EBAO. It establishes the framework for Air Force components to function and fight as part of a larger joint and multinational team. Specific guidance on particular types of Air Force operations can be found in other operational-level doctrine as well as Air Force tactics, techniques, and procedures documents. This publication conveys basic understanding of key design and planning processes and how they are interrelated.

The US' national security and national military strategies establish the ends, goals, and conditions the armed forces are tasked to attain in concert with non-military instruments of national power. Joint force commanders (JFCs), in turn, employ strategy to determine and assign military objectives, and associated tasks and effects, to obtain the ends, goals, and conditions stipulated by higher guidance in an effort to produce enduring advantage for the US, its allies, and its interests. Strategy is a prudent idea or set of ideas for employing the instruments of national power in a synchronized and integrated fashion to achieve theater, national, and multinational objectives. Airmen should follow a disciplined, repeatable approach to strategy development in order to maximize airpower's contribution to overarching national aims.

Today, the United States faces many security challenges including an ongoing conflict against implacable extremists, engagement with regimes that support terrorism, and the need to support international partners. Against this backdrop, US military forces may be called upon to conduct a full range of operations in a variety of conflicts and security situations, including major operations and campaigns, irregular warfare , information operation, homeland defense, humanitarian assistance/disaster relief efforts, building partnerships with other nations, and others.

The operational environments in which airpower is employed may be characterized by simultaneous action by Air Force forces against more than one adversary at a time–including the potential for near-peer and peer competitors–who may attempt to achieve objectives against US interests by using asymmetric advantages across all instruments of power: diplomatic, informational, military, and economic. Conflicts may occur with little or no warning and they may stretch the Air Force as it works with JFCs to provide support for the joint force while simultaneously addressing Air Force-unique missions.

Airpower Overview (Doctrine Annexes)

1. Counterair Operations (Annex 3-01) *See pp. 4-7 to 4-16.*

The Air Force defines counterair as a mission that integrates offensive and defensive operations to attain and maintain a desired degree of control of the air and protection by neutralizing or destroying enemy aircraft and missiles, including cruise and ballistic missiles, both before and after launch

2. Counterland Operations (Annex 3-03) *See pp. 4-17 to 4-32.*

Counterland operations are defined as "airpower operations against enemy land force capabilities to create effects that achieve joint force commander (JFC) objectives."

3. Countersea Operations (Annex 3-04) *See pp. 4-33 to 4-36.*

Countersea operations are those operations conducted to attain and maintain a desired degree of maritime superiority by the destruction, disruption, delay, diversion, or other neutralization of threats in the maritime environment.

4. Counterspace Operations (Annex 3-14) *See pp. 4-37 to 4-40.*

The Air Force uses four space operations functions to clearly delineate the capabilities required for successful global joint operations and supersede the space mission areas: space situational awareness (SSA); counterspace operations; space support to operations; and space service support.

5. Air Mobility Operations (Annex 3-17) *See pp. 4-41 to 4-56.*

Joint doctrine defines air mobility as "the rapid movement of personnel, materiel, and forces to and from or within a theater by air." The foundational components of air mobility operations—airlift, air refueling, air mobility support, and aeromedical evacuation—work with other combat forces to achieve national and joint force commander objectives.

6. Global Integrated Intelligence, Surveillance, and Reconnaissance (Annex 2-0) *See pp. 4-57 to 4-64.*

The Air Force defines global integrated ISR as "cross-domain synchronization and integration of the planning and operation of ISR assets; sensors; processing, exploitation and dissemination systems; and, analysis and production capabilities across the globe to enable current and future operations."

7. Strategic Attack (Annex 3-70) *See pp. 4-65 to 4-70.*

Strategic Attack (SA) is offensive action specifically selected to achieve national strategic objectives. These attacks seek to weaken the adversary's ability or will to engage in conflict, and may achieve strategic objectives without necessarily having to achieve operational objectives as a precondition.

8. Nuclear Operations (Annex 3-72) *See pp. 4-71 to 4-78.*

The Air Force's responsibilities in nuclear operations are to organize, train, equip, and sustain forces with the capability to support the national security goal of deterring nuclear attack on the United States, our allies, and partners.

9. Personnel Recovery (Annex 3-50) *See pp. 4-79 to 4-82.*

Our adversaries clearly understand there is great intelligence and propaganda value to be leveraged from captured Americans that can influence our national and political will and negatively impact our strategic objectives. For these reasons, the Air Force maintains a robust and well trained force to locate and recover personnel who have become "isolated" from friendly forces. Personnel recovery (PR) is an overarching term that describes this process, and the capability it represents.

10. Special Operations (Annex 3-05) *See pp. 4-83 to 4-88.*
Special operations are operations requiring unique modes of employment, tactical techniques, equipment and training often conducted in hostile, denied, or politically sensitive environments and characterized by one or more of the following: time sensitive, clandestine, low visibility, conducted with and/or through indigenous forces, requiring regional expertise, and/or a high degree of risk.

11. Cyberspace Operations (Annex 3-12) *See pp. 4-89 to 4-94.*
Cyberspace is a global domain within the information environment consisting of the interdependent network of information technology infrastructures, including the Internet, telecommunications networks, computer systems, and embedded processors and controllers.

12. Information Operations (Annex 3-13) *See pp. 4-95 to 4-102.*
The purpose of information operations (IO) is to affect adversary and potential adversary decision making with the intent to ultimately affect their behavior in ways that help achieve friendly objectives. Information operations is defined as "the integrated employment, during military operations, of information-related capabilities [IRCs] in concert with other lines of operation to influence, disrupt, corrupt, or usurp the decision making of adversaries and potential adversaries while protecting our own."

13. Electronic Warfare (Annex 3-51) *See pp. 4-103 to 4-108.*
Electronic Warfare (EW) is waged to secure and maintain freedom of action in the electromagnetic spectrum (EMS). Military forces rely heavily on the EMS to sense, communicate, strike, and dominate offensively and defensively across all warfighting domains. EW is essential for protecting friendly operations and denying adversary operations within the EMS.

14. Public Affairs Operations (Annex 3-61) *See pp. 4-109 to 4-112.*
Air Force PA "advances Air Force priorities and achieve mission objectives through integrated planning, execution, and assessment of communication capabilities. Through strategic and responsive release of accurate and useful information, imagery, and musical products to Air Force, domestic, and international audiences, PA puts operational actions into context; facilitates the development of informed perceptions about Air Force operations; helps undermine adversarial propaganda efforts; and contributes to the achievement of national, strategic, and operational objectives".

15. Combat Support (Annex 4-0) *See pp. 8-1 to 8-4.*
The Air Force defines combat support (CS) as the foundational and crosscutting capability to field, base, protect, support, and sustain Air Force forces across the range of military operations. The nation's ability to project and sustain airpower depends on effective CS.

16. Engineering Operations (Annex 3-34) *See pp. 8-5 to 8-8.*
Air Force civil engineer forces establish, operate, sustain, and protect installations as power projection platforms that enable Air Force and other supported commanders core capabilities through engineering and emergency response services across the full mission spectrum.

17. Medical Operations (Annex 4-02) *See pp. 8-9 to 8-12.*
The Air Force is increasingly called upon to deliver medical capabilities throughout the range of military operations. Diverse medical missions may consist of civil-military operations, global health engagement, or humanitarian assistance/disaster relief as part of joint or multinational operations.

18. Force Protection (Annex 3-10) *See pp. 8-13 to 8-16.*
Joint doctrine defines FP as "preventive measures taken to mitigate hostile actions against Department of Defense personnel (to include family members), resources, facilities, and critical information" (Joint Publication 3-0, Joint Operations). FP is a fundamental principle of all military operations as a way to ensure the survivability of a commander's forces.

I. Applying Airpower

Airpower entails the use of military power and influence to achieve objectives at all levels by controlling and exploiting air, space, and cyberspace. It encompasses military, civil, and commercial capabilities, the industrial infrastructure, and a doctrine of employment. Airpower is an indivisible, unitary construct—one that unifies Airmen, rather than portraying them as a collection of "tribes" broken into technological or organizational "stovepipes." Other doctrine publications deal with specific aspects of airpower or specific types of Air Force operations, but in all cases readers should remember that airpower accomplishes or contributes to achieving national objectives across all domains via operations in and through air, space, and cyberspace.

Due to speed, range, and its multidimensional perspective, airpower operates in ways that are fundamentally different from other forms of military power; thus, the various aspects of airpower are more akin to each other than to the other forms of military power. Airpower is the product, not the sum, of air, space, and cyberspace operations. Each depends on the others to such a degree that the loss of freedom of action in one may mean loss of advantage in all other domains. Airpower has the ability to create effects across an entire theater and the entire globe, while surface forces, by their nature, are constrained to divide the battlespace into discrete operating areas. Airmen view operations, including the application of force, more from a functional than a geographic perspective, and usually classify actions taken against targets (including nondestructive and nonkinetic actions) by the effects created rather than the targets' physical locations within the battlespace.

A. Airpower as Maneuver in Warfare

The multidimensional nature of airpower provides distinct advantages. Traditionally, the physical structure of ground maneuver forces has consisted of fronts, flanks, and rears. While these concepts do not apply as readily to airpower, it can be useful to make an analogy in surface terms in order to convey the Air Force's contribution to joint warfare. In such terms, airpower adds flanks in other dimensions that make the vertical and virtual battle as important as the horizontal battle. Using a metaphor from surface warfare, the airspace above the battlespace is like an additional flank in the third dimension, which can be exploited to achieve a relative advantage. Thus, as with surface flanks, commanders should seek to gain positions of advantage by turning an enemy's vertical flank, while trying not to expose their own vertical flank(s). Through cross-domain effects (effects created in one or more domains through operations in another), airpower can also create virtual "flanks" or "rears" in other dimensions, such as time and cyberspace (or assist the joint force in doing so).

Integrated with surface forces, airpower can reduce the need for operations like surface probing actions through such capabilities as wide-ranging intelligence, surveillance, and reconnaissance (ISR), information exploitation, and comprehensive situational awareness and understanding. This enables freedom of action for surface forces, greatly enhancing their effectiveness and that of the entire joint force.

B. Parallel and Asymmetric Operations

Air Force capabilities are usually employed to greatest effect in parallel, asymmetric operations.

Parallel Operations

Parallel operations are those that apply pressure at many points across an enemy's system in a short period of time to cause maximum shock and dislocation effects across that system. Sequential, or serial, operations, in contrast, are those that apply pressure in sequence, imposing one effect after another, usually over a significant period of time. Parallel operations limit an enemy's ability to react and adapt and thus place as much stress as possible on the enemy system as a whole. For example, in Operation DESERT STORM, the Iraqi command and control structure was severely degraded through parallel attacks on the electric grid, communications nodes, and command facilities. In the past, target sets were often prioritized and attacked sequentially, and thus it usually took considerable time for effects to be felt across an enemy system. While focusing on one node in a system, the enemy was often able to adapt to losses or compensate with other resources, thus slowing or even negating desired effects. Today, airpower often enables a truly parallel approach.

Asymmetric Operations

"Asymmetric," in this context, refers to any capability that confers an advantage for which the adversary cannot directly compensate. Asymmetric operations can confer disproportionate advantage on those conducting them by using some capability the adversary cannot use, will not use, or cannot effectively defend against. Conversely, symmetric operations are those in which a capability is countered by the same or similar capability. For example, tank-on-tank battles, like the battle of Kursk during WW II, are symmetric, as was the Allied battle for air superiority over Germany in that same war. The use of Coalition air power to immobilize and defeat Iraqi armored forces in Operations DESERT STORM and IRAQI FREEDOM was asymmetric, since the Iraqis could not counter this coalition strength. Similarly, al Qaeda's use of airliners as terror weapons against the United States on 11 September 2001 was asymmetric, since a direct counter would not be used by the United States to prevent the attacks and the US had no effective defense in place at the time. Asymmetric warfare pits friendly strengths against the adversary's weaknesses and maximizes our capabilities while minimizing those of the enemy to achieve rapid, decisive effects.

C. Additional Considerations

In some situations, airpower may be the only force immediately available and capable of providing an initial response. Due to the speed at which Air Force capabilities can be employed, this may occur early in a crisis, before significant friendly surface forces can build up in theater. In such cases, airpower can be brought to bear against the enemy system to directly reduce the enemy's ability to achieve immediate war aims, often through strategic attack.

II. Airpower & the Range of Military Operations

The Air Force conducts operations along a varying scale of military involvement and violence, referred to as the range of military operations (ROMO). They range from continuous and recurring operations such as military engagement, security cooperation, and deterrence; through smaller-scale contingencies and crisis response operations, as well as irregular warfare; to major operations and campaigns such as declared wars. Conflicts may escalate or de-escalate from one form to another. Warfighters may find that military activities like security cooperation and engagement take place simultaneously with major combat operations and irregular warfare. No two operations are alike: scope, duration, tempo, and cultural/political context vary widely. Military leaders should carefully assess the nature of their assigned missions to determine the appropriate mix of forces and discern implied missions and requirements. As military professionals, Airmen should possess the skills and apply airpower doctrine to design, plan, execute, and assess military operations across the ROMO. As an institution, the Air Force organizes, trains, and equips to conduct operations across the ROMO.

Military operations take place in and through the air, land, maritime, space, and cyberspace domains and the information environment. The Air Force exploits advantages in the air, space, and cyberspace domains to achieve joint force commander (JFC) and national objectives in all domains and the information environment. In either a supporting or supported role, these functions can be conducted independently from, or in concert with, land and maritime operations.

The Range of Military Operations and the Conflict Continuum

Airpower is a vital component of successful military operations and can often provide for decisive, rapid, and more efficient attainment of enduring advantage. It has been an asymmetric advantage for the United States in many operations. Defeating enemy forces has traditionally been the most important of the tasks assigned to the military, and while that remains vitally important, national strategic guidance increasingly emphasizes the importance of preventing conflict, deterring adversaries, and shaping the operational environment so as to obtain continuing strategic advantage for the US and its allies. The Department of Defense (DOD) refers to the ongoing and recurring operations intended to accomplish this apart from the realm of war and other major operations as the steady state, and can design, plan, execute, and assess steady-state operations and activities as part of geographically-aligned theater campaigns. The strategies created to accomplish this are called theater campaign plans (TCPs). From a Service perspective, preparation is a foremost priority during the steady state, as success in a crisis depends upon preparedness and readiness at the beginning of that crisis.

See pp. 1-10 to 1-11 for further discussion.

Homeland Operations

The Air Force plays a significant role in homeland operations. It employs airpower to assist federal, state, and local governments, as well as other branches of the Department of Defense (DOD) and non-governmental organizations (NGOs) in detecting, helping preempt, responding to, mitigating, and recovering from a full spectrum of threats and incidents, man-made and natural, within the United States and its territories and possessions. Homeland operations consist of two major mission areas: homeland defense and defense support of civil authorities (DSCA), along with the integral subset mission of emergency preparedness.

While homeland operations may arguably be considered a subset within the ROMO previously described, Air Force doctrine considers these activities important enough to warrant separate discussion.

See p. 1-16 and 2-14 to 2-15 for further discussion.

Chap 4

I. Counterair Operations

Ref: Annex 3-01, Counterair Operations (27 Oct '15).

The US Air Force flies, fights, and wins in the domains of air, space, and cyberspace. Control of the air provides the joint force with freedom of action while reducing vulnerability to enemy detection, attack, and other effects.

Joint doctrine provides broad guidance for countering air and missile threats (Joint Publication [JP] 3-01, Countering Air and Missile Threats), but does not describe the full spectrum of air control, as Annex 3-01 does.

The Air Domain

The Air Force brings specific capabilities to a joint force to achieve various levels of air control by operating in the air domain. Clearly defined domains help identify the conditions and capabilities under which systems and personnel conduct operations, but do not mandate or imply command relationships. The air domain is the area, beginning at the Earth's surface, where the atmosphere has a major effect on the movement, maneuver, and employment of joint forces.

Control of the air is normally one of the first priorities of the joint force. This is especially so whenever the enemy is capable of threatening friendly forces from the air or inhibiting a joint force commander's (JFC's) ability to conduct operations. Counterair is a mission that integrates offensive and defensive operations to attain and maintain a desired degree of air superiority (JP 1-02) Counterair missions are designed to destroy or negate enemy aircraft and missiles, both before and after launch. Counterair helps ensure freedom to maneuver, freedom to attack, and freedom from attack.

Counterair is directed at enemy forces and other target sets that directly (e.g., aircraft, surface-to-air missiles) or indirectly (e.g., airfields, fuel, command and control facilities, network links) challenge control of the air. Airmen integrate capabilities from all components to conduct intensive and continuous counterair operations aimed at gaining varying degrees of air control at the time and place of their choosing.

Air Control

The Air Force defines counterair as a mission that integrates offensive and defensive operations to attain and maintain a desired degree of control of the air and protection by neutralizing or destroying enemy aircraft and missiles, including cruise and ballistic missiles, both before and after launch. Counterair operations are conducted across all domains and determine the level or degree of control of the air. Control of the air describes a level of influence in the air domain relative to that of an adversary, and is typically categorized as parity, superiority, or supremacy. The degree of control lies within a spectrum that can be enjoyed by any combatant. This can range from a parity (or neutral) situation, where neither adversary can claim control over the other, to air superiority, to air supremacy over an entire operational area.

Normally, counterair operations are classified as offensive or defensive. However, airpower's inherent flexibility allows missions and aircraft to shift from defensive to offensive (or vice versa) to adapt to changing conditions in the operational environment. Counterair operations can be conducted across the tactical, operational, and strategic levels of war by any component of the joint force. Operations are conducted over and in enemy, friendly, and neutral territory. The JFC's objectives and desired effects determine when, where, and how these operations are conducted to gain the desired degree of air control.

The Counterair Framework

Like other air, space, and cyberspace operations, counterair is fundamentally effects based. This means that counterair operations are designed, planned, executed, assessed, and adapted in order to influence or change system behavior to achieve desired outcomes (AFDD 2). Effective counterair operations should be part of a larger, coherent plan that logically ties the overall operation's end state to all objectives and effects and tasks. This plan should guide execution and the means of gaining feedback and measuring success must be planned for and evaluated throughout and after execution. This approach should consider all potential instruments of power and all available means to achieve desired effects, and must consider the entire operational environment. The operational environment is a composite of the conditions, circumstances, and influences that affect the employment of capabilities and bear on the decisions of the commander (JP 1-02). Non-military instruments of national power may not seem relevant to counterair operations, but they can be decisively important in certain circumstances, as when diplomatic efforts permit or deny basing or overflight rights that critically impact counterair efforts. Conversely, counterair capability can help deter hostile adversary action by providing a credible military threat to enemy maneuver and freedom to attack.

In an effects-based framework, effects fall into two broad categories: direct effects, or those immediate outcomes created by "blue" (friendly) actions, and indirect effects, higher-order effects created upon "red" (adversary) or "gray" (neutral) actors within the operational environment.

The counterair framework, illustrated on the following pages (pp. 4-10 to 4-11), shows typical "blue" actions taken to create effects in support of counterair operations.

Air Refueling Requirements

Air refueling is an essential enabler of counterair operations. Many air assets that perform the counterair mission have relatively short on-station times or operate from bases far removed from their intended targets. These assets rely on air refueling to extend range, on-station time, and tactical flexibility. Strategists and planners should build needed refueling support into the air component's planning products. Refueling coordination also requires constant management by planners. Detailed refueling instructions should be included in the air tasking order (ATO) special instructions (SPINS) and the airspace control order (ACO).

See pp. 4-50 to 4-51 for discussion of air refueling operations from Annex 3-17.

Intelligence, Surveillance, and Reconnaissance (ISR) Requirements

Effective counterair operations require timely, reliable, and accurate intelligence, so proper joint intelligence preparation of the operational environment (JIPOE) can be crucial to counterair operations. Near-real time information from air, surface, and space-based sensors may provide warning, situational awareness, targeting, and assessment. ISR is also needed to identify and attack or exploit emerging targets that pose a substantial threat to friendly operations. Timely target detection, development, and geolocation, as well as weapon selection, mission planning, and assessment all depend on integrated collection and analysis. Effective integration of ISR assets is often as crucial to successful counterair operations as are traditional lethal effects.

The ISR, combat plans, and strategy divisions within the air operations center (AOC) determine and prioritize the measures and indicators used to assess counterair operations. These measures and indicators help evaluate whether friendly actions have been accomplished and desired counterair effects within the operational environment have been created.

See pp. 4-54 to 4-67 for discussion of ISR from Annex 2-0, Global Integrated ISR Operations.

I. Air Control Relationships

Ref: Annex 3-01, Counterair Operations (27 Oct '15).

Counterair operations can be conducted across the tactical, operational, and strategic levels of war by any component or element of the joint force. Operations may be conducted over and in enemy, friendly, and international airspace, land, and waters; as well as space and cyberspace. They range from seeking out and destroying the enemy's aircraft (manned and unmanned) and missiles (air-to-air, surface-to-air, cruise, and ballistic), through taking measures to minimize the effectiveness of those systems, to countering efforts to contest control of the air through other domains such as cyberspace. The joint force commander's (JFC's) objectives and desired effects determine when, where, and how these operations are conducted to gain the desired degree of control of the air.

Air Control Relationship Elements

Air Parity

Air Superiority

Air Supremacy

A. Air Parity

A condition in which no force has control of the air. This represents a situation in which both friendly and adversary land, maritime, and air operations may encounter significant interference by the opposing force. Parity is not a "standoff," nor does it mean aerial maneuver or ballistic missile operations have halted. On the contrary, parity may be typified by fleeting, intensely contested battles at critical points during an operation with maximum effort exerted between combatants in their attempt to achieve some level of favorable control.

B. Air Superiority

Joint doctrine defines air superiority as, "that degree of dominance in the air battle by one force that permits the conduct of its operations at a given time and place without prohibitive interference from air and missile threats" (JP 3-01). For conceptual clarity, Air Force doctrine further defines air superiority as "that degree of control of the air by one force that permits the conduct of its operations at a given time and place without prohibitive interference from air and missile threats, including cruise and ballistic missiles." Air superiority may be localized in space (horizontally and vertically) and in time, or it may be broad and enduring.

C. Air Supremacy

Joint doctrine defines air supremacy as "that degree of air superiority wherein the opposing force is incapable of effective interference within the operational area using air and missile threats" (JP 3-01). For conceptual clarity, Air Force doctrine further defines air supremacy as "that degree of control of the air by one force that permits the conduct of its operations at a given time and place without effective interference from air and missile threats, including cruise and ballistic missiles." Air supremacy may be localized in space (horizontally and vertically) and in time, or it may be broad and enduring. This is normally the highest level of control of the air that air forces can pursue.

Control of the air hinges on the idea of preventing prohibitive or effective interference to joint forces in the air domain from enemy forces, which would prevent joint forces from creating their desired effects.

II. Counterair Operations

Ref: Annex 3-01, Counterair Operations (27 Oct '15).

Counterair is directed at enemy forces and other target sets that directly (e.g., aircraft, surface-to-air missiles) or indirectly (e.g., airfields, fuel, command and control facilities, network links) challenge control of the air. Normally, counterair operations are classified as offensive or defensive. However, airpower's inherent flexibility allows missions and aircraft to shift from defensive to offensive (or vice versa) to adapt to changing conditions in the operational environment. Counterair operations can be conducted across the tactical, operational, and strategic levels of war by any component of the joint force. Operations are conducted over and in enemy, friendly, and neutral territory. They range from seeking out and destroying the enemy's ability to conduct airborne attacks with both aircraft and missiles, to taking measures to minimize the effectiveness of those attacks. The JFC's objectives and desired effects determine when, where, and how these operations are conducted to gain the desired degree of air control.

The Counterair Framework

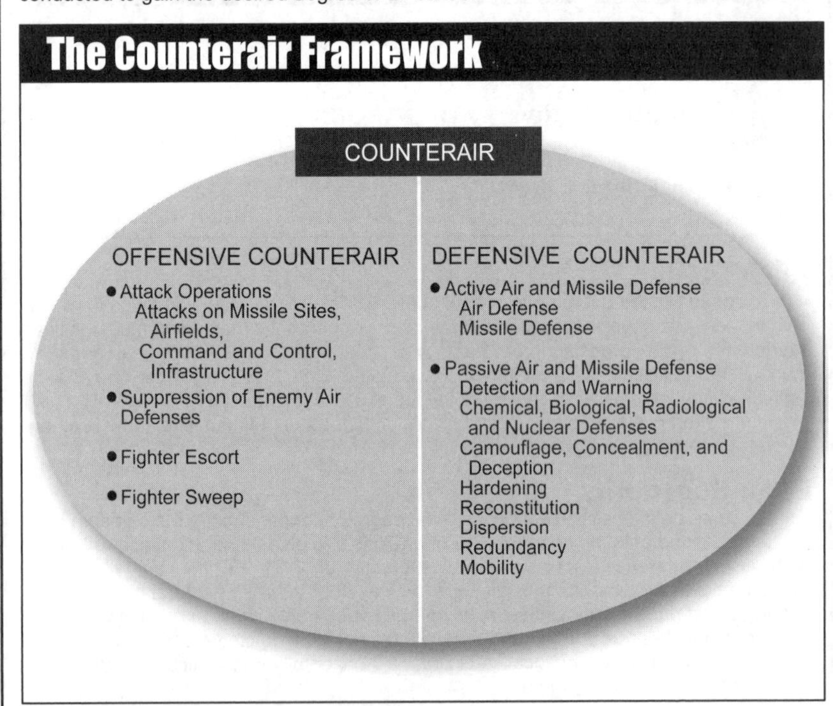

A. Offensive Counterair (OCA)

The objective of offensive counterair (OCA) is to destroy, disrupt, or degrade enemy air capabilities by engaging them as close to their source as possible, ideally before they are launched against friendly forces. Otherwise, OCA operations seek out and destroy these targets as close to their launch locations as possible. These operations may range throughout enemy, friendly, and international airspace and waters and are generally conducted at the initiative of friendly forces. OCA targets may include but are not limited to: enemy air defense systems, theater missile systems, airfields, airfield support infrastructure, C2 nodes, multi-domain launch platforms, and launch platform supporting infrastructure. OCA operations enable friendly use of contested airspace and reduce the threat of airborne attacks against friendly forces.

- **Attack operations**. Attack operations are intended to destroy, disrupt, or degrade counterair targets on the ground. These missions are directed against enemy air and missile threats, their C2, and their support infrastructure (e.g., airfields, launch sites, launchers, fuel, supplies, and runways). The main goal is to prevent enemy employment of air and missile assets.
- **Suppression of enemy air defenses**. SEAD is an OCA mission designed to neutralize, destroy, or degrade enemy surface-based air defenses by destructive or disruptive means. SEAD requirements may vary according to mission requirements, system capabilities, and threat complexity. SEAD planners should coordinate with ISR operators to ensure collection and exploitation opportunities are considered prior to destroying or disrupting emitters. SEAD operations fall into three categories:

  - **Area of responsibility (AOR)/joint operating area (JOA) air defense suppression**: Operations conducted against specific enemy air defense systems to destroy, disrupt, or degrade their effectiveness. It targets high payoff air defense assets, resulting in the greatest degradation of the enemy's total system and enabling effective friendly operations.

  - **Localized suppression**: Operations normally confined to geographical areas associated with specific ground targets or friendly transit routes, contributing to local air superiority.

  - **Opportune suppression**: Usually unplanned, including aircrew self-defense and attack against targets of opportunity. The JFC or JFACC normally establishes specific ROE to permit airborne assets the ability to conduct opportune suppression.
- **Fighter escort**. Escorts are aircraft assigned to protect other aircraft during a mission (JP 1-02). Escort missions are flown over enemy territory to target and engage enemy aircraft and air defense systems. Friendly aircraft en route to or from a target area may be assigned escort aircraft to protect them from enemy air-to-air and surface-to-air threats. Typically, escort to low-observable ("stealth") aircraft requires special consideration and planning at the air operations center (AOC) level.
- **Fighter sweep**. An offensive mission by fighter aircraft to seek out and destroy enemy aircraft or targets of opportunity in a designated area.

B. Defensive Counterair (DCA)

The objective of defensive counterair (DCA) is to protect friendly forces and vital interests from enemy airborne attacks and is synonymous with air defense. DCA consists of active and passive air defense operations including all defensive measures designed to destroy attacking enemy airborne threats or to nullify or reduce the effectiveness of such threats should they escape destruction. The basic active defense criteria to detect, identify, intercept, and destroy remain the same for any airborne threat. DCA forces generally react to the initiative of the enemy and are subject to the weapons control procedures of the area air defense commander (AADC).

Several types of DCA tasks also help to provide a permissive environment for friendly air action.

- **Active air and missile defense**. Active air and missile defense is defensive action taken to destroy, nullify, or reduce the effectiveness of air and missile threats against friendly forces and assets. It includes actions to counter enemy manned and unmanned aircraft, cruise missiles, air-to-surface missiles, and ballistic missiles. These actions are closely integrated to form essential DCA capabilities, but may involve different defensive weapon systems or tactics, techniques and procedures (TTP).
- **Passive air and missile defense.** Passive defense includes all measures, other than active defense, taken to minimize the effectiveness of hostile air and missile threats against friendly forces and assets. It consists of several categories of activities.

Integrated Air and Missile Defense (IAMD)

IAMD is the integration of capabilities and overlapping operations to defend the homeland and United States national interests, protect the joint force, and enable freedom of action by negating an adversary's ability to achieve adverse effects from their air and missile capabilities. At the theater level, IAMD is a subset of counterair and an approach which combines OCA attack operations and DCA operations (see The Counterair-IAMD Relationship figure below) to achieve the joint force commander's desired effects. Within the IAMD approach, OCA attack operations are commanded by the JFACC and DCA is commanded by the AADC. The JFACC is responsible for integration between the offensive and defensive components of IAMD. The OCA attack operations component of IAMD will, in all likelihood, not be planned and executed in isolation but rather will be a part of a wider offensive effort against a variety of adversary targets.

III. Planning Considerations

Counterair planning may be conducted at every echelon of command and across the range of military operations. Counterair planning should take into account the capabilities of all the Services, joint force components, and interagency and multinational partners. Counterair planning is conducted using the joint operation planning process for air.

See pp. 6-5 to 6-18 for discussion of the joint operation planning process for air (JOPPA).

During joint intelligence preparation of the operational environment (JIPOE), planners should determine the adversary's active and passive counterair capabilities, as well as his intent to contest control of the air with those capabilities, if possible. This, in turn, should inform the joint force air component commander's (JFACC's) and joint force commander's (JFC's) decision-making efforts during mission analysis and course of action (COA) development.

Normally, the JFACC's first priority should be to define—in both time and space—that level of air control needed to achieve the JFC's objectives. Once defined, the JFACC should identify the current level of control in the air (parity, superiority, or supremacy) and what actions are required to reach the desired level of control. This determination will drive the priorities for AOC planners. AOC planners and the JFACC must inform the JFC as to which level of air control is realistically achievable given current capabilities and allocation of assets. When analyzing forces available, it is important to consider the capabilities of other joint force components and multinational partners.

A. Offensive Counterair (OCA)

Offensive counterair (OCA) may be the highest payoff air component mission when the enemy has the capability to significantly threaten friendly forces with air and missile assets. Given finite resources, the JFACC should judiciously allocate them in order to meet the JFC's objectives. Successful OCA results in greater freedom from attack, enabling increased freedom of action, and freeing assets for other operations against the enemy. In other words, the initial investment in OCA operations to achieve the desired level of control of the air may pay significant dividends toward overall mission accomplishment. Determining which enemy capabilities hinder control of the air is fundamental to successful OCA operations. For instance, it may not be necessary to completely destroy a given capability, but only temporarily degrade it in order to achieve desired effects. The latter may require much less effort, thereby freeing up assets for other missions. This type of analysis may vary from one operation to another but often results in an effective set of target priorities and an efficient use of assets to achieve the desired effects.

The nature of airpower is such that offensive combat power can frequently be "massed" by distributing forces. In fact, the most effective OCA efforts may be achieved as part of a broader, parallel attack on the adversary as a system-of-systems with all available assets, to include cyberspace and space capabilities.

Planners should utilize intelligence to determine the adversaries' capabilities and expect at a minimum that adversaries will have at least a rudimentary integrated air defense system (IADS), consisting of both active and passive defenses, even if they do not possess any significant offensive air potential.

OCA Targeting Priorities and Methods

The following considerations are important for determining OCA targeting priorities and methods:

- **Threat**. The threat posed by specific enemy capabilities (aircraft, theater missiles, etc.) includes an assessment of the urgency or the need to counter that threat. A WMD-capable missile launcher would normally merit diversion of assets from a less immediate threat, such as a SAM site.
- **Direct effects.** First-order results of actions with no intervening effects between action and outcome. These are usually immediate, physical, and readily recognizable (e.g., weapon employment results). These are important in determining whether friendly tasks were accomplished. Planning for them must also consider such factors as collateral damage potential and rules of engagement restrictions.
- **Indirect effects.** Second, third, or higher-order effects created through intermediate effects or causal linkages following causal actions. These may be physical, psychological, functional, or systemic in nature. They may be created in a cumulative, cascading, sequential, or parallel manner. They are often delayed and typically are more difficult to recognize and assess than direct effects.
- **Forces available.** The forces available are assessed against the number, types, and priority of targets that can be attacked. Sufficient and capable forces should be provided to ensure the desired results are obtained.
- **Time available and time required.** Time constraints are integral to prioritization and planning. The time allowed to achieve the direct and indirect effects as well as the duration required of those effects will influence the number and type of forces required.
- **Risk.** Risk calculation involves weighing the risk to friendly forces against expected gains from target attack. Risk calculation should also consider the risks entailed in not taking planned actions. Different objectives and circumstances drive different acceptable levels of risk.
- **Measures and indicators.** These are the essential component parts of assessment; the means of evaluating progress toward creating effects and achieving objectives. They should be determined during planning.

The types of resources available to perform OCA tasks (listed under "execution," below) are only "tools" in a planner's "toolkit." Desired effects should drive planning efforts and there may be many ways to impose a particular effect. The means may be chosen based on a number of criteria, including desired higher-order indirect effects.

Planning for OCA usually takes place in the AOC as part of the joint operation planning process for air. In early stages of planning, the JFACC, along with the AOC's strategy and intelligence, surveillance, and reconnaissance (ISR) divisions, will determine objectives, desired effects, and relative priorities. Planners in the strategy, combat plans, and ISR divisions will determine enemy systems, capabilities, and assets that can be used to contest control of the air. Combat plans and combat operations personnel will use this information to match desired effects to targets provided by the ISR division, and match targets with friendly forces to create tactical tasks.

OCA Planning at the AOC and Unit Mission Planning Levels

The following considerations are important at AOC and unit mission planning levels:

- **Enemy threat, location, and capabilities.** The enemy threat to air operations needs proper consideration in the planning, positioning, and timing of OCA mission details. Specific threats to the OCA effort (aircraft, missiles, AAA, electronic attack) may require substantial emphasis be placed on their disruption prior to striking intended targets.
- **Friendly C2 capabilities.** Theater C2 assets such as AWACS and JSTARS, are tasked by numerous units and agencies. As such, OCA planners should not assume that complete C2 capabilities will be available for every OCA mission. In all cases, C2 instructions should be carefully monitored, because this is the avenue through which higher-priority re-tasking will come.
- **Rules of engagement.** ROE (and related special instructions [SPINS]) found in tasking orders, as well as rules for use of force, often used in situations such as homeland defense and civil support missions) may critically affect how missions are performed. All levels, from the JFACC down to individual aircrews, should understand the ROE that apply to the accomplishment of their missions.
- **Weaponeering.** Assigning the correct weapons and platforms to a specific target set is a critical job. Accurate weaponeering increases the chances of achieving desired effects.
- **Deconfliction.** The sheer number of airborne assets demands that planners deconflict to protect friendly forces from unnecessary risk.
- **Environmental conditions.** The significance of environmental conditions on counterair cannot be overstated. Weather can limit sensor or seeker sensitivity and ultimately limit the planner's munitions selection. Likewise, varying terrain can be a challenge to pilots or offer refuge to an adversary. Terrain will often limit munitions selection.
- **Distance, timing, and refueling.** OCA and DCA air assets typically require refueling support for sustained presence. Refueling coordination requires constant management by planners, and details need to be stated in ATO SPINS.

B. Defensive Counterair (DCA)

While OCA seeks to affect enemy counterair systems close to their source, DCA seeks to affect those same systems closer to their intended targets. In some cases, DCA may also be the only allowed means of countering air and missile threats due to constraints imposed by the political situation. Effective OCA greatly reduces the DCA requirement, freeing assets for more offensive operations, but some degree of DCA is normally necessary in every operation. DCA operations defend friendly lines of communication, protect friendly forces and assets by denying the enemy the freedom to carry out offensive attacks from the air, and provide a secure area from which all elements of the joint force can operate effectively. DCA operations can be conducted in conjunction with or independent of OCA operations and generally fall into one of two categories: Active or passive defense.

Just as in OCA operations, DCA planners prioritize which assets and capabilities to defend. Planners at all levels identify enemy targets and capabilities to defend against, while matching available forces against the threat. They use many of the same OCA planning considerations. Planners determine which mission-critical assets and capabilities to protect, which will vary from operation to operation.

Active Air Defense

Active air defense is direct defensive action taken to destroy, nullify, or reduce the effectiveness of hostile air and missile threats against friendly forces and assets (JP 1-02). Active air defense operations are conducted using a mix of weapon and sensor systems, supported by secure and highly responsive C2 systems, to find, fix track, target, and destroy or reduce the effectiveness of hostile airborne threats. These operations attempt to neutralize or degrade the effectiveness of enemy attacks and protect friendly forces and interests through the direct employment of weapons systems. Active air defense targets include any airborne threat that negatively impacts friendly operations.

Integrated employment of air-to-air and surface-to-air defense systems through coordinated detection, identification, engagement, and assessment of enemy forces is necessary to defeat enemy attacks and protect friendly forces. Planners should keep in mind the complexities of airspace control in a DCA environment. Airspace control in an active air defense environment is extremely difficult and becoming more complicated with the proliferation of UAS. Rapid, reliable, and secure means of identification are critical to the survival of friendly aircraft and to facilitate an effective defense against enemy air and missile attacks.

The efficient execution of air defense operations requires the ability to quickly detect a potential air defense threat, identify it, target and track it, and attack it. DCA engagements may occur inside friendly airspace, requiring careful deconfliction between friendly assets, such as fighters in the DCA role and friendly SAMs. An agile ISR capability is essential to provide continuous surveillance and reporting of real time and near-real time target track data. To maximize damage to the enemy force, the engagement process is continuous throughout the threat's approach, entry into, and departure from the friendly operational area. Target track production is a sequential process that begins with the surveillance function.

Near-real time surveillance and threat analysis depends on the ability to fuse all-source sensor data (ground, air, sea, and space-based sensors) into an accurate theater attack assessment. As a track is detected, it is identified and labeled; this information is then disseminated as rapidly as possible. The track data provided should be sufficiently detailed and timely to permit the C2 system to evaluate the track, determine the significance of the threat, and designate air defense forces for interception. The optimum employment of air defense weapon systems involves the earliest possible discrimination of friend from foe to maximize beyond-visual-range engagement. To prevent fratricide, great caution should be exercised when employing autonomous CID in DCA operations.

If no IADS is established, procedural means should be used to permit the safe passage of friendly aircraft while still allowing for the use of air defense weapons (fighter engagement zones, missile engagement zones, and joint engagement zones).

Passive Air Defense

Unlike active air defense measures, passive air defense does not involve the employment of lethal weapons. Rather, these measures improve the survivability of friendly forces by reducing the potential effects of enemy attacks. Passive air defense measures are designed to provide protection for friendly forces and assets by complicating the enemy's identification, surveillance, and targeting processes and by countering the enemy's planned effects.

The first step of passive air defense is to hide valuable assets from the enemy or to encourage him to attack decoys. Like active air defense measures, a thorough passive defense should include layered defense in depth. Passive measures can work concurrently to achieve this goal. These measures include camouflage, concealment, and deception; hardening; reconstitution; dispersal; electronic and infrared countermeasures; and low observable (LO) or stealth technologies. Passive air defenses are often an additional means of defense should active air defense efforts fail.

IV. Combat Identification (CID)

Ref: JP 3-01, Countering Air and Missile Threats () pp. III-14 to III-15.

CID is the process of attaining an accurate characterization of detected objects in the operational environment sufficient to support an engagement decision. The CID process uses all available PID methods to achieve the highest confidence possible for that decision, because it is normally one of the most critical decisions to be made. The CID process applies to all joint forces for both defensive and offensive actions.

The CID process is determined by the JFC (commonly in coordination with subordinate commanders), supported by the ROE, and may be situational dependent and/or time sensitive. CID allows the commander to balance the level of confidence in the ID method against the risk associated with an erroneous ID. While high confidence-low risk is always desired, the commander may face situations when the absence of PID requires procedural ID be used with a recognized increase in the risk of friendly fire or to misidentify an enemy (i.e., low confidence-high risk). This remains a commander's decision. For example, during DCA operations against numerous simultaneous attacks by enemy aircraft and CMs, potentially with WMD, it may be necessary to accept lower confidence ID methods for hostiles and increased risk of friendly fire to minimize the risk of a "leaker" getting through to the target. Unambiguous lines of command and clarity of ROE are particularly important to the CID process, especially when delegating authority for engagement decisions during decentralized execution.

ID Matrix

The ID matrix is a logic tree for categorizing a track (e.g., friendly, hostile, unknown, or neutral) and following it throughout its life in the AOR/JOA. If not identified as friendly, an object being tracked may require further assessment based on position, the ROE, and weapons control status (WCS).

For a detailed discussion of CID refer to ATP 3-01.15[FM 3-01.15]/MCRP 3-25E/ NTTP 3-01.8/AFTTP 3-2.31, Multi-Service Tactics, Techniques, and Procedures for an Integrated Air Defense System.

Chap 4

II. Counterland Operations (AI & CAS)

Ref: Annex 3-03, Counterland Operations (17 Mar '17) and JP 3-09.3 Close Air Support (Jul '09).

Counterland operations are defined as "airpower operations against enemy land force capabilities to create effects that achieve joint force commander (JFC) objectives." The aim of counterland operations is to dominate the surface environment using airpower. By dominating the surface environment, counterland operations can assist friendly land maneuver while denying the enemy the ability to resist. Although most frequently associated with support to friendly surface forces, counterland operations may also be conducted independent of friendly surface force objectives or in regions where no friendly land forces are present. For example, recent conflicts in the Balkans, Afghanistan, and Iraq illustrate situations where counterland operations have been used absent significant friendly land forces or with small numbers of special operations forces (SOF) providing target cueing. This independent attack of adversary land operations by airpower often provides the key to success when seizing the initiative, especially in the opening phase of an operation.

Counterland Operations

Air Interdiction (AI)

Close Air Support (CAS)

Counterland operations provide the JFC two distinct types of operations for engaging enemy land forces. The first is air interdiction (AI), which is defined as "air operations conducted to divert, disrupt, delay, or destroy the enemy's military surface capabilities before it can be brought to bear effectively against friendly forces, or to otherwise achieve objectives that are conducted at such distances from friendly forces that detailed integration of each air mission with the fire and movement of friendly forces is not required." Air Interdiction indirectly supports land forces and directly supports JFC objectives in the absence of friendly land forces. The second distinct type of air operations is close air support (CAS) which is defined as "air action by fixed- and rotary-wing aircraft against hostile targets that are in close proximity to friendly forces and that require detailed integration of each air mission with the fire and movement of those forces." While AI can support either the JFC or the land component, CAS directly supports land maneuver forces. Whether destroying enemy surface forces, interdicting supply routes, or providing CAS to friendly troops, counterland operations

Counterland operations can serve as the main attack and be the decisive means for achieving JFC objectives. Although often associated with support to friendly surface forces, counterland operations also include operations that directly support the JFC's theater strategy rather than exclusively supporting a surface component. In some cases, counterland operations can provide the sole US effort against the enemy.

Counterland Operations

Ref: Annex 3-03, Counterland Operations (17 Mar '17), pp. 9-15.

The commander, Air Force forces (COMAFFOR), executes counterland operations by conducting air interdiction (AI) as the supported or supporting commander or by supporting land forces with close air support (CAS). AI and CAS missions can function under an overall theater posture of offense or defense and are typically coordinated with a ground scheme of maneuver to maximize the effect on the enemy.

I. Air Interdiction (AI) *See pp. 4-20 to 4-24.*

The purpose of interdiction operations is to divert, disrupt, delay, and destroy, by either lethal or nonlethal means in order to achieve objectives. Actions associated with one desired effect may also support the others. Air interdiction (AI) is defined as "air operations conducted to divert, disrupt, delay, or destroy the enemy's military potential before it can be brought to bear effectively against friendly forces, or to otherwise achieve objectives that are conducted at such distance from friendly forces that detailed integration of each air mission with the fire and movement of friendly forces is not required." 5 AI targets may include fielded enemy forces or supporting components such as operational command and control (C2) nodes, communications networks, transportation systems, supply depots, military resources, and other vital infrastructure. When conducted as part of a joint campaign, AI needs the direction of a single commander who can exploit and coordinate all the forces involved.

The commander, Air Force forces (COMAFFOR) is normally the supported commander for the joint force commander's (JFC's) overall AI effort. When designated as the supported commander, the COMAFFOR will conduct theater-wide or joint operations area-(JOA-) wide AI to support the JFC's overall theater objectives. With the preponderance of AI assets and the ability to plan, task, and control joint air operations, the COMAFFOR can best plan and execute AI. The COMAFFOR recommends theater and/or JOA-wide targeting priorities and, in coordination with other component commanders, forwards the air apportionment recommendation to the JFC. The COMAFFOR plans and executes the interdiction effort in accordance with the JFC's guidance.

II. Close Air Support (CAS) *See pp. 4-24 to 4-30.*

Close air support (CAS) is defined as "air action by fixed- and rotary-winged aircraft against hostile targets that are in close proximity to friendly forces and that require detailed integration of each air mission with the fire and movement of those forces." CAS provides supporting firepower in offensive and defensive operations to destroy, disrupt, suppress, fix, harass, neutralize, or delay enemy targets as an element of joint fire support. The speed, range, and maneuverability of airpower allows CAS assets to attack targets that other supporting arms may not be able to engage effectively. When conditions for air operations are permissive, CAS can be conducted at any place and time friendly forces are in close proximity to enemy forces and, at times, may be the best means to exploit tactical opportunities.

Although in isolation CAS rarely achieves campaign-level objectives, at times it may be the more critical mission due to its contribution to a specific operation or battle. CAS should be planned to prepare the conditions for success or reinforce successful attacks of surface forces. CAS can halt enemy attacks, help create breakthroughs, destroy targets of opportunity, cover retreats, and guard flanks. To be most effective, CAS should be used at decisive points in a battle and should normally be massed to apply concentrated combat power and saturate defenses. Elements of the theater air control system (TACS) must be in place to enable command and control and clearance to attack in response to rapidly changing tactical circumstances. In fluid, high-intensity warfare, the need for terminal attack control, the unpredictability of the tactical situation, the risk of collateral damage and friendly fire incidents, and the proliferation of lethal ground-based air defenses make CAS especially challenging.

CAS requires a significant level of coordination between air and surface forces to produce desired effects, avoid excessive collateral damage and prevent friendly fire incidents. CAS employment should create effects that support the ground scheme of maneuver. The fluidity of the ground situation that exists within close proximity usually requires real-time direction from a joint terminal attack controller (JTAC) to ensure that targets of highest priority to the ground commander are struck.

Types of AI and CAS

Counterland missions are either scheduled or on-call. Scheduled missions result from preplanned requests during the normal air tasking order (ATO) cycle and allow for detailed coordination between the tactical units involved. Additionally, preplanned requests may result in counterland sorties in an on-call status (either airborne or ground alert) to cover periods of expected enemy action, respond to immediate requests, or attack dynamic targets. Scheduled air interdiction (AI) missions use detailed intelligence to attack known or anticipated targets in an operational area to generate effects that achieve joint force commander (JFC) objectives. Scheduled close air support (CAS) missions are normally allocated to a specific ground unit or operation. Air planners attach a "G" or "X" prefix to the ATO mission identifier to designate either ground or airborne alert, respectively.

- **AI** is a mission scheduled to strike particular targets in response to JFC or component target nominations.
- **GAI** is the AI term used to identify an on-call mission placed on ground alert to provide responsive AI throughout the theater in response to emerging targets.
- **XAI** is the AI term used to identify an airborne alert AI mission tasked for on-call targets that may be retasked during execution for targets of opportunity (also referred to as armed reconnaissance).
- **SCAR** (Strike Coordination and Reconnaissance) missions use aircraft to detect targets for dedicated AI missions in a specified geographic zone. The area may be defined by a box or grid where worthwhile potential targets are known or suspected to exist, or where mobile enemy surface units have relocated because of ground fighting.
- **CAS** is a mission scheduled to provide air support in response to preplanned CAS requests.
- **GCAS** is the CAS term used to identify an on-call mission placed on ground alert status to provide responsive air support to ground forces that encounter substantial ene**my resistance. CAS assets located close to the supported ground forces normally provide faster response times. GCAS missions may be changed to XCAS as the situation dictates.**

Derivative Missions Associated with Counterland

Derivative mission-types are frequently tasked to complement and support counterland operations. The following discussion briefly describes the two most common missions that are associated with, and facilitate the effective accomplishment of, CAS and AI.

- **Forward Air Controller (Airborne) (FAC[A]).** FAC(A) missions provide joint terminal attack control for CAS aircraft operating in close proximity to friendly ground forces. Because of the risk of fratricide, FAC(A)s are specially trained aviation officers qualified to provide delivery clearance to CAS aircraft. The FAC(A) is the only person cleared to perform such control from the air, and can be especially useful in controlling CAS against targets that are beyond the visual range of friendly ground forces.
- **Strike Coordination and Reconnaissance (SCAR).** SCAR missions use aircraft to detect targets for dedicated AI missions in a specified geographic zone. The area may be defined by a box or grid where worthwhile potential targets are known or suspected to exist, or where mobile enemy surface units have relocated because of ground fighting.

I. Air Interdiction (AI)

Air interdiction (AI) represents a flexible and lethal form of airpower that can be used in various ways to prosecute the joint operation. However employed, certain principles such as centralized control/decentralized execution should be followed to achieve maximum effectiveness with minimum losses. AI can channel enemy movement, constrain logistics, disrupt communications, or force urgent movement to put the enemy in a favorable position for friendly forces to exploit. To be most effective, AI requires persistence, concentration, joint integration, and intelligence that is both timely and accurate. Whether supporting the ground offensive by attacking ground-nominated targets or decisively halting an enemy advance with theater-wide interdiction, AI provides a powerful tool for defeating the enemy ground force.

AI increases airpower's efficiency because it does not require detailed integration with friendly forces. Detailed integration requires extensive communications, comprehensive deconfliction procedures, and meticulous planning. AI is inherently simpler to execute in this regard. Therefore, if the enemy surface force presents a lucrative target, AI conducted before friendly land forces make contact can significantly degrade the enemy's fighting ability and limit the need for close air support (CAS) when the two forces meet in close combat.

The air component often conducts theater-wide air attacks against enemy land forces and their resources to achieve joint force commander (JFC) objectives. This autonomous use of AI usually occurs outside of a surface component's area of operations (AO). Special operations forces (SOF) air and ground assets may play a significant supporting role during AI with their ability to seamlessly integrate into the find, fix, track, target, engage and assess (F2T2EA) process.

Using JFC priorities and understanding the surface component's scheme of maneuver, the commander, Air Force forces (COMAFFOR) can employ AI to provide effects that facilitate and support the maneuver. The COMAFFOR may support a land scheme of maneuver by conducting AI within a surface commander's AO. After coordinating priorities, effects, timing, and targets with surface components, the COMAFFOR directs responsive AI across the joint operations area (JOA) against enemy military capabilities that contribute directly to, or are maneuvering to reinforce, the conflict. US ground commanders often consider AI synonymous with what they express as "shaping" operations within the ground commander's AO. From an Airman's perspective, shaping may be regarded as preparing the operational environment with AI to assist the surface component's scheme of maneuver.

A. Effects of Air Interdiction

Air interdiction (AI) effects differ with every situation and can significantly affect the course of an operation. AI against an enemy with minimal logistics requirements, a simple force structure, and primitive logistics systems differs from AI conducted against a highly mechanized, modern force possessing intensive logistics requirements. Interdiction conducted against enemy forces and logistics, without regard to the overall theater situation, may be largely ineffective; therefore planning for interdiction should be closely integrated in the joint force commander's (JFC's) overall planning process.

The effectiveness of AI is dependent on a number of variables. The time required for AI to affect the enemy, and the duration and depth of those effects, depends on several factors. These factors include, but are not limited to, the distance between interdiction operations and the location of intended effects; the means and rate of enemy movement (ships, trains, aircraft, trucks), the physical target (forces, supplies, fuel, munitions, infrastructure), the level of enemy activity, enemy tactics, and the resilience of the targeted force or system AI will have a more robust effect in linear combat against a modern, mobile, conventional force utilizing significant resources. The timing and magnitude of effects will vary depending upon where AI is conducted

B. Interdiction Objectives

Ref: Annex 3-03, Counterland Operations (17 Mar '17), pp. 21-22.

It is not necessary for an air interdiction (AI) operation to focus solely on a single objective; in fact, AI typically inflicts multiple effects on the enemy. The enemy army traveling to the front while under air attack will suffer some level of destruction. The remaining force will likely be delayed in getting to its destination and will suffer some level of physical and psychological disruption. The following describes the objectives for interdiction:

1. Divert

AI diverts enemy fielded forces from areas where critically needed, to a location more favorable to the joint force commander (JFC), or around established lines of communications (LOCs). It may divert resources en route to repair and recover damaged equipment and facilities as well as forces tasked to keep existing LOCs open.

2. Disrupt

AI planners should focus on the enemy critical vulnerabilities that result in disruptive effects on command and control (C2), intelligence collection, and transportation and supply lines (e.g. ammunition or petroleum, oil, and lubricants [POL]). Planners should consider the psychological effect on the enemy's morale and will, historically an airpower strongpoint. When analyzing the enemy considerations include the enemy's strategy, current operational situation, what reserves or workarounds are available to the enemy, and time before the enemy is affected by friendly actions.

3. Delay

Delaying the enemy allows friendly forces to gain time and momentum. While its purpose is to improve the JFC's operational environment, for delay to have a major impact on combat operations, the enemy must face urgent movement requirements in support of its own operations or in countering friendly maneuver, or enhance the effect of a planned friendly maneuver. Ideally, by the air component maintaining the initiative, the opponent is forced to make unplanned urgent movements at times and places that maximize their exposure to additional friendly targeting. Delay payoffs include prolonging the time of risk to attack to land or naval forces, vehicles amassed behind a damaged route segment, or ships trapped in harbor due to mines rendering them ineffective and placing them at risk to lethal action.

4. Destroy

Destruction of the enemy surface force, supporting elements, and supplies is the most direct of the four objectives of AI but the act (actual or perceived) may also provide synergy among the four. The enemy's perception of its imminent destruction can achieve substantial delay and diversion of enemy resources being as effective as physically destroying target systems, if it causes the enemy to react in a way upon which friendly forces can capitalize. Destroying transportation systems may cause the enemy to move only at night or to mass air defense assets (which may be useful elsewhere) around critical transportation nodes. The actual or perceived destruction of LOCs may divert engineering resources from other tasks to prepare alternate routes in anticipation of possible attacks. This may be true when transportation systems remain largely undamaged. Planners should be cognizant that destruction may also inhibit friendly freedom of action. For example, destruction of key transportation targets could hinder future surface operations that intend to use the same infrastructure. Appropriate coordination of AI with other joint force components helps preserve friendly freedom of action.

C. Types of Air Interdiction Requests

Ref: Annex 3-03, Counterland Operations (17 Mar '17), pp. 29-32.

Air interdiction (AI) requests fall into two categories: preplanned and immediate. Each type of request is influenced by a variety of factors. Unless time constraints dictate otherwise, preplanned requests should always be accomplished to allow for proper weapon-target combination, target area tactics planning, threat avoidance, weather study, and other variables, to maximize the probability of target destruction with minimum losses and minimization of collateral damage. Attacking mobile or short-notice targets provides a more flexible response that can capitalize on opportunities, but lack of mission planning can reduce effectiveness, increase the risk of causing collateral damage and higher friendly losses may be expected. Real-time information technology and digital cockpit imagery reduce, but do not eliminate, these factors. Kill box operations can also add a flexible response option, enabling timely and effective coordination and control as well as facilitating rapid attacks. Combining the traditional aspects of both an airspace coordinating measure (ACM) and fire support coordination line (FSCM) enables expeditious air-to-surface attack of targets that can also be augmented by or integrated with surface-to-surface indirect fires.

Preplanned Requests

Preplanned AI is the normal method of operation in which aircraft attack prearranged or planned targets. This mode is used to hit specific targets that are known in advance, and detailed intelligence information is available to support strike planning. Preplanned attacks are normally flown against fixed targets or against mobile targets that are not expected to move in the interval between planning and execution (e.g., revetted tanks). Target information for scheduled AI can come from sources that vary from overhead reconnaissance to ground-based special operations forces (SOF). Preplanned AI is conducted within the normal air tasking cycle and provides enough time for close coordination with other joint force components. It is crucial for component liaisons to communicate and work together to facilitate centralized planning and effective integration, and avoid duplicating effort. Preplanned AI requests evolve into scheduled and on-call missions.

- **Scheduled missions** are planned against targets on which air attacks are delivered at a specific time.
- **On-call missions** are planned against targets other than scheduled missions for which a need can be anticipated but which will be delivered upon request rather than a specific time. On-call AI missions can produce responsive, flexible effects. In cases where a specific area to search for enemy AI targets cannot be predetermined, these missions are designated as airborne air interdiction (XAI) or ground-based alert air interdiction (GAI) on the air tasking order (ATO) and may be put on an airborne alert status.

Immediate Requests

Immediate AI meets specific requests which arise during the course of a battle and which by their sudden nature are not planned in accordance with the normal ATO process. Immediate AI requests can respond to unplanned or unanticipated targets that require urgent, time-sensitive attention. It should be noted that many immediate requests for AI allow sufficient time for in-depth planning prior to execution even if those requests fall inside of the normal 72-hour air tasking cycle that defines "immediate." Immediate AI often responds to attack requests against dynamic and time-sensitive targets (TSTs).

Dynamic Targeting

Dynamic targeting prosecutes targets identified too late, or not selected for action in time to be included in deliberate targeting. It is the active process of identifying, prosecuting, and effectively engaging emerging targets. Dynamic targeting includes prosecution of several categories of targets:

- JFC-designated TST—targets or target set of such high importance to the accomplishment of the joint force commander's (JFC's) mission and objectives, or one that presents such a significant strategic or operational threat to friendly forces or allies, that the JFC dedicates intelligence collection and attack assets, or is willing to divert assets away from other targets in order to engage it. The dynamic targeting process is referred to as find, fix, track, target, engage and assess (F2T2EA).
- Targets that are considered crucial for success of friendly component commanders' missions, but are not JFC-approved TSTs. Component commanders may nominate targets to the JFC for consideration as TSTs. If not approved as TSTs by the JFC, these component-critical targets may still require dynamic execution with cross-component coordination and assistance in a time-compressed fashion
- Targets that are scheduled to be struck on the ATO being executed but have changed status in some way (such as fire support coordination measures changes)
- Other targets that emerge during execution that friendly commanders deem worthy of targeting, prosecution of which may not divert resources from higher-priority targets

Time-Sensitive Targets (TST)

A TST is a JFC-validated target or set of targets requiring immediate response because it is a highly lucrative, fleeting target of opportunity or it poses (or will soon pose) a danger to friendly forces. The commander, Air Force forces may recommend TSTs to the JFC. TSTs are prosecuted using the dynamic targeting process described above, but are of higher priority and may require additional coordination with other components or the joint task force. The destruction of these high payoff targets is considered critical for achieving JFC objectives. The JFC is ultimately responsible for TST prosecution and relies upon the component commanders for conducting TST operations.

When using on-call or dynamically re-tasked assets, immediate AI often relies on an offboard sensor such as Joint Surveillance Target Attack Radar System (JSTARS) to provide initial target detection and attack targeting information. Using real-time target information via data-link, response times can be as short as a few minutes, depending on the distances and C2 arrangements involved. Immediate AI requests allow assets to exploit enemy vulnerability that may be of limited duration. It can work particularly well when attacking enemy ground forces on the move in the enemy rear area and provide a responsive use of counterland attack when supporting the surface component. The air support operations center (ASOC) normally coordinates and directs immediate AI requests flown short of the fire support coordination line (FSCL).

The same quick-responsive nature of immediate AI that allows it to take advantage of fleeting opportunities can also have a negative impact on individual mission success. Scheduled missions allow aircrews more time to study the target imagery and to align attack axes to optimize weapons effects. Detailed study can reduce threat exposure and allow mission planners to optimize the weapon's fusing for maximum effect. Preplanning allows better packaging of strike and support assets when required. The bottom line for dynamic targeting of airborne assets is that it should be used in those cases when the need for a short reaction time outweighs the reduced effectiveness that may result when compared with preplanned operations. Moreover, opportunity costs should be considered. Commanders should ensure the benefits of diverting airpower away from a preplanned target outweigh the costs by pondering several variables. Is it affordable to delay striking a preplanned target? What are the priorities? Will diverting airpower to an unplanned target create greater effects or is it less efficient? In short, the payoff of striking a dynamic target should be worth the cost of diverting preplanned assets.

To increase situational awareness during dynamic targeting, C2 elements should ensure that aircrews have the most current information pertaining to the location of SOF, friendly ground forces, and no-strike target lists and the nature of the enemy. AI deep in the operational area will usually produce extensive, protracted effects that take longer to occur while AI conducted near the front lines typically produces immediate, but geographically limited, effects. During major operations and campaigns the effects of AI are typically more apparent by influencing an enemy's ability to command, mass, maneuver, supply, and reinforce available conventional combat forces. AI may have negligible effects against an insurrection during stability operations where the enemy employs a shadowy force structure, a simple logistics net and unconventional tactics. Timely, accurate intelligence and persistent operations, allows AI to disrupt enemy supply operations, destroy weapons caches, or deny sanctuary to insurgents. To maximize the influence AI has on an enemy, commanders need to understand how its effects will differ depending on the nature of the conflict being fought.

Whether the Air Force is involved in major operations and campaigns or smaller scale contingencies, AI can:

- Channel movements
- Constrict logistics systems
- Disrupt communications
- Force urgent movement
- Attrit enemy fielded forces

II. Close Air Support (CAS)

Close air support (CAS) is defined as "air action by fixed- and rotary-wing aircraft against hostile targets which are in close proximity to friendly forces and which require detailed integration of each air mission with the fire and movement of those forces. Employing ordnance within close proximity of ground troops and the requirement for detailed integration are two characteristics that distinguish CAS from other types of air warfare.

Close Proximity

Close proximity does not represent a specific distance. Instead, the word "close" is situational and requires detailed integration and terminal attack control (TAC) based on friendly force proximity to enemy targets. Detailed integration and TAC help ensure engagement of correct targets and mitigation of friendly fire incidents and collateral damage. Thus, CAS is not defined by a specific region of an operation, it can be conducted at any place and time friendly surface forces are in close proximity to enemy forces. For example, special operations forces (SOF) operating anywhere in the joint operations area (JOA) may require CAS support if there are friendly troops within close proximity to the enemy forces being attacked.

Detailed Integration

The requirement for detailed integration because of fires, proximity, or movement is the determining factor for CAS. Detailed integration describes a level of coordination required to achieve desired effects while minimizing the risk of a friendly fire incident—from either surface fires or air-delivered weapons. Because of this level of integration, each element should be controlled in real time to prevent friendly fire incidents with ground or air forces. Procedures should be flexible enough so that CAS, surface fires, and the ground scheme of maneuver are not overly restricted. The range at which the preponderance of effects against the enemy shifts from surface fires to airpower is the prime factor (among several) used to define the maximum range requiring detailed integration and a good depth for commanders to consider delineating between CAS and air interdiction (AI).

A. CAS Responsibilities

The joint force commander establishes the guidance and priorities for CAS in the concept of operations, operation plan or campaign plan, air apportionment decision, and by making capabilities and forces available to the components.

The commander, Air Force forces (COMAFFOR) is given the authority necessary to accomplish missions and tasks assigned by the establishing commander. For CAS, these responsibilities normally include recommending air apportionment, allocating forces/capabilities made available from the JFC and components including command and control elements of the theater air control system (TACS), creating and executing the air tasking order (ATO), and other applicable actions associated with CAS execution. The COMAFFOR maintains close coordination with the other component commanders to ensure CAS requirements are being met in accordance with JFC guidance.

B. CAS Objectives

Close air support (CAS) provides firepower in offensive and defensive operations, day or night, to destroy, suppress, neutralize, disrupt, fix, or delay enemy forces in close proximity to friendly ground forces. For CAS to be employed effectively, it should be prioritized against targets that present the greatest threat to the supported friendly surface force. Moreover, CAS assets should arrive in a timely manner. CAS that arrives late may be ineffective due to the fluid nature of ground battle.

Almost any enemy threat in close proximity to friendly forces on the modern battlefield is suitable for CAS targeting. However, indiscriminate CAS application against inappropriate targets decreases mission effectiveness, increases the risk of friendly fire incidents, and dilutes availability of CAS aircraft to an unacceptable level. Although there is no single category of targets most suitable for CAS application, mobile targets and their supporting firepower (in general) present the most immediate threat to friendly surface forces and thus are prime candidates for consideration. This is especially true when supporting light forces, such as airborne or amphibious units, since they are not able to bring as much organic heavy firepower into battle as heavier mechanized or armored units. CAS provides the surface commander with highly mobile, responsive, and concentrated firepower. It enhances the element of surprise , is capable of employing munitions with great precision, and is able to attack targets that are inaccessible or invulnerable to available surface fire.

The success of CAS during both offensive and defensive operations in contiguous, linear warfare may depend on massing effects at decisive points —not diluting them across the entire battlefield. During large-scale ground operations, there are often more requests for CAS than can be attacked by the available air assets. The centralized command and control of CAS employment is essential to allow the massing of its effects where needed most. This may often be beyond the troops-in-contact range, as CAS missions operating there will have reduced risk of friendly fire incidents, and enemy forces destroyed or delayed there are often kept from engaging friendly surface forces. Surface commanders should properly prioritize and focus the firepower of apportioned and allocated CAS at decisive places and times to achieve their objectives. Distributing CAS among many competing requests dilutes the effects of those assets and may result in less, rather than more effective air support to ground forces.

C. CAS Effects

When it is necessary to provide troops in contact with supporting fires, close air support (CAS) can devastate enemy forces while spearheading offensive operations or covering retrograde operations. CAS can also be used for harassment, suppression, and neutralization. However, because those effects are typically assigned to surface fire support assets, such use may represent a less efficient use of limited CAS mis-

D. CAS Planning

Ref: Adapted from JP 3-09.3 Close Air Support (Jul '09), p. III-4 to II-12.

1. Receipt of Mission/ Prepare for Mission Analysis

As integral parts of the planning team, the action officers and ALOs should "gather the tools" and be prepared to provide pertinent information from the following to the ground force commander's staff:

- **Air order of battle (apportionment, allocation, and distribution decision)**
- **ATO**
- **ACO**
- **SPINS**
- **OPORD**
- **Standard operating procedure (SOP)**

2. Mission Analysis

CAS planner responsibilities for mission analysis actually begin before the new mission is received. As part of the ongoing staff estimate, they must continuously monitor and track the status of fire support systems to include available air support. Specifically, during mission analysis CAS planners perform the following actions:

- **Update latest products** (ATO, ACO, SPINS, etc.)
- Estimate **air combat capability** to support the operations
- Determine **capabilities and limitations** of assigned personnel and equipment. (# of JTACs, systems, equipment status, communications status, etc.)
- Provide input to the **ground commander's initial guidance**
- Determine **specified, implied, and mission essential tasks**
- Consider **mission, enemy, terrain and weather, troops and support available-time available** (METT-T)
- Assist in developing the **mission statement**
- Anticipate **air power required** to support the mission based on:

1. **HHQ priorities of fires**
2. **Facts and assumptions**
3. **Weight of effort decisions**

- Provide the following products:

1. **AO/ALO estimate**
2. **Available CAS assets**
3. **CAS constraints and restraints** (ground alert CAS and airborne alert CAS response times, weather limitations, ROE, etc.)
4. **Warning order(s)** to subordinate units
5. **Verification** that subordinate TACP elements understand the warning order and have the ability to support the mission

- **Key Considerations**. During the mission analysis step, CAS planners should be familiar with the following elements of the HHQ order:

1. **CONOPs/Scheme of Maneuver.** What is the commander's intent? Is this an offensive or defensive operation? What type of offensive or defensive operation (deliberate attack, hasty defense, etc.)? How does ROE impact CAS?
2. **Concept of fires/essential fire support tasks (EFSTs).** What are the commander's desired task and purpose for fires? How can CAS contribute? What other joint functions (C2, intelligence, fires, movement and maneuver, protection, sustainment) are affected? Have all CAS assets been properly integrated with JAAT operations?
3. **JIPOE.** What is the enemy order of battle? What effects will time of day, terrain, and weather have on CAS operations? What are the likely enemy avenues of approach?
4. **Intelligence, Reconnaissance, and Surveillance.** What ISR assets are available? Where are ISR assets positioned? How can CAS operators communicate directly/indirectly with ISR assets? What are the commander's critical information requirements (CCIRs)? Can CAS assets satisfy CCIRs?
5. **Observation Plan.** How can CAS take advantage of available "eyes" on the battlefield? Are terminal attack control methods (i.e., types of CAS) considered? Where will JTACs/JFOs/FAC(A)s be required?
6. **Communications Plan.** How will maneuver elements, fire support, and TACP personnel communicate? Are JTACs integrated into the ground force communications plan? Are communications plans reliable and redundant?

- **Preplanned Air Support Request**. Once CAS planners have analyzed the mission and are familiar with CAS requirements, initial CAS requests should be drafted and submitted. See Appendix A, "Joint Tactical Air Strike Request." Further refinements to these initial requests can be forwarded as details become available. Adherence to ATO cycle time constraints is critical.

3. COA Development

After receiving guidance, the staff develops COAs for analysis and comparison. Guidance and intent focuses staff creativity toward producing a comprehensive, flexible plan within available time constraints. During this step, CAS planners:

- **Update latest products** (ATO, ACO, SPINS, etc.)
- **Analyze relative combat power.** This is typically accomplished by weighing the individual effectiveness of air platforms against anticipated enemy surface forces to include air defense threats.
- **Generate options used to develop possible COAs**. Options should be suitable, feasible, acceptable, distinguishable, and complete.
- **Array initial forces** to determine CAS requirements
- Develop **fire support/ACMs**
- Develop the **CAS integration plan** by examining opportunities for the best use of air power including the placement of TACP assets
- The AO/ALO assists in developing **engagement areas,** target areas of interest (TAIs), triggers, objective areas, obstacle plan, and movement plan
- Prepare **COA statements and sketches** (battle graphics). This part involves brainstorming to mass the most effective combat power against the enemy (CAS, EW, ISR, and surface fire support).
- **Key Considerations.** During COA development (for each COA), CAS planners must consider:

  1. Commander's Intent. How does the commander intend to use CAS? What are his objectives? Does CAS facilitate the commander's ability to achieve his mission objective?

  2. CCIRs. What CCIR can CAS assets provide? Will TACPs, JFOs, and/or FAC(A)s be able to provide critical battlefield information? How will this information be relayed to the maneuver unit?

  3. Enemy Situation. Where is the enemy and how does he fight (enemy order of battle)? Where is he going? Where can I kill him? When will he be there? What can he do to kill me? How am I going to kill him?

  4. Statements and Sketches. Once COA development has started, sketches of each COA should be made with notes for the staff to better understand what each can offer the unit. How will CAS aircraft enter/exit the operational area? Does the CAS overlay reflect artillery positioning areas and azimuths of fire (AOFs)? Does the plan promote simultaneous engagement of targets by CAS and surface fires? Has the CAS overlay been shared with all battlefield operating system elements? Where will JTACs/JFOs be positioned on the battlefield? What ACMs and FSCMs are needed to support the COA?

  5. Priority of CAS Fires. Priority of fires (POF) for each COA must be identified. As part of the POF, priority of CAS fires must also be identified. The ground maneuver commander establishes which element will receive POF and priority of CAS. It is also important to make the commander and his staff aware of their unit's priority for CAS relative to other units in the operational area. Does the element with priority of CAS fires have a designated JTAC? What if priorities change or CAS is unavailable for the planned COA? How will changes in priority be communicated with forward elements and JTACs? Does the priority of CAS support the commander's intent for each COA?

- **TACP**: The TACP provides the following inputs during COA development:

  1. Specific TACP portions of the following plans:

  a. Observation plan (to include target area, aircraft, and BDA)

  b. Employment plan (i.e., ACAs)

  c. Communications plan

  2. Evaluation of overall TACP capabilities/limitations:

  a. Personnel

  b. Equipment

  3. Consideration of the most effective TAC procedures

  4. Update initial or submit new JTARs with all information currently available

  5. Current geospatial products and overlays

Continued on next page

CAS Planning (Cont)

Ref: Adapted from JP 3-09.3 Close Air Support (Jul '09), p. III-4 to II-12.

Continued from previous page

4. COA Analysis/War Game

The planning staff "fights the battle" to determine the advantages and disadvantages of each COA and to identify which COA best accomplishes the commander's intent. CAS planners should:

- Identify **strengths and weaknesses** for CAS in each COA
- Conduct an **initial tactical risk assessment** for each COA
- Recommend **terminal attack control criteria** for commander approval. Type of control to use where and under what conditions

  1. Determine best locations for certified JTACs/FAC(A)s

  2. Plan use of JFOs/observers and assess communications requirements
- **Evaluate CAS integration** with other fire support assets
- **Assess effectiveness** of ACA and other FSCMs/ACMs
- Gather **war gaming tools**

  1. Updated ATO/SPINS information

  2. Decision-making matrices/devices

  3. Briefing cards/CAS briefs

  4. Standard conventional load listings

  5. Aircraft and weapons capabilities information
- List all **friendly forces**

  1. CAS aircraft

  2. FAC(A)

  3. Airborne C2

  4. Ground forces, including fire support assets

  5. JTACs

  6. JFOs/other observers/ISR assets

  7. Other aviation and support assets
- List **assumptions**

  1. Aircraft operating altitudes

  2. Enemy surface to air threat posture

  3. CAS tactics

  4. JTAC procedures in effect

  5. How terrain and weather affects CAS
- List known **critical events and decision points (DPs)**

  1. Line of departure or defend no later than times

  2. CAS triggers (named areas of interest [NAIs]/TAIs)

  3. ACM/FSCM requirements

  4. SEAD/marking round requirements
- Determine **evaluation criteria**

  1. Timeliness

  2. Accuracy

  3. Flexibility

  4. Mass

  5. Desired effects
- Select the **war game method**

  1. Rehearsal of Concept (ROC)/Terrain Model/Sand Table. Commanders and staffs may use a form of rehearsal called a "ROC drill." A ROC drill is a leader and staff rehearsal that usually uses a sand table or similar training aid. Its primary purpose is to synchronize the actions of all six joint functions (C2, intelligence, fires, movement and maneuver, protection, and sustainment).

  2. Map

  3. Radio

  4. Other
- Select a **method to record and display results**

  1. Event logs

  2. Timetables

  3. Reaction times, etc.
- **War game the battle and assess the results.** Did CAS support the commander's intent for fires? Was CAS effectively integrated with ground scheme of maneuver? Was C2 of CAS reliable and effective? Were FSCMs and ACMs effective in supporting the COA?
- **Fires Paragraph**. CAS and other fire support planners begin to refine the fires paragraph to the OPORD by further developing specific tasks, purpose, methods, and desired effects of fires. The resulting list of tasks becomes the CAS EFSTs. EFSTs have four distinct components: task, purpose, method, and effects (TPME):

Continued from previous page

1. Task. Describes the targeting objectives fires must achieve against a specific enemy formation's function or capability. Examples include:

a. "Disrupt movement of 3rd Guards Tank Regiment."

b. "Delay Advanced Guard Main Body movement by 2 hours."

c. "Limit advance of 32nd Motorized Rifle Regiment."

d. "Destroy lead elements of the Forward Security Element."

2. Purpose. Describes the maneuver or operational purpose for the task. Examples include:

a. "To allow 2nd BN to advance to phase line Smith."

b. "To seize and hold Objective Panther."

c. "To enable Task Force 2-69 Armor to secure access to Brown's Pass."

3. Method. Describe how the task and purpose will be achieved. Examples include:

a. CAS engages armored targets vicinity of Brown Pass not later than 1400L.

b. CAS attacks defensive positions at point of penetration at 1300Z.

c. CAS available to engage targets of opportunity entering the main defensive belt.

4. Effects of Fires. Attempts to quantify the successful accomplishment of the task. Examples:

a. CAS destroys 8–10 vehicles vicinity Brown's Pass; 2-69 Armor secured Brown's Pass.

b. CAS disables enemy engineer platoon at point of penetration; 2nd BN advanced to phase line Smith, seized and held Objective Panther.

c. CAS destroys 10 T-80s/T-72s in main defensive belt; 2nd BN advanced to phase line Smith, seized and held Objective Panther.

5. Orders Production

The staff prepares the order or plan to implement the selected COA and provides a clear, concise CONOPs, a scheme of maneuver, and concept of fires. Orders and plans provide all necessary information that subordinates require for execution, but without unnecessary constraints that would inhibit subordinate initiative. TACPs should produce the CAS specific appendix to the fire support annex as required.

- **Fire Support Annex**. Fire support and CAS planners will also produce a fire support annex. This annex is necessary to expand upon the fire support information in paragraph 3 of the OPORD. A fire support execution matrix (FSEM) may also be developed as part of or used in place of a standard fire support annex. Regardless of format, further expansion of fire support information includes:

1. Purpose. Addresses exactly what is to be accomplished by fire support during each phase of the battle. It should be specific in addressing attack guidance and engagement criteria. This is the most important part of the fires paragraph. The fire support annex must articulate how fires, as a joint function, will be synchronized with the other five joint functions (C2, intelligence, movement and maneuver, protection, and sustainment).

2. Priority. Designates POF and when or if it shifts for each phase. Include all fire support systems to include CAS when assigning POFs.

3. Allocation. Designates the allocation of fire support assets to include the following: targets allocated to units for planning; CAS sorties for planning; smoke, expressed in minutes and size; priority targets, final protective fires (FPFs), and special munition priority targets; and laser equipped observation teams.

4. Restrictions. Addresses FSCMs and the use of specific munitions. Some examples are critical FSCMs and specific munition restrictions such as those placed on the employment of illumination, smoke, dual-purpose improved conventional munitions, family of scatterable mines, and cluster bomb units (CBUs).

- **Airspace Coordinating Measures Annex**. This addresses ACMs required to support the CAS and fire support plans.

sions. Ground commanders should use their organic firepower when better suited for the task before calling in requests for CAS. However, a ground commander's organic firepower—particularly longer range systems—may not always be the most appropriate fire support asset. Thus, when planned and integrated well, CAS provides desired effects that can be exploited by the maneuver commander. Ultimately, each of the different CAS applications should be weighed against other, potentially more effective, uses for CAS-capable assets such as air interdiction or even strategic attack. CAS generates the following benefits:

- Facilitate Ground Action
- Induce Shock, Disruption, and Disorder
- Support Stability Operations

III. Terminal Attack Control

A need for flexible, real-time targeting guidance, collateral damage minimization and friendly fire incident avoidance are critical considerations when conducting close air support (CAS). To integrate air-ground operations safely and effectively, either a joint terminal attack controller (JTAC) or a forward air controller–airborne (FAC [A]) provides terminal attack control (TAC) for CAS missions. Terminal attack control is defined as "the authority to control the maneuver of and grant weapons release clearance to attacking aircraft" (JP 3-09.3, Close Air Support).

Joint Terminal Attack Controller (JTAC)

A JTAC is defined as "a qualified (certified) Service member who, from a forward position, directs the action of combat aircraft engaged in CAS and other offensive air operations. A qualified and current JTAC will be recognized across the Department of Defense as capable and authorized to perform terminal attack control" (JP 3-09.3). The JTAC provides recommendations on the integration of CAS with the ground commander's scheme of maneuver. A JTAC should be trained to:

- Know the enemy situation and location of friendly units and civilians.
- Know the supported commander's target priority, desired effects, and timing of fires.
- Know the commanders intent and applicable rules of engagement (ROE).
- Validate targets of opportunity.
- Advise the commander on proper employment of air assets.
- Submit immediate requests for CAS.
- Control CAS with supported commander's approval.
- Deconflict aircraft and fires from CAS sorties.
- Perform battle damage assessment (BDA).

Terminal Attack Control roles and responsibilities are outlined in Table 34 of AFTTP 3-2.6 Multi-Service Tactics, Techniques, and Procedures for Joint Application of Firepower.

Forward Air Controller FAC(A)

The FAC(A) is a specifically trained and qualified aviation officer who exercises control from the air of aircraft engaged in CAS of ground troops. The FAC(A) is normally an airborne extension of the tactical air control party (TACP). "A qualified and current forward air controller (airborne) will be recognized across the Department of Defense as capable and authorized to perform terminal attack control" (JP 3-09.3). Only specially trained and certified aircrews are authorized to perform this duty, as it requires detailed knowledge of friendly and target locations, artillery, available aircraft weapons and fuel states, the ability to conduct all three types of terminal attack control, and the flexibility to prioritize and adjust in a dynamic environment. At the request of the JTAC/TACP, a FAC(A) can assume the tasks of Brief, Stack,

Types of Terminal Control

Ref: Annex 3-03, Counterland Operations (17 Mar '17), pp. 44-48.

There are three types of terminal attack control (TAC) designated as Types 1, 2, and 3 (discussed below). Each type is characterized by a specific set of procedures outlined in JP 3-09.3, Close Air Support. The ground commander considers the situation and issues guidance to the joint terminal attack controller (JTAC) based on the associated risks identified in the tactical risk assessment. The intent is to offer the lowest level supported commander the latitude to determine which type of TAC best accomplishes the mission. Risk level is not directly tied to a given type of TAC. The three types of control are not ordnance-specific and the tactical situation will define the risk level.

Type 1 Control

Type 1 control will be used when the JTAC requires control of individual attacks and must visually acquire the attacking aircraft and the target for each attack (JP 3-09.3). "Visually acquire" is literally eyes-on or via optics such as binoculars, without the use of third party devices such as laptops or other digital imagery. Analysis of attacking aircraft geometry is required to reduce the risk of collateral damage or the attack affecting friendly forces. Language barriers when controlling coalition aircraft, lack of confidence in a particular platform, ability to operate in adverse weather, or aircrew capability are all examples where visual means of TAC may be the method of choice.

Type 2 Control

Type 2 control will be used when the JTAC requires control of individual attacks and any or all of the conditions exist: JTAC is unable to visually acquire the attacking aircraft at weapons release; JTAC is unable to visually acquire the target; and/or the attacking aircraft is unable to acquire the mark/target prior to weapons release (JP 3-09.3). The JTAC must acquire the target visually or utilize targeting data from a scout, fire support team (FIST), joint fires observer (JFO), unmanned aircraft (UA), special operations forces (SOF), CAS aircrew, or other asset with accurate real-time targeting information. Type 2 control may be applicable during certain conditions, such as night, adverse weather, and high altitude or standoff weapons employment. Type 2 control is also applicable when using configured UA or targeting pod sensor aimpoint via remotely operated video enhanced receiver. A JTAC, who can see a laser spot on the target or a real-time feed from a targeting pod, may be better able to minimize collateral damage and deconflict an attack from friendly forces than one relying on visual contact with an attacking aircraft at high altitude. Currently fielded technology has the capability to improve the flow of information between the JTAC and pilot. These tools are an additional means to ensure the destruction of the enemy, minimization of collateral damage and prevent friendly fire incidents, and in many cases are a more reliable means of aimpoint verification.

Type 3 Control

Type 3 control is used when the JTAC requires the ability to provide clearance for multiple attacks within a single engagement subject to specific attack restrictions. Type 3 control does not require the JTAC to visually acquire the aircraft or the target; however, all targeting data must be coordinated through the supported commander's battle staff (JP 3-09.3). During Type 3 control, JTACs provide attacking aircraft targeting restrictions (e.g., time, geographic boundaries, final attack heading, specific target set, etc.) and then grant a "blanket" weapons release clearance to meet the prescribed restrictions. The JTAC will monitor radio transmissions and other available digital information to maintain control of the engagement. The JTAC maintains abort authority. Observers may be utilized to provide targeting data and the target mark during Type 3 control. Type 3 is a CAS TAC procedure and should not be confused with TGO or AI. Missions attacking targets not in close proximity to friendly forces, and beyond the range requiring detailed integration with surface fires and maneuver, should be conducted using air interdiction (AI) procedures vice CAS.

Mark and/or Control. Each of these tasks has a specific responsibility associated to it, understanding that the absence or misidentification of the tasks and duties for the FAC(A) during planning and/or execution will likely result in delayed CAS operations. FAC(A)s should receive land maneuver commander clearance, normally through the TACP, before expending or authorizing other aircraft to expend ordnance. The FAC(A) may provide TAC, relay CAS briefings, provide immediate target and threat reconnaissance, and mark targets for attacking aircraft. Threats and weather permitting, the FAC(A) may see well beyond the visual range of ground-based JTACs. The FAC(A) can perform tactical battle management by cycling aircraft through the target area while prioritizing targets in coordination with a JTAC. In this role, the FAC(A) is operating as a tactical air coordinator (airborne) (TAC[A]). The FAC(A) may provide positive identification (PID), collateral damage estimation (CDE), and immediate BDA.

The TAC(A) is an extension of the theater air control system (TACS) air support control agencies. In the absence of Joint Surveillance Target Attack Radar System (JSTARS) or a FAC(A), a TAC(A) may provide communications relay between the TACP and attack aircraft. A two-ship FAC(A) flight, especially in higher threat environments, may divide responsibilities so one aircraft fills the normal FAC(A) role while the second becomes a TAC(A). The TAC(A) expedites CAS aircraft-to-JTAC handoff during "heavy traffic" CAS operations.

Joint Fires Observer (JFO)

A JFO can request, adjust, and control surface-to-surface fires, provide targeting information in support of CAS, and perform terminal guidance operations (TGO). TGO are those actions that provide electronic, mechanical, voice or visual communications that provide approaching aircraft and/or weapons additional information regarding a specific target location. The JFO adds joint warfighting capability but cannot provide TAC during CAS operations. Unless qualified as a JTAC or FAC(A), personnel conducting TGO do not have the authority to control the maneuver of or grant weapons release to attacking aircraft. JFOs provide the capability to exploit those opportunities that exist in the operational environment where a trained observer could be used to efficiently support air delivered fires, surface-to-surface fires, and facilitate targeting. The JFO is not an additional person provided to a team but rather an existing team member who has received the supplemental proper training and certification. The intent of a JFO is to add joint warfighting capability, not circumvent the need for qualified JTACs. JFOs expand the target set available to ground commanders by passing accurate targeting information to both the JTAC and aircrew.

Special Tactics Team (STT)

Air Force STTs are composed primarily of special operations combat control and pararescue personnel. Combat control personnel support SOF ground elements by providing air-ground interface, fire support, target designation, C2 communications, and airfield/helicopter landing zone/drop zone surveys. Some combat controllers are JTAC-qualified.

CAS Execution With Non-JTAC Personnel

In certain circumstances, the ground commander might require air support when a joint terminal attack controller (JTAC) or forward air controller (airborne) (FAC[A]) is not available but detailed integration with friendly forces fire and movement is still required. Aircrew executing close air support (CAS) under these circumstances bear increased responsibility for the detailed integration required to minimize friendly fire incidents and collateral damage normally done by a JTAC/FAC(A). Non-JTAC personnel must clearly state to strike aircraft that they are not a JTAC. In these circumstances, CAS aircrew should assist these personnel/units to the greatest extent possible in order to bring fires to bear.

Refer to AFTTP (I) 3-2.6, Multi-Service Procedures for the Joint Application of Firepower (JFIRE), for additional discussion.

Chap 4

III. Countersea Operations

Ref: Annex 3-04, Countersea Operations (7 Nov '14).

Our nation depends on assured access to the world's waterways and coastal regions for global economic trade, as well as providing a stabilizing military presence abroad. These waterways, along with our maritime fleet, provide the means for projecting the bulk of our heavy forces forward, sustaining them over the long term, and projecting force ashore from the seas. Where airpower is the key to rapid forward presence and striking power over long distances, sea power is key to extended forward presence, maritime power projection, mass force deployment, and sustainment through sealift. Protecting sea lanes, littorals, and our maritime assets operating within them are vital to US defense posture, economic prosperity, and national security.

I. Countersea Operations

Countersea operations are those operations conducted to attain and maintain a desired degree of maritime superiority by the destruction, disruption, delay, diversion, or other neutralization of threats in the maritime environment. The main objective of countersea operations is to secure and dominate the maritime domain and prevent opponents from doing the same.

The countersea function entails Air Force operations in the maritime domain to achieve, or aid in the achievement of, superiority in that medium. This function fulfills Department of Defense (DOD) requirements for the use of Air Force forces to counter adversary air, surface, and subsurface threats, ensuring the security of vital sea and coastal areas, and enhancing the maritime scheme of maneuver. More importantly, it demonstrates the teamwork required of Service forces working together in a joint environment. Air Force forces achieve effects in the maritime domain through the integrated employment of airpower. The overarching effect of countersea operations is maritime superiority—denial of this domain to the adversary while assuring access and freedom of maneuver for US and allied maritime forces. To this end, Air Force operations can make significant contributions to maritime components in support of joint force objectives.

II. The Maritime Domain

From a military perspective, the maritime domain is not limited to the open seas. The DOD Dictionary of Military and Associated Terms (Joint Publication [JP] 1-02) defines the maritime domain as "the oceans, seas, bays, estuaries, islands, coastal areas, and the airspace above these, including the littorals." Littoral comprises two segments of the operational environment: 1. Seaward: the area from the open ocean to the shore, which must be controlled to support operations ashore. 2. Landward:

Refer to the Naval Operations and Planning SMARTbook for further discussion of the Maritime Domain. The Naval Operations & Planning SMARTbook covers essential Navy keystone warfighting doctrine and maritime operations at the JFMCC/CFMCC, Fleet and JTF levels -- to include maritime forces, organization and capabilities; maritime operations; maritime headquarters (MHQ) and the maritime operations center (MOC); the maritime operations process; naval planning; naval logistics; and naval theater security cooperation.

III. (Air Force) Countersea Operations

Ref: Annex 3-04, Countersea Operations (7 Nov '14), pp. 34-

Countersea operations can be used in various ways to support the joint force commander's (JFC's) campaign. Conducted independently, or in conjunction with other military operations, countersea operations may be used for the following purposes:

- At the initial phase of a campaign or major operation where the objective is to establish a military lodgment to support subsequent phases
- Serve as a supporting operation during a campaign in order to deny use of an area or facilities to the enemy, or to fix enemy forces' attention in support of other combat operations
- Support stability operations in order to deter war, resolve conflict, promote peace and stability, or support civil authorities in response to crises that require controlling the surrounding maritime domain
- Support military operations for homeland defense, by controlling use of the maritime domain along US coastal waters to prevent enemies from attacking civilian population centers, disrupting sea lines of communication (SLOC), or committing terrorism on US sovereign soil
- As an independent operation without other Service forces present, to achieve operational or strategic objectives in the maritime domain

Maritime Surveillance and Reconnaissance

In the maritime domain, control must be achieved in the air, on the surface, and under the surface as part of battlespace dominance. Air Force forces help enable control of air and surface maritime areas through intelligence, surveillance, and reconnaissance (ISR) coverage and their significant abilities to collect data. Air Force forces provide rapid and large area surveillance and reconnaissance coverage, often arriving on station prior to other forces. This coverage can be used to observe the maritime domain in a homeland security role or overseas.

Planning and employing this capacity could occur as a single Service or jointly. Operations may involve interfacing with multinational forces, Navy forces, the Coast Guard, or other agencies responsible for homeland security. Preparation and execution of ISR should include coordination through liaison officers (LNOs) working in the air operations center (AOC) or with other agencies.

Antisurface Ship Warfare (Surface Warfare)

Commanders may employ Air Force forces to interdict enemy maritime surface forces. These operations are conducted to destroy or neutralize enemy naval surface forces and merchant vessels. Planning should address and define marshalling areas; area of attack; rules of engagement; required coordination and deconfliction with friendly vessels in or near the area of operation; fighter, joint, missile, and self-defense engagement zones; vessel identification; and other factors that may influence platform choices, weapons mix, tactics, and support requirements.

Antisubmarine Warfare (ASW)

Air Force forces successfully performed antisubmarine warfare (ASW) during World War II. Currently, Air Force assets could perform ASW in an ISR and interdiction role by monitoring and, if needed, attacking surfaced enemy submarines under way or in port, as well as the port itself, or locations used for refueling or supply. Additionally, currently fielded Air Force assets have sensors and weapons required to detect and engage surfaced submarines, in support of the joint force maritime component commander's (JFMCC) undersea warfare efforts. However, extensive planning and training would be required for Air Force forces to effectively attack deployed, submerged submarines.

Airpower

Aerial Minelaying Operations

Mine warfare is defined as "the strategic, operational, and tactical use of mines and mine countermeasures either by emplacing mines to degrade the enemy's capabilities to wage land, air, and maritime warfare or by countering of enemy-emplaced mines to permit friendly maneuver or use of selected land or sea areas." (JP 1-02, DOD Dictionary of Military and Associated Terms). Mine warfare is divided into two basic subdivisions: mine laying for area denial and countering enemy–laid mines. The most expeditious mine laying operations are accomplished by aircraft, including Air Force bombers, which might be the best suited for mining in threat areas.

Counterair Operations

Counterair is defined as "a mission that integrates offensive and defensive operations to attain and maintain a desired degree of air superiority and protection by neutralizing or destroying enemy aircraft and missiles, both before and after launch." (JP 1-02). "Counterair" and the US Navy/US Marine Corps term "air warfare" (AW), are virtually synonymous. The Navy employs an air defense commander (ADC) as part of its composite warfare commander structure to enable air and ship platforms to engage the enemy in much the same way Air Force assets perform counterair.

Depending upon the proximity of a forward operating location to an objective area and the availability of air-to-air refueling support, commanders may employ Air Force fighter aircraft in the maritime domain to gain air superiority. Counterair is divided into offensive counterair (OCA) and defensive counterair (DCA). Suppression of enemy air defenses is a component of OCA.

Air-to-Air Refueling

Planning air–to–air refueling in support of maritime operations should ensure refueling compatibility between tankers and aircraft receiving fuel. Because maritime support aircraft missions generally begin from locations outside the AO, determination of air refueling tracks and offload requirements should account for operating radius of aircraft, distance to and from the AOR, and threat reaction requirements.

Amphibious Operations

Amphibious operations may require Air Force forces to perform functions such as counterair to provide air superiority, counterland for interdiction and/or joint close air support, airlift for air assault or resupply, and ISR from air and space assets. The commander, Air Force forces should plan with the JFMCC, commander, amphibious task force (CATF), and commander, landing force (CLF) to ensure functional integration and to accomplish

Close Air Support (CAS)

Amphibious operations may entail CAS in the littoral environment. However, there are significant differences that make this type of CAS operation more difficult than traditional CAS. Amphibious operations involve many fire support elements creating deconfliction challenges and increased potential for fratricide. Air, sea surface and sub-surface, and land elements operate and converge in one confined area to support the LF.

Space Capability

The JFACC, as the space coordinating authority, will coordinate space operations, integrate space capabilities, and have primary responsibility for in-theater joint space operations planning.

Other Air Force Countersea Operations

Other Air Force operations such as airlift, cyberspace operations, information operations, special operations, command and control, personnel recovery operations, and weather services may also provide support to countersea operations.

the area inland from the shore that can be supported and defended directly from the sea. (Joint Publication [JP] 2-01.3, Joint Intelligence Preparation of the Operational Environment). Countersea operations are equally relevant to "brown" water (navigable rivers, lakes, bays and their estuaries), "green" water (coastal waters, ports and harbors) and "blue" water (high seas and open oceans) environments.

The inclusion of "the airspace above these" in the domain definition indicates the decisiveness of air operations within the maritime domain. Although the "airspace above" could be considered the air domain, nothing in the definition of that domain implies or mandates exclusivity, primacy, or command and control of that domain. Command and control is established through command relationships within the various operational areas as described in JP 1, Doctrine for the Armed Forces of the United States, and is the authority of the joint force commander based upon most effective use of available resources to accomplish assigned missions.

With the potential emergence of a credible naval opponent, maritime operations are once again focusing on defeating enemy naval forces while retaining a focus on the role of power projection ashore from the littorals. Airpower provides a rapid, maneuverable, and flexible element in this environment. Air Force capabilities can extend the reach and increase the flexibility of naval surface, subsurface, and aviation assets, playing a key role in controlling the maritime domain. Air Force and Navy capabilities synergistically employed enable the joint force to control the maritime domain.

IV. International Law

To effectively conduct countersea operations, commanders, planners, and aircrews must be aware of the legal issues that can impact such operations. National policy and legal requirements dictate that countersea operations be conducted in compliance with international law. The law relating to countersea operations is particularly complex in that much of the law is customary international law developed throughout naval history. In addition, commanders, planners, and aircrews must have knowledge of the air navigation regimes that dictate where aircraft can lawfully fly. Part of the preparation for countersea operations must be a review of the law of armed conflict (LOAC) and law of the sea requirements, which affect these operations.

Portions of the United Nations Law of the Sea Convention of 1982 are consistent with customary international law concerning maritime navigation and overflight rights. Air Force members involved in countersea operations must be aware of the rights of aircraft over the various maritime zones. These zones include the high seas, exclusive economic zones, contiguous zones, territorial seas, internal waters, archipelagic waters, international straits, and archipelagic sea lanes. These zones are important because they determine the amount of control that a coastal state may exercise over foreign aircraft and ships. All of these zones are measured from national baselines, hence knowledge of where these baselines are located is essential if aircraft are to be able to assert and exercise their lawful rights in conducting countersea operations.

Some nations assert security zones beyond the limits of their territorial sea but international law does not recognize any such zone. Military aircraft generally have freedom of navigation rights outside of territorial seas. Any nation may declare a temporary warning zone including over areas of the high seas. These zones do not restrict the right of navigation but advise ships and aircraft of hazardous (but lawful) activities. These may include missile testing, gunnery practice, and space vehicle recovery operations. In the exercise of their inherent right of self-defense under the United Nations Charter, nations may declare various forms of maritime control areas. These may include air or maritime exclusion zones, or other types of defensive sea areas in which a measure of control is exercised over foreign ships and aircraft. During times of conflict, Air Force units must be particularly aware of the rights of neutral nations. These rights protect the sovereignty of neutral nations, which includes national ships and aircraft.

IV. Counterspace Operations

Ref: Annex 3-14, Counterspace Operations (27 Aug '18) and JP 3-14, Space Operations (Apr '18).

The Air Force uses four space operations functions to clearly delineate the capabilities required for successful global joint operations and supersede the space mission areas listed in previous versions of Annex 3-14: space situational awareness (SSA); counterspace operations; space support to operations; and space service support. Taken together, these functions provide the ability to understand, operate and exploit the space domain. It is necessary to understand the domain (SSA) to conduct effective command and control, in turn enabling friendly forces to operate effectively in the domain (counterspace). These actions permit the conduct of operations at a given time and place without prohibitive interference by opposing forces, and are the basis of space superiority, a necessary step to enable exploitation of the domain (space support to operations) in order to provide battlefield advantages to joint warfighters. Space service support and SSA capabilities enable space operations across each of the space operations functions.

Space Operations Functions

Space Situational Awareness (SSA)

Counterspace Operations

Space Support to Operations

Space Service Support

I. Space Superiority

Achieving space superiority is of primary concern to Airmen as it enables the continuous provision and advantages of space-enabled capabilities to joint warfighting operations. Space superiority is, "the degree of control in space of one force over any others that permits the conduct of its operations at a given time and place without prohibitive interference from terrestrial or space-based threats"

Space supremacy is the degree of control in space by one force over another that permits the conduct of operations at a given time and place without effective interference from opposing forces. The concept of space superiority / supremacy is similar to air superiority / supremacy; however, the desired control may not always be achievable, particularly against a peer or near-peer adversary. Additionally, "place" does not refer to controlling physical space. It refers to specific terrestrial areas that may be impacted by space operations. Space superiority / supremacy may be localized in time and space, or it may be broad and enduring.

II. Space Operations Functions

Ref: Annex 3-14, Counterspace Operations (27 Aug '18), pp. 7 to 15.

A. Space Situational Awareness (SSA)

Space situational awareness (SSA) is foundational and fundamental to the conduct of all space operations functions and is especially critical to the effective conduct of counter-space operations. Joint Publication 3-14, Space Operations, defines SSA as "the requisite foundational, current, and predictive knowledge and characterization of space objects and the operational environment upon which space operations depend – including physical, virtual, information, and human dimensions – as well as all factors, activities, and events of all entities conducting, or preparing to conduct, space operations." SSA makes it possible to understand the space domain, allowing effective command and control of counterspace missions, leading to the desired control of space. SSA is divided into four functional capabilities:

- **Detect / Track / Identify.** Detect / track / identify (D/T/ID) is the ability to search, discover, and track space objects in order to maintain custody of objects and events; distinguish objects from others; and recognize objects as belonging to certain types, missions, etc.
- **Threat Warning and Assessment.** Threat warning and assessment (TW&A) is the ability to predict and differentiate between potential or actual attacks,space weather environment effects, and space system anomalies, as well as provide timely friendly force status.
- **Characterization.** Characterization is the ability to determine strategy, tactics, intent, and activity, including characteristics and operating parameters of all space capabilities (ground, link, and space segments) and threats posed by those capabilities.
- **Data Integration and Exploitation.** DI&E is the ability to fuse, correlate and integrate multi-source data into a UDOP and enable decision-making for space operations.

B. Counterspace Operations

Counterspace* is a mission, like counterair, that integrates offensive and defensive operations to attain and maintain the desired control and protection in and through space. These operations may be conducted across the tactical, operational, and strategic levels in all domains (air, space, land, maritime,and cyberspace), and are dependent on robust space situational awareness (SSA) and timely command and control (C2). Counterspace operations include both offensive counterspace (OCS) and defensive counterspace (DCS) operations. *(* Counterspace is referred to as "space control" in Joint Publication 3-14, Space Operations.)*

- **Offensive Counterspace (OCS)**. OCS operations are undertaken to negate an adversary's use of space capabilities, reducing the effectiveness of adversary forces in all domains. These operations target an adversary's space capabilities (space, link, and ground segments, or services provided by third parties), using a variety of reversible and non-reversible means. These actions may include strikes against adversary counterspace capabilities before they are used against friendly forces. OCS operations may occur in multiple domains and may result in a variety of desired effects including deception, disruption, denial, degradation,or destruction.

  OCS operations may occur in multiple domains and may result in a variety of desired effects including deception, disruption, denial, degradation,or destruction.

  - **Deceive**. Measures designed to mislead an adversary by manipulation, distortion, or falsification of evidence or information into a system to induce the adversary to react in a manner prejudicial to their interests.

- **Disrupt**. Measures designed to temporarily impair an adversary's use or access of a system for a period of time, usually without physical damage to the affected system.
- **Deny**. Measures designed to temporarily eliminate an adversary's use, access, or operation of a system for a period of time, usually without physical damage to the affected system.
- **Degrade**. Measures designed to permanently impair (either partially or totally) the adversary's use of a system, usually with some physical damage to the affected system.
- **Destroy**. Measures designed to permanently eliminate the adversary's use of a system, usually with physical damage to the affected system.

• **Defensive Counterspace (DCS)**. DCS operations protect friendly space capabilities from attack, interference, and unintentional hazards, in order to preserve US and friendly ability to exploit space for military advantage. Space capabilities include the space segment (e.g., on-orbit satellites), ground segment (e.g., space operations centers and telemetry, tracking, and commanding stations), and the link segment (the electromagnetic spectrum).

• **Navigation Warfare (NAVWAR)**. NAVWAR contributes to counterspace operations by preventing adversary use of PNT information while protecting the unimpeded use of the information by forces and preserving peaceful use of this information outside the area of operations.

Airpower

C. Space Support to Operations

The space support to operations function provides capabilities to aid, protect, enhance and complement the activities of other military forces, as well as intelligence, civil, and commercial users. These capabilities improve the integration and availability of space capabilities to increase the effectiveness of military operations and achieve national and homeland security objectives. Space support to operations capabilities contribute to counterspace operations, incorporate both active and passive measures for self-protection, and benefit from defensive counterspace (DCS) actions to suppress attacks, as required, in all domains. Space support to operations capabilities include:

• Intelligence, Surveillance and Reconnaissance (ISR)

• Launch Detection

• Missile Tracking

• Environmental Monitoring

• Satellite Communications

• Positioning, Navigation, and Timing (PNT)

D. Space Service Support

Space service support capabilities ensure access to, transport through,operations in, and, as appropriate, return from space through reliable, flexible, resilient, responsive, and safe launch and satellite operations. Space service support consists of spacelift, range, and satellite operations. Space service support capabilities contribute to counterspace operations, incorporate both active and passive measures for self-protection, reconstitute capabilities lost due to enemy attack, and benefit from defensive counterspace actions to suppress attacks,as required, in all domains:

• Spacelift Operations

• Range Operations

• Satellite Operations (On-Orbit Reconstitution, Disposal of Space Vehicles, and Rendezvous and Proximity Operations (RPO)

III. Threats to Space Operations

Ref: Annex 3-14, Counterspace Operations (27 Aug '18), pp. 4 to 5.

Potential adversaries see increasing value in the ability to attack US and allied space capabilities. Adversaries may employ multiple means, developed organically or acquired from third parties. Near- and long-term threats include the following:

Terrestrial Attack. Kinetic attack or sabotage against terrestrial nodes and supporting infrastructure. Examples of terrestrial nodes include operations centers, command and control nodes, and communications relays.

Electronic Attack (EA). Electromagnetic (EM) energy used to attack a link segment, to include uplink, downlink, and crosslink signals.

Directed Energy (DE). Directed-energy threats include laser, radio frequency (RF), and particle-beam weapons. Laser systems may be used to temporarily disrupt or deny capabilities or to permanently degrade or destroy satellite subsystems. RF weapons concepts include ground and space-based RF emitters that fire high-power bursts of EM energy at a satellite, imparting disruptive EM fields into the wiring and electrical components in order to upset and possibly damage the computer processing subsystems. Particle-beam weapons could be used to fire beams of charged particles at a satellite, superheating and destroying structural materials and mission components.

High Altitude Nuclear Detonation. A nuclear explosion can potentially affect all three segments of several space systems at the same time. Since the effects of nuclear detonation move out rapidly and permeate all space, no satellites have to be targeted directly. An electromagnetic pulse will induce damaging voltages and currents into unprotected electronic circuits and components of affected satellites and terrestrial nodes. The radiation generated by the detonation could damage satellite components and shorten their effective operational lives from years to days.

Anti-Satellite (ASAT) Weapons. Weapons capable of destroying or degrading spacecraft and spacecraft components and/or denying or disrupting their capabilities. There are two basic types. Direct ascent systems are best visualized as being "surface-to-space missiles," while on-orbit ASAT systems are also possible. ASATs may cause structural damage by impacting the target. Even small projectiles can inflict substantial damage or destroy a satellite.

Offensive Cyberspace Operations. Cyberspace attacks may disrupt or deny space-based or terrestrial-based computing functions used to conduct or support satellite operations and to collect, process, and disseminate mission data.

Environment. Neutral and environmental threats include weather, space debris, and unintentional EM interference. While not intended to do harm, this category of neutral and environmental threats causes increasing concern due to the potential impact to space operations.

Weather. Just as weather affects air operations, space and terrestrial weather can impact satellites, their communications links, and ground segments.

Debris. The space domain is becoming more congested with active satellites and debris. This congestion increases the satellite collision probability, which could damage satellites and even result in additional debris. The resulting debris would likely continue to accumulate and congest the most valuable orbits for the foreseeable future.

Electromagnetic Interference. The demand placed on the electromagnetic spectrum continues to grow as the number of satellites, satellite services,and users increases. Increased congestion limits the available spectrum and increases the potential for unintentional interference on friendly signals. To complicate the issue further, international spectrum management practices create uncertainty in gaining access to the required spectrum and impose strict limitations on power, bandwidth, and coverage.

Chap 4

V. Air Mobility Operations

Ref: Annex 3-17, Air Mobility Operations (5 Apr '16) and JP 3-17, Air Mobility Operations (Sept '13).

Air mobility operations support all of the geographic combatant commanders and functional combatant commanders. The foundational components of air mobility operations—airlift, air refueling, air mobility support, and aeromedical evacuation—work with other combat forces to achieve national and joint force commander objectives.

Joint doctrine defines air mobility as "the rapid movement of personnel, materiel, and forces to and from or within a theater by air." The Department of Defense (DOD) transportation mission involves many transportation communities and assets, services, and systems owned by, contracted for, or controlled by the DOD. US Transportation Command serves as the manager of the transportation community and is supported by the Air Force's Air Mobility Command, the Army's Surface Deployment and Distribution Command, and the US Navy's Military Sealift Command.

Airpower

Air Mobility Operations

Airlift

Air Refueling (AR)

Air Mobility Support

Aeromedical Evacuation (AE)

See p. 4-43 for an overview and further discussion of the four types of air mobility operations.

I. Global Mobility Enterprise

The global mobility enterprise is an integrated series of nodes that support air mobility operations. The four components of the enterprise consist of Airmen, equipment, infrastructure, and command and control (C2). In a dynamic, complex, or contested environment, the enterprise requires global situational awareness through collaboration, coordinated operations, and adherence to processes and support disciplines.

Specifically, the airfields or nodes that are part of this enterprise have the four components (Airmen, equipment, infrastructure, and C2). When contingencies arise, planners identify key nodes and components. Mobility Airmen label these nodes as aerial ports of embarkation, aerial ports of debarkation/hubs, intermediate staging bases, and forward operating bases. Through mission analysis, planners adjust the nodes to drive greater velocity and thus effectiveness throughout the global mobility enterprise. Most importantly, restricting any component or failing to protect all lines of communication from physical or cyberspace attacks within the enterprise can jeopardize its ability to support air mobility operations.

II. Command Relationships

Air mobility serves all combatant commanders as well as other government agencies and is thus normally optimized functionally across geographic commands. In some instances the nature of the operation may require a transfer of mobility air forces (MAF) to the geographic combatant commander (GCC) to enhance unity of effort and responsiveness of air mobility forces in theater. When functional forces participate in operations across multiple area of responsibilities (AOR), a supporting/supported relationship normally exists between the functional component commander (FCC [supporting]) and the GCC (supported). To advise on the FCC's supporting relationship during operations, the commander, Air Force forces (COMAFFOR) may request a director of mobility forces (DIRMOBFOR). The DIRMOBFOR is under the command of the COMAFFOR. Unless the COMAFFOR requests a DIRMOBFOR, the theater's air operations center (AOC) air mobility division Chief fulfills the DIRMOBFOR duties during daily operations.

Air Mobility Command (AMC)

Air Mobility Command (AMC) is the air component to US Transportation Command (USTRANSCOM) and is also its component major command, making the AMC/CC the COMAFFOR to the commander, USTRANSCOM (CDRUSTRANSCOM). CDRUSTRANSCOM normally delegates operational control (OPCON) to AMC/CC, who normally further delegates OPCON to the commander, Eighteenth Air Force (18 AF/CC) for day-to-day execution. 18 AF is USTRANSCOM's designated component numbered Air Force and is known as Air Forces Transportation (AFTRANS). 18 AF (AFTRANS)/CC delegates tactical control to the 618th Air Operations Center (AOC) (Tanker Airlift Control Center [TACC]) commander for day-to-day operations.

Air Mobility Command (AMC) is the Air Force major command primarily responsible for providing intertheater airlift, air refueling, air mobility support, and aeromedical evacuation capability. AMC organizes, trains, equips, and employs its assigned and attached forces to meet worldwide air mobility requirements. As the air component to US Transportation Command (USTRANSCOM), AMC prepares those forces to meet intertheater air mobility taskings. AMC also plans, coordinates, and manages the Civil Reserve Air Fleet (CRAF) program that provides a pool of civil airlift capability made available to the Department of Defense in times of crises.

AMC is the designated lead command for Air Force air mobility issues and works closely with theater air component commands from each combatant command to establish appropriate standards enabling a smooth transition to contingency operations. In this capacity, AMC develops weapon system standards and integrates command and control processes for the entire mobility air forces (MAF) enterprise. Standardization of processes and procedures is crucial to ensure consistent capability across the MAF.

US Transportation Command

US Transportation Command (USTRANSCOM) provides the air, land, and sea transportation for the Department of Defense (DOD), as well as other government agencies. Commander, USTRANSCOM serves as the single manager of the Defense Transportation System (DTS) and is designated by the Secretary of Defense as the global synchronization and distribution process owner. The DTS includes USTRANSCOM's three Service components: Air Mobility Command (AMC), Surface Deployment and Distribution Command, and Military Sealift Command. USTRANSCOM provides common-user airlift for the entire DOD.

The USTRANSCOM/J-3 fusion center receives, processes, and sources all transportation requests. The fusion center determines the best modes and nodes to meet mission requirements. Requirements that must move by air, based on mission timing or security, are tasked to their air component command, AMC.

III. Types of Air Mobility Operations

Ref: Annex 3-17, Air Mobility Operations (5 Apr '16), pp. 7 to 9.

Intertheater / Intratheater

Air mobility operations are described as either intertheater (operations between two or more geographic combatant commands) or intratheater (operations exclusively within one geographic combatant command). Differences exist between intertheater and intratheater airlift operations. Effective integration and synchronization of intertheater and intratheater air mobility operations is crucial to air mobility support to the warfighter. A combination of intertheater and intratheater air mobility operations requires close coordination and cooperation between the 618th Air Operations Center (AOC) (Tanker Airlift Control Center [TACC]) and the respective geographic AOC.

Air mobility allows forces to reach destinations quickly, thereby opening opportunities for seizing the initiative via speed and surprise, and by providing follow-on sustainment of critical materiel. The four types of air mobility operations are:

A. Airlift *See pp. 4-48 to 4-49.*

Airlift is "the movement of personnel and materiel via air mobility forces to support of strategic, operational, or tactical objectives." Airlift provides rapid, flexible, and secure transportation. Because airlift is a high demand asset, it should be used carefully when satisfying warfighter requirements.

B. Air Refueling (AR) *See pp. 4-50 to 4-51.*

AR is defined as "the refueling of an aircraft in flight by another aircraft."2 AR extends presence, increases range, and serves as a force multiplier. AR significantly expands the options available to a commander by increasing the range, payload, persistence, and flexibility of receiver aircraft.

C. Air Mobility Support *See pp. 4-52 to 4-53.*

Air mobility support provides command and control (C2), aerial port, and maintenance for mobility air forces. Air mobility support is part of the global air mobility support system (GAMSS). The GAMSS consists of a limited number of permanent en route support locations plus deployable forces that deploy according to a global reach laydown strategy.

D. Aeromedical Evacuation (AE) *See pp. 4-54 to 4-55.*

AE provides time-sensitive en route care of regulated casualties to and between medical treatment facilities using organic and/or contracted aircraft with medical aircrew trained explicitly for that mission. AE forces can operate as far forward as aircraft are able to conduct air operations, across the full range of military operations, and in all operating environments. Specialty medical teams may be assigned to work with the AE aircrew to support patients requiring more intensive en route care. The Air Force description supplements the joint definition in JP 3-17: "AE is the movement of patients under medical supervision to and between medical treatment facilities by air transportation."

Air mobility operations are described as either intertheater (operations between two or more geographic combatant commands) or intratheater (operations exclusively within one geographic combatant command). Differences exist between intertheater and intratheater airlift operations. Effective integration and synchronization of intertheater and intratheater air mobility operations is crucial to air mobility support to the warfighter. A combination of intertheater and intratheater air mobility operations requires close coordination and cooperation between the 618th Air Operations Center (AOC) (Tanker Airlift Control Center [TACC]) and the respective geographic AOC.

A. Geographic Organization and Control

A geographic combatant commander (GCC) exercises operational control (OPCON) over assigned and attached forces and normally delegates OPCON of assigned and attached mobility air forces (MAF) to the theater commander, Air Force forces (COMAFFOR). For example, Commander, US Pacific Command (CDRUSPACOM) delegates OPCON of assigned and attached MAF to the commander, Pacific Air Forces (PACAF), who acts as the theater COMAFFOR to CDRUSPACOM.

The COMAFFOR executes control of assigned and attached Air Force mobility forces through the air operations center (AOC). One of the AOC divisions, the air mobility division (AMD), plans, coordinates, tasks, and executes intratheater air mobility operations and, when required, plans, coordinates, tasks, and executes intertheater operations to meet requirements established by the GCC. The AOC coordinates intertheater air mobility support operations with the 618 AOC (Tanker Airlift Control Center [TACC]). A theater COMAFFOR may designate a director of mobility forces (DIRMOBFOR) as a coordinating authority between the 618 AOC (TACC), the geographic AOC's AMD, and joint task force (JTF)-specified command and control nodes to meet all validated air mobility requirements. The COMAFFOR and DIRMOBFOR should ensure intratheater MAF are organized to properly interact with other intratheater and intertheater forces. When air mobility forces are attached to a subordinate JTF, they become part of that air expeditionary task force commanded by the GCC's COMAFFOR.

Airpower

B. Air Operations Center (AOC)

A geographic air operations center (AOC) plans, tasks, and schedules attached and assigned aircraft within and outside an area of responsibility to meet geographic combatant commander (GCC) requirements. The AOC publishes an air tasking order (ATO) on a predetermined cycle (daily, weekly, etc.) to meet its mission requirements during normal operations, and may publish an ATO more frequently during wartime and other contingency operations.

Establishing a routine battle rhythm for air operations is essential to create successful measures of effect, especially in rapidly changing conditions. These situations often require a high level of cargo and personnel throughput and the necessity of quickly constructing a routine command and control (C2) battle rhythm for air assets becomes even more critical to rapid global mobility.

Air Mobility Division (AMD)

The air mobility division (AMD) plans, coordinates, tasks, and executes theater air mobility missions. The AMD tasks intratheater mobility air forces (MAF) through wing and unit command posts and through applicable C2 nodes deployed forward. The AMD works for the AOC commander and coordinates closely with the director, mobility forces (DIRMOBFOR). The AMD coordinates with the theater deployment and distribution operations center (DDOC) and the 618 AOC (Tanker Airlift Control Center [TACC]).1 The DIRMOBFOR should be collocated with the AOC to facilitate coordination with the AMD and the other AOC divisions as applicable.

See facing page for further discussion.

C. Director Of Mobility Forces (DIRMOBFOR)

The director of mobility forces (DIRMOBFOR) is a senior mobility officer who is familiar with the area of responsibility and possesses extensive background in air mobility operations. The commander, Air Force forces (COMAFFOR) may appoint a DIRMOBFOR to function as a coordinating authority between the joint task force (JTF), 618 Air Operations Center (AOC) Tanker Airlift Control Center (TACC), theater AOC, the geographic combatant commander's (GCC) J-4, and the joint mobility center (JMC)/joint deployment distribution operations center (JDDOC) to ensure the appro-

Air Mobility Division (AMD)

Ref: Annex 3-17, Air Mobility Operations (5 Apr '16), pp. 7 to 9.

The air mobility division (AMD) plans, coordinates, tasks, and executes theater air mobility missions. The AMD tasks intratheater mobility air forces (MAF) through wing and unit command posts and through applicable C2 nodes deployed forward. The AMD works for the AOC commander and coordinates closely with the director, mobility forces (DIRMOBFOR). The AMD coordinates with the theater deployment and distribution operations center (DDOC) and the 618 AOC (Tanker Airlift Control Center [TACC]).1 The DIRMOBFOR should be collocated with the AOC to facilitate coordination with the AMD and the other AOC divisions as applicable.

The AMD is normally comprised of four core teams: the airlift control team (ALCT); the air refueling (AR) control team (ARCT); the air mobility control team (AMCT); and the aeromedical evacuation control team (AECT). A fifth team, the air mobility support team (AMST) may also be established if required. Major products include airlift apportionment plans and AR inputs to the AOC's master air attack plan, ATO, airspace control order, and special instructions.

Airlift Control Team (ALCT)

The ALCT provides intratheater airlift functional expertise to plan, coordinate, manage and execute intratheater airlift operations in support of the COMAFFOR.

Air Refueling Control Team (ARCT)

The ARCT coordinates AR to support combat air operations or to support a strategic air bridge.

Air Mobility Control Team (AMCT)

The AMCT directs or redirects air mobility forces in response to requirements changes, higher priorities, or immediate execution requirements.

Aeromedical Evacuation Control Team (AECT)

The AECT provides mission planning, scheduling and execution of theater aeromedical evacuation missions and position of aeromedical evacuation ground forces.

Air Mobility Support Team (AMST)

The AMST may be established to facilitate reports, briefs and analysis to the AMD Chief and provide support to the four AMD teams.

Air Mobility C2 Structures

Ref: JP 3-17, Air Mobility Operations (Sept '13), chap. II.

The Air Force air and space operations center (AOC) is the senior C2 element of TACS and includes personnel and equipment of the necessary disciplines to ensure effective control of air operations (e.g., communications, operations, intelligence).

The following structures consist of fixed and mobile units and facilities that provide the JAOC with the information and communications required to monitor the ongoing air operation and control Air Force aircraft in theater air operations:

- **Control and Reporting Center (CRC).** The CRC is directly subordinate to the JAOC and is charged with broad air defense, surveillance, and control functions. The CRC provides the means to flight-follow, direct, and coordinate the support and defense of theater air mobility aircraft operating in the operational area.
- **Tactical Air Control Party (TACP).** TACPs consist of personnel equipped and trained to assist US ground commanders to plan and request tactical air support.
- **Emergency Operating Center (EOC).** As the C2 facility of wings, EOCs provide control and communications facilities to link wing commanders to the JAOC and enable them to command their forces. To facilitate joint operations, Army ground liaison officers or other component representatives may be assigned to an EOC.
- **Contingency Response Group (CRG), CRE, Contingency Response Team (CRT), and Contingency Load Planning Teams (CLPTs).** CRGs include contingency support elements that deploy to initiate airbase opening operations. CRG capabilities include airfield assessment, C2, aerial port passenger and cargo processing, limited aircraft servicing and maintenance, threat assessment, contracting/finance, limited force protection, limited fuels/logistics support, limited physical security/defense, and limited medical care.
- **Special Tactics Team (STT).** Commander, United States Special Operations Command (USSOCOM) exercises COCOM over all CONUS-based special operations forces (SOF), including STTs. STTs are small, rapidly deployable, task-organized combat control teams (CCTs), pararescue jumper, and combat weather (special operations weather team) personnel. They are uniquely organized, trained, and equipped to facilitate the air-ground interface during joint special operations and sensitive recovery missions.
- **An air mobility liaison officer (AMLO)** is a rated air mobility officer specifically trained to advise the supported Army/Marine Corps unit commander and staff on the optimum and safe use of air mobility assets. They support units at the corps, division, separate regiment, and selected brigade echelons, but may be aligned with echelons above corps as required. Air liaison officers are organized and empowered to serve as single authoritative voices representing and advising the ground commanders they support.
- **AMC liaison officers (LNOs)** are normally assigned to a Marine expeditionary force command element. The AMC LNOs perform similar functions as the AMLOs, but are not designated as AMLOs.
- **Airborne Elements.** As airborne C2 nodes, the E-2C HAWKEYE and the E-3A Airborne Warning and Control System (AWACS) may perform limited C2 functions in support of theater air mobility operations.
- **The medical support team** provides the personnel and equipment required to administer medical care for injuries and illness, and to administer preventive medical care reducing the risk of a catastrophic or detrimental event that could impact on mission effectiveness. The team also recommends strategies to CRG, CRE, and CRT commanders for countermeasures against environmental and physiological stressors, in order to enhance mission effectiveness.
- **Battlefield Coordination Detachment (BCD).** The airlift section of the BCDs will be located within the JAOC and will consist of intelligence and operations personnel organized into airlift, air defense, fire support, and airspace control elements. Overall, the BCD monitors and interprets the land battle situation and provides the necessary interface for the exchange of current intelligence and operational data. The airlift section is collocated with the AMD and is responsible for monitoring movements on joint airlift operations supporting Army forces (ARFOR) and providing feedback to ARFOR operations and logistics staff officers.

- **Army Tactical Operations Centers (TOCs).** TOCs are found in Army units down to maneuver battalion however, AMLOs will normally be located at the division level and above. Intratheater airlift requests will be validated and prioritized by the Army service component commander.
- **Ground Liaison Officers (GLOs).** Army units may assign GLOs to the JAOC/AOC and theater airlift EOCs. In those positions, they monitor and report on the current airlift situation to their parent units. They also advise Air Force mission commanders and staffs on Army component air movement requirements, priorities, and other matters affecting the airlift situation. GLOs assigned to the JAOC/AOC report through the BCD. They are also the principal points of contact between the Air Force CRGs and Army arrival/departure airfield control groups (A/DACGs) for controlling Army theater airlift movements.
- **Army Arrival/Departure Airfield Control Group (A/DACG).** The A/DACG is a provisional organization designed to assist AMC and the deploying unit in receiving, processing and loading or unloading personnel and equipment. A/DACGs are designed to coordinate and control the movement of personnel and materiel through air terminals. The capabilities of the A/DACGs are tailored based on the mission and military units performing aerial port operations.
- **Army Movement Control Teams (MCTs).** MCTs are responsible for coordinating the movement of personnel and materiel from air terminals to their designated destinations. MCTs operate independently of the A/DACG and are responsible for controlling movement on an area basis.
- **Army Long-Range Surveillance Teams (LRSTs).** LRSTs can support airlift by conducting reconnaissance and surveillance operations of named areas of interest around terminal areas. LRSTs, which are organized from long-range surveillance detachments and companies, are organic to each Army division. Typically, one to six LRSTs support an airborne or air assault operation. If required, LRSTs can also mark DZs and LZs and direct fire support for airlift operations.
- **Army Drop Zone Support Teams (DZSTs).** In the absence of, or in conjunction with, an Air Force STT, DZSTs provide Army units with limited organic capabilities to support airdrop operations. DZSTs direct airdrop operations on DZs and consist of at least two personnel, including an airborne jumpmaster- or pathfinder-qualified leader. They can support airdrops (up to three aircraft) of personnel, equipment, and CDS bundles.
- **Army Tactical Aviation Control Teams (TACTs).** Composed of air traffic control or pathfinder-qualified personnel, TACTs locate, identify, and establish DZs and LZs. They install and operate navigational aids and communications around the terminal, control air traffic in that vicinity and, to a limited degree, gather and transmit weather information.
- **Joint Patient Movement Requirement Center.** A joint activity established to coordinate the joint patient movement requirements function for a joint task force operating within a unified command area of responsibility. It coordinates with the theater patient movement requirements center for intratheater patient movement and the Global Patient Movement Requirements Center for intertheater patient movement.
- **Joint Task Force – Port Opening (JTF-PO).** USTRANSCOM also provides a JTF-PO to rapidly open and operate ports of debarkation and initial distribution networks for joint distribution operations supporting humanitarian, disaster relief, and contingency operations. The JTF-PO (APOD), consists of an air element for airfield operations and a surface element for cargo transfer and movement control. The surface element operates a forward distribution node for clearance of cargo from the APOD. The JTF-PO (APOD) is designed to arrive early at an airfield to establish single port management and provide in-transit visibility from the beginning of an operation. The JTF-PO deploys under the authority of the CDRUSTRANSCOM, in direct support of the CCDR; it is designed to operate for 45-60 days and be relieved by follow-on forces.
- **Joint Deployment and Distribution Operations Center.** The integration of intertheater and intratheater movement control is the responsibility of the supported combatant command and USTRANSCOM. Subsequently, each GCC has a JDDOC. The JDDOC is a CCDR's movement control organization designed to synchronize and optimize national and theater multimodal resources for deployment, distribution, and sustainment. The JDDOC is normally placed under the control and direction of the combatant command J-4, but may also be placed under other command or staff organizations.

IV. Airlift

Ref: JP 3-17, Air Mobility Operations (Sept '13), chap. IV.

Airlift operations transport and deliver forces and materiel through the air in support of strategic, operational, and/or tactical objectives. Airlift offers its customers a high degree of speed, range, and flexibility. Airlift enables commanders to respond and operate in a wide variety of circumstances and time frames that would be impractical through other modes of transportation.

Airlift supports the US National Military Strategy by rapidly transporting personnel and materiel to and from or within a theater. Airlift is a cornerstone of global force projection. It provides the means to rapidly deploy and redeploy forces, on short notice, to any location worldwide. Within a theater, airlift employment missions can be used to transport forces directly into combat. To maintain a force's level of effectiveness, airlift sustainment missions provide resupply of equipment, personnel, and supplies. Finally, airlift supports the movement of patients to treatment facilities and noncombatants to safe havens. Airlift's characteristics — speed, flexibility, range, and responsiveness — complement other US mobility assets.

Airlift operations are defined by the nature of the mission rather than the airframe used. Most aircraft are not exclusively assigned to one operational classification. In fact, the vast majority of the air mobility force is capable of accomplishing any classification of airlift. Intertheater and intratheater capabilities are available to all users of Air Force airlift.

Intertheater Airlift

Intertheater airlift provides the critical link between theaters. A key strength of airlift is its ability to quickly redeploy forces from one theater to another.

Intratheater Airlift

Intratheater airlift provides air movement of forces, personnel, and materiel within a GCC's AOR. Typically, aircraft capable of accomplishing a wide range of operational and tactical level missions conduct these operations. Intratheater operations provide both general support, usually through common-user airlift in response to the JFC's movement priorities, and direct support, normally using Service-organic airlift assets or with assets provided by another Service to responsively satisfy Service component commander's priorities. Intratheater airlift requirements include TPFDD force movements and the continuation of sustainment movements arriving in the theater, as well as on-demand movements and routinely scheduled airlift missions for the movement of non-unit related cargo and personnel.

Airlift Missions

1. Passenger and Cargo Movement
2. Combat Employment and Sustainment
3. Aeromedical Evacuation (AE)
4. Special Operations Support
5. Operational Support Airlift (OSA)

Airlift Missions

1. Passenger and Cargo Movement

The basic mission of airlift is passenger and cargo movement. This includes combat employment and sustainment, AE, special operations support, and operational support airlift (OSA). USAF airlift forces perform these missions to achieve strategic-, operational-, and tactical-level objectives across the range of military operations. Normally, movement requirements are fulfilled through regularly scheduled channel missions over fixed route structures with personnel and cargo capacity available to all customers. These regularly scheduled requirements are validated through the appropriate Service organization to USTRANSCOM or GCC, and then tasked by the 618 AOC (TACC), an AMD, or another appropriate C2 node. Depending on user requirements, requests not supportable through the channel structure can be fulfilled through use of other mission categories such as SAAM, exercise, and contingency missions. Requests that cannot be satisfied by any of the above missions may be referred to other transportation modes of the DTS. The airlift system has the flexibility to surge and meet requirements that exceed routine, peacetime demands for passenger and cargo movement.

2. Combat Employment and Sustainment

Combat airlift missions are missions that rapidly move forces, equipment and supplies from one area to another in response to changing battle conditions. Combat employment missions allow a commander to insert surface forces directly and quickly into battle and to sustain combat operations.

3. Aeromedical Evacuation (AE)

AE is the movement of regulated patients under medical supervision to and between MTFs by air transportation. AE specifically refers to USAF provided movement of regulated patients using organic and/or contracted mobility airframes with AE aircrew trained explicitly for this mission. Movement of patients requires special ATC considerations to comply with patient-driven altitude and pressurization restrictions as well as medical equipment approved for use with aircraft systems.

4. Special Operations Support

Specified airlift forces provide unique airland and airdrop support to SOF. Since there are a limited number of airlift assets dedicated to this mission, the principle of economy of force is particularly applicable. When performing special operations missions, highly trained airlift and AR crews normally act as an integral member of a larger joint package. Because these airlift missions routinely operate under adverse conditions in a hostile environment, extensive planning, coordination, and training are required to enhance mission success. Airlift and AR used in a special operations role provides commanders the capability to achieve specific campaign objectives, which may not be attainable through more conventional airlift practices.

5. Operational Support Airlift (OSA)

OSA is the movement of high-priority passengers and cargo with time, place, or mission-sensitive requirements. OSA missions are a special classification of airlift mission support to provide for the timely movement of limited numbers of priority personnel or cargo. The OSA aircraft fleet consists of executive and non-executive aircraft. The executive fleet is dedicated to the airlift of DOD and federal senior officials and DOD-approved senior officials. Non-executive aircraft support passenger and cargo airlift during peacetime, but also support CCMD wartime requirements during conflict. USTRANSCOM is responsible for the scheduling and tasking of OSA operations regarding CONUS-based assets while the Services validate OSA requests. Theaters with their own OSA fleets are responsible for scheduling and execution tasking of OSA operations within their AORs. Within a theater, OSA assets and their scheduling should reside with their respective Service component, and may be made available for tasking at the CCDRs direction.

V. Air Refueling (AR)

Ref: JP 3-17, Air Mobility Operations (Sept '13), chap. V.

AR allows air assets to rapidly reach any trouble spot around the world with less dependence on forward staging bases. Furthermore, AR significantly expands the force options available to a commander by increasing the range, payload, loiter time, and flexibility of other aircraft.

Because AR increases the range of other aircraft, many types of aircraft may be based at locations well outside the range of an adversary threat. AR allows some aircraft to participate in contingency operations without having to forward-deploy. CONUS-based operations reduce the theater logistics requirements, thereby simplifying sustainment efforts. Positioning forces outside the adversary's reach permits a greater portion of combat assets to concentrate on offensive rather than defensive action. As a result of the reduced need to forward-deploy forces, AR reduces force protection requirements as well.

Although other Services and nations maintain some organic AR capability, the Air Force possesses the overwhelming preponderance of common-user AR assets. With boom and drogue capability, these assets are capable of refueling most Air Force, Navy, and Marine Corps aircraft, and can accommodate most foreign aircraft. Additionally, all USAF tanker aircraft are capable of performing an airlift role and are used to augment core airlift assets.

Air Refueling Missions

AR is a critical force multiplier across the full range of global and theater employment scenarios. Tankers directly enhance the operational flexibility of US and allied/coalition strike, support, and surveillance aircraft. In the same manner, the nearly unlimited flight endurance provided by tanker assets is an indispensable component of the US strategic airborne command post concept.

- **Global Strike Support.** AR assets are a critical enabler for global strike operations (conventional or nuclear). For example, AR significantly increases the range and endurance of bomber aircraft, directly enhancing their flexibility to strike at distant targets and maximizing their operational utility for warfighter mission requirements.
- **Air Bridge Support.** An air bridge creates an ALOC linking CONUS and a theater, or any two theaters. AR makes possible accelerated air bridge operations since en route refueling stops for receivers are reduced or eliminated. It reduces reliance on forward staging bases, minimizes potential en route maintenance delays, and enables airlift assets to maximize their payloads.
- **Aircraft Deployment Support.** AR assets can extend the range of deploying combat and combat support aircraft, allowing them to fly nonstop to an AOR or JOA. This capability increases the deterrent effect of CONUS-based forces and allows a rapid response to regional crises.
- **Theater Support to Combat Air Forces.** Intratheater AR enables fighter aircraft to increase their range, endurance, and flexibility. During a combat operation, the highest priority for intratheater AR forces is normally supporting combat and combat support aircraft executing air operations. This is especially true during the initial phases of a conflict.
- **Special Operations Support.** AR enables SOF to maintain a long-range operating capability. The Air Force maintains AR crews who are trained to air refuel fixed- and rotary-wing special operations aircraft. Successful mission completion requires special equipment, specialized crew training, and modified operational procedures.

Force Extension

Force extension is the AR of one tanker by another and is the most efficient means to provide deployment support, given a limited number of tanker aircraft. This capability can be used whenever the fuel requirements of the escorting tanker and its receivers exceed the tanker's takeoff fuel capacity. Since takeoff fuel is limited by the amount of payload carried, dual-role tankers may require force extension.

Air Refueling Operations

AR's contribution to air power is based on the force enabling and force multiplying effects of increased range, payload, and endurance provided to refueled aircraft. AR forces conduct both intertheater and intratheater AR operations.

1. Intertheater AR

Intertheater AR supports the long-range movement of combat and combat support aircraft between theaters, or between theaters and JOAs. Intertheater AR operations also support global strike missions and airlift assets in an air bridge. AR enables deploying aircraft to fly nonstop to their destination, reducing closure time.

2. Intratheater AR

Intratheater AR supports operations within a CCDR's AOR by extending the range, payload, and endurance of combat and combat support assets. Both theater-assigned and CDRUSTRANSCOM-assigned AR aircraft can perform these operations. When CDRUSTRANSCOM-assigned AR forces participate in these operations, they are typically attached to the GCC who exercises OPCON over these forces through the COMAFFOR. Although the primary purpose is to refuel combat air forces operating within the theater, consideration should be given to the best utilization of the tanker fleet to meet the President's and SecDef's objectives.

Anchor Areas and AR Tracks

AR is normally conducted in one of two ways: in an anchor area or along an AR track. While AR is normally conducted in friendly airspace, missions may require operations over hostile territory and in contested airspace. Anchor areas and tracks may place tankers in an extremely vulnerable position and should be limited to friendly airspace when possible. AR missions over hostile territory should be conducted only after careful risk considerations and when at least regional air superiority is achieved.

- In **anchor areas**, the tanker flies a racetrack pattern within defined airspace while waiting for receiver aircraft to arrive. Once joined with the receiver, the tanker then flies in an expanded racetrack pattern while refueling the receiver. Anchor AR is normally used for intratheater operations where airspace is confined or where receivers operate in a central location. Anchor areas are best suited for small, highly maneuverable aircraft, especially in marginal weather conditions.
- An **AR track** is a published track or precoordinated series of navigation points which can be located anywhere throughout the world. To maximize effectiveness, AR tracks will normally be placed along the receiver's route of flight. However, AR track location(s) must sometimes balanced with tanker availability and basing to develop an integrated AR plan making the best use of limited receiver and tanker assets overall. AR along an AR track is the preferred method for intertheater operations.
- The **tanker rendezvous (RV)** can be accomplished in multiple ways. *For more information about RV procedures, refer to Allied Tactical Publication (ATP)-56, Air to Air Refueling.*

VI. Air Mobility Support/GAMSS

Ref: Annex 3-17, Air Mobility Operations (5 Apr '16), pp. 62 to 67 and JP 3-17, Air Mobility Operations (Sept '13), chap. VI.

Within the range of combat support capabilities are three functional areas: air mobility command and control (C2), aerial port, and air mobility maintenance, which collectively comprise air mobility support.

Air mobility support forces are divided between Air Mobility Command (AMC), which controls the majority of assets in support of USTRANSCOM's functional role, and the geographic combatant commanders (GCC) who control sufficient assets to meet their specific regional needs. These forces, combined with operating locations and the interrelated processes and systems that move information, cargo, and passengers, make up the global air mobility support system (GAMSS). This structure consists of a number of continental United States (CONUS) and en route locations, as well as deployable forces capable of augmenting the fixed en route locations or establishing operating locations where none exists.

Air mobility operations may dictate the use of contingency response (CR) forces, especially at austere locations or during a rapidly developing crisis. CR forces provide the three core air mobility support functions but also include additional combat support functions, enabling them to operate the airfield and to sustain themselves. These additional functions include weather, civil engineering, security forces, medical, contracting, finance, communications, logistics, air traffic control, public affairs, intelligence, legal and airfield operations. They can be tailored to meet the specific requirements of the operation. AMC owns the preponderance of the CR forces, which can be tasked to respond to contingencies globally. Additionally, there are limited CR forces assigned to GCCs.

Functions of Air Mobility Support

Air mobility support—command and control (C2), aerial port, and maintenance as well as forces at en route locations—are tasked to provide these services, but can also be augmented with additional functions (such as combat support, aircrew flight equipment, and intelligence) to create a more robust throughput and support capability. The level of support throughout the global air mobility support system (GAMSS) can be tailored to match the mission requirements. Additionally, deployable air mobility support forces can expand the GAMSS at existing locations or establish capabilities where none exists. Deployable air mobility support forces are designed for short-term deployments.

Command and Control Systems

Whether operational control is exercised by Commander, US Transportation Command (USTRANSCOM) or a theater commander, Air Force forces, air mobility support forces provide their own C2 systems to plan, flow, and track air movements and provide in-transit visibility (ITV) of equipment and passengers. Communication requirements may include various radio and satellite communications systems, and mobility mission planning and execution systems supporting their airfield operations as well as those of supported air mobility aircrews that may transit or operate from their location. USTRANSCOM-assigned mobility support forces normally use this capability to report to the 618th Air Operations Center (AOC) (Tanker Airlift Control Center [TACC]), while theater-assigned support forces normally report to their theater AOC.

Among the most important services that GAMSS provides are ITV and flight following. Commanders depend on accurate, timely ITV of assets to efficiently manage those assets and associated supporting operations. Consequently, the effectiveness of the GAMSS relies significantly on the integration of C2 data into a comprehensive ITV picture. (NOTE: In selected cases, Air Force Special Operations Command special tactics teams can provide limited initial C2 capability for both air traffic control and aircraft reporting.)

Aerial Port

An aerial port is an airfield that has been designated for the sustained air movement of personnel and materiel. The GAMSS possesses a robust aerial port capability. In order to be responsive as a throughput network, fixed en route aerial port operations are sized to ensure a minimum throughput capacity is maintained at all times, based not on steady state workload, but on established planning factors. Deployed aerial port operations, on the other hand, are usually sized to meet the forecasted workload requirements of the operation they are supporting. GAMSS units are designed to establish and operate air mobility terminals and have the ability to onload and offload a set number of aircraft based on forecast workload requirements. In addition, GAMSS aerial port specialists establish marshalling yards and traffic routing for cargo, aircraft servicing, passenger manifesting, and air terminal operations center services.

Aircraft Maintenance

The ability to provide basic maintenance at all times, particularly for airlift aircraft, is critical to global mobility. Designed primarily to support air mobility aircraft operations, en route maintenance units are not intended to provide sustainment maintenance. In addition, the contingency response wing provides mobile GAMSS maintenance capability comprised of mostly cross-functional maintenance specialties designed to provide aircraft marshalling, parking, refueling, and limited aircraft repair capability. When specialized aircraft repair capability is required at a forward location that exceeds the core capacity at the site, a maintenance recovery team can be deployed to accomplish the repair. As a rule, planners and units receiving maintenance augmentation from GAMSS forces should consider supplementing maintenance capability as soon as practical to ensure continued operations.

Airbase Opening

Contingency response (CR) forces are normally the first Air Force presence on an expeditionary airbase regardless of how the base is gained (e.g., base seizure or acceptance from a host nation) or which follow-on US entity operates the base. CR forces are eventually replaced by follow-on forces (see figure "Airbase Opening Force Module Construct"). When opening a base, CR forces normally coordinate actions with theater command elements to ensure theater-specific responsibilities such as force protection (FP) meet requirements. All deployed CR forces should integrate with the host for FP and communications. Defined operational areas and responsibilities for CR personnel should be specified during planning of seizure and airbase opening operations. Additional issues that should be considered during planning may include the handoff of the airfield from any seizure force to the contingency response group (CRG) or other GAMSS element, CRG/GAMSS element to follow-on unit, and redeployment and reconstitution of the CRG/GAMSS units once other expeditionary support forces are in place (normally not later than D+45 days).

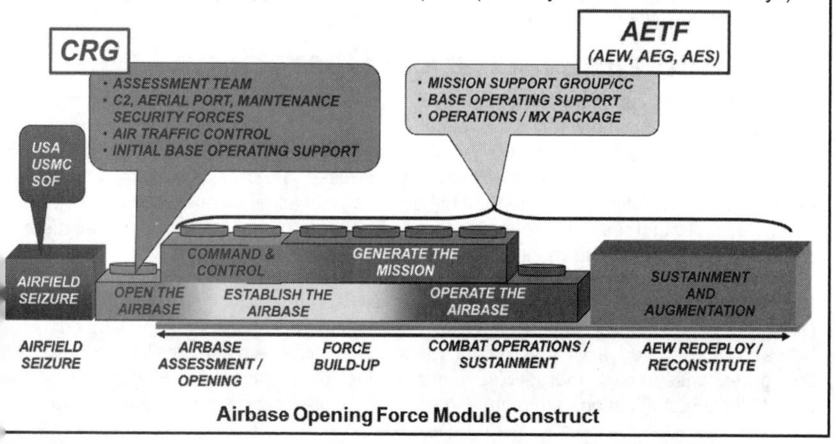

Airbase Opening Force Module Construct

VII. Aeromedical Evacuation (AE)

Ref: Annex 3-17, Air Mobility Operations (5 Apr '16), pp. 71 to 77.

The aeromedical evacuation (AE) system provides time-sensitive mission critical en route care to patients to and between levels of care (e.g. Role I-V). The Air Force's AE capability comprises a system of systems including AE liaison teams, AE crew stages, AE crews, critical care air transport teams, tactical critical care evacuation teams, other specialty teams and en route patient staging systems. These forces execute patient movement predominately on mobility air forces aircraft, as well as aboard sister Service, contracted, and international partner airframes. AE forces operate as far forward as air operations occur. The system is designed to be flexible to operate across the spectrum of potential scenarios and interface with joint, multinational, and special operations forces.

See also pp. 8-9 to 8-12 for an overview and discussion of medical operations from Annex 4-02.

Aeromedical evacuation (AE) forces may be tasked across the range of military operations. In certain circumstances, AE forces may also be tasked to evacuate injured or ill host nation personnel, enemy prisoners of war, detainees, and coalition forces in patient status. AE improves casualty recovery rates by providing timely and effective en route medical care of sick and wounded patients to medical facilities offering appropriate definitive medical care. The AE system provides patient movement by air, clinical specialty teams, specific patient movement items (PMI) equipment for in-flight care, patient staging facilities, command and control (C2) of AE forces and AE operations, and support to the communication network between airlift C2 agencies.

The Air Force is responsible for the AE mission. Air Mobility Command (AMC) is the Air Force's lead command for AE. AMC is charged with the responsibility to operate the common-user AE force and to procure and execute commercial augmentation (i.e., civilian air ambulance). AMC/Surgeon General (SG) is the US Transportation Command (USTRANSCOM)/SG's execution agency responsible for resourcing, maintaining and recycling PMI medical equipment to support Department of Defense (DOD) patient movement. It oversees the global patient movement requirement center. 18 AF (Air Forces Transportation [AFTRANS]) manages and operates the AE intertheater and hub and spoke operations, and provides AE elements and planning assistance to all theaters of operation. United States Air Forces in Europe (USAFE) and Pacific Air Forces (PACAF) are responsible for their theater-assigned AE units and associated airlift units. During contingencies where requirements exceed theater AE capabilities, AMC normally provides tailored augmentation forces to support increased intratheater requirements and expands or establishes the intertheater capability to support movement between theaters of operation or to the continental United States (CONUS), as required.

Aeromedical Evacuation Operations

During support of operations, aeromedical evacuation (AE) employs its full capability, to include staging, AE aircrew members, specialty teams, and integrated communications. During expeditionary operations, AE includes the movement of military casualties from forward operating bases (FOB) to definitive care facilities. The AE system may also be tasked to provide patient movement for noncombatant evacuation operations (NEO), injured US combat forces, repatriated American citizens, allied prisoners of war, detainees, coalition forces, and Department of Defense (DOD) civilian contractors.

Defense Support of Civil Authorities

Defense Support of Civil Authorities (DSCA) enables mutual assistance and support between DOD and any civil government agency. This includes planning and preparation for response to civil emergencies or attacks, including national security emergencies. Most DSCA situations are managed within the state. In a natural disaster the state normally declares when the situation is beyond the state's response capability and then requests federal support for the state emergency management agency from the Federal Emergency Management Agency (FEMA). The Director of patient stage operations is the senior AE DOD representative responsible for coordinating AE efforts at the aerial port of embarkation (APOE) and coordinating resource requirements with DOD, state, and federal units and agencies at the APOE. This person is responsible for all aspects of patient care and operations affecting patient care at the APOE.

When the DOD is asked to provide support, most FEMA-requested, regulated patient evacuations requiring air transportation are accomplished by AE. US Transportation Command (USTRANSCOM) is the authority that validates AE requirements in support of civilian authorities. Once validated, the requirement is tasked to 18 AF (Air Forces Transportation [AFTRANS]) or the geographic combatant commander (GCC)'s air mobility division (AMD) for execution. 618th Air Operations Center (AOC) (Tanker Airlift Control Center [TACC]) is the 18 AF (AFTRANS) lead agency for patient movement planning, coordinating, and, when directed, executing DSCA support. Air Mobility Command (AMC) also provides trained AE coordinating officers and coordinating elements for DSCA from existing active and reserve component forces in execution of the National Response Framework. AE assets required depend on the size and scope dictated by the disaster or contingency and may be supported by in-place AE infrastructure or the deployment of AE assets to the disaster area.

See also p. 1-16. Refer to Annex 3-27, Homeland Operations, for additional information.

AE Interface with Special Operations and Personnel Recovery Operations

Some expeditionary forward deployed forces, such as special operations forces (SOF), Marine expeditionary forces, and personnel recovery operations forces, do not possess organic patient evacuation capability and should identify requirements for, and obtain patient evacuation support at forward airbases. See Annex 3-50, Personnel Recovery Operations, for more information about personnel recovery.

Evacuation of casualties within a joint special operations area (JSOA) can be particularly complex since SOF often operate with small, widely dispersed teams, and in locations not easily accessible. SOF are responsible for care and evacuation of casualties from the forward location to the secure airfield where AE forces may be prepositioned to support the operation. SOF conduct the evacuation of patients with their organic capabilities. At the secured airfield patient evacuation and specialty care teams (e.g., critical care air transport team [CCATT]) assume responsibility for the casualties, freeing special operations medical assets to return to forward locations. Patient evacuation assets provide the support required to move patients through the en route care system.

Detainee Missions and AE

AE personnel are not normally used for providing care to detainees unless they require in-flight medical care. Security of detainees is not a responsibility of the en route care system. Strict adherence to detainee handling guidelines is required.

Inter-fly Agreements with Services and Coalition AE Support

The Air Force employs aircraft for the movement of patients and uses AE crew members and specialty teams (e.g., CCATT, tactical critical care evacuation team, etc.) to provide in-flight patient care. Other Services and coalition forces use various ground transport and a variety of aircraft for patient movement. Air Force AE aircrew members may perform appropriate duties in non-Air Force aircraft in the interest of the US government and approved by the appropriate Air Force component, the affected GCC, and the controlling aircraft authority. Conversely, coalition forces may also integrate with Air Force AE forces.

priate prioritization of intertheater requirements in support of intratheather air mobility taskings. The DIRMOBFOR may be sourced from within the geographic combatant commander's (GCC's) organizations or US Transportation Command (USTRANSCOM). Normally, the DIRMOBFOR is attached to the COMAFFOR's special staff and should be given appropriate liaison authority. The DIRMOBFOR provides advice to the air mobility division (AMD) on air mobility matters that should be responsive to the timing and tempo of air operations center (AOC) operations. The AMD remains under the control of the AOC commander who manages the execution of operations for the COMAFFOR.

D. Command and Control of Other Air Mobility Operations

This section discusses command and control of other air mobility-related operations including tanker, homeland, and nuclear operations.

Tanker Command and Control

Normally, operational control (OPCON) of the continental US (CONUS)-based tanker force for operational missions remains with US Transportation Command (USTRANSCOM). However, tanker assets, when authorized by the Secretary of Defense (SecDef), are transferred to a geographic combatant commander (GCC) with specification of OPCON for intratheater operations. OPCON for US Air Forces in Europe (USAFE) and Pacific Air Forces (PACAF) assigned tanker assets rests with their parent GCC and can be provided in support to another combatant command. Typically, the command and control (C2) agency is the AOC for the respective combatant command (CCMD) (e.g., the 601th Air Operations Center (AOC) for US Northern Command [USNORTHCOM] missions).

Homeland Operations

The C2 of mobility air forces (MAF) during homeland operations is the same as the functional and geographic operations discussed earlier. During homeland operations, USNORTHCOM is the geographic combatant command, except for Hawaii and other Pacific territories in the US Pacific Command area of responsibility. First Air Force (Air Forces North [AFNORTH]) supports USNORTHCOM, and uses its AOC for CONUS-based homeland operations, with air mobility expertise and operations obtained from a director of mobility forces and air mobility division, linked back to the 618 AOC (Tanker Airlift Control Center [TACC]).

Domestic emergencies often call for the use of air mobility assets to support civil authorities. The air mobility capabilities of the Department of Defense (DOD) far exceed that of state and local resources, and therefore are a crucial piece of the operational planning in response to civil crises. DOD can provide specialized skills and assets that can rapidly stabilize and improve a situation until civil authorities can effectively respond to the needs of the populace. The focus of defense support to civil authorities (DSCA) is to save lives, prevent human suffering, and mitigate property damage.

Nuclear Operations

For nuclear operations, Air Mobility Command, as USTRANSCOM's air mobility component, supports nuclear operations via change in OPCON of air refueling and airlift forces to the commander, US Strategic Command (CDRUSSTRATCOM). For theater nuclear operations, C2 of USTRANSCOM forces mimics conventional theater operations to the greatest extent possible. However, if political considerations warrant that all tanker support to theater nuclear operations must originate from CONUS, CDRUSTRANSCOM maintains OPCON of tankers and provides AR in a supporting role to CDRUSSTRATCOM.

VI. Global Integrated ISR Operations

Ref: Annex 2-0, Global Integrated ISR Operations (29 Jan '15).

Intelligence, surveillance, and reconnaissance (ISR) from the air dates back to the use of balloons to observe adversary positions during the French Revolution. Today's knowledge-based environment, provides the opportunity to observe and analyze the meaning and impact of a wide variety of events and convey useful, timely intelligence on adversaries' capabilities and intentions to decision-makers. However, in this "Age of Information," ISR capabilities have expanded to operate from and through air, land, maritime, space, and cyberspace domains to achieve desired effects across the range of military operations in support of US national security objectives.

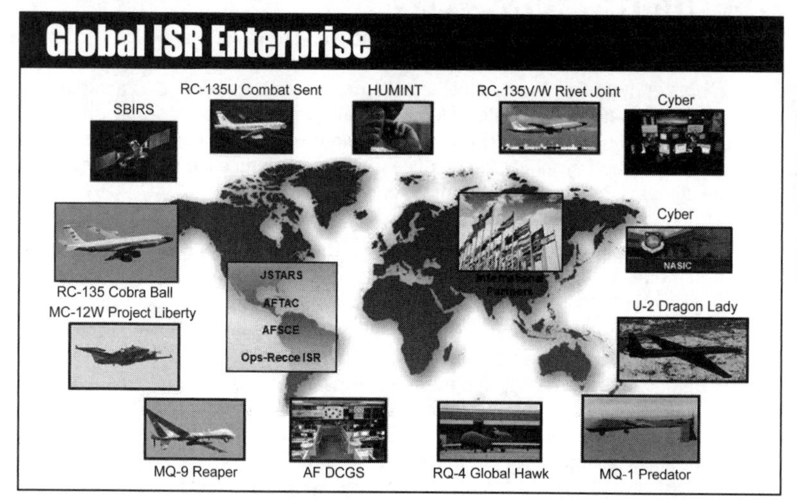

ISR Defined

ISR is defined as "an activity that synchronizes and integrates the planning and operations of sensors, assets, and processing, exploitation, and dissemination systems in direct support of current and future operations. This is an integrated intelligence and operations function." ISR consists of separate elements but requires treatment as an integrated whole in order to be optimized.

Global Integrated ISR Defined

The Air Force defines global integrated ISR as "cross-domain synchronization and integration of the planning and operation of ISR assets; sensors; processing, exploitation and dissemination systems; and, analysis and production capabilities across the globe to enable current and future operations."

Global integrated ISR enables use of multiple assets from multiple geographic commands and leverages national capabilities in support of Service-specific requirements to collect data across all domains to satisfy strategic, operational and tactical requirements. The data may be used by national, joint, service specific, coalition, or allied personnel or any combination thereof. Global integrated ISR enables the integration of this collected information to deliver intelligence to the right person at the right time, anywhere on the globe.

An example of how the ISR enterprise projects global presence and battlespace awareness is through distributed operations. A global integrated Predator mission includes the aircraft, a CONUS-based or forward-deployed pilot and sensor operator team, the datalinks that allow it to be flown remotely from a location outside of theater, and all of the networks that allow its data to be streamed in near real-time to many locations around the world. It also includes the analytic capability being leveraged outside of the area of responsibility, possibly from multiple locations, that allows global collaboration to exploit the collected data, plus the dissemination capability that allows finished intelligence to flow back to multiple end users and be stored for future reference.

The ultimate goal of global integrated ISR operations is to support national security objectives through enhanced understanding of the operating environment and adversary intentions. JP 2-0, Joint Intelligence, states that "information is of greatest value when it contributes to the commander's decision-making process by providing reasoned insight into future conditions or situations." Global integrated ISR operations provide actionable intelligence to the commander in the fastest way possible.

I. Airman's Perspective

Airpower

Global integrated intelligence, surveillance, and reconnaissance (ISR) operations are domain, Service and platform neutral. The focus is on meeting information requirements and providing actionable intelligence to commanders. Global integrated ISR is further enhanced when integrated with joint, Departmental, national, and multinational ISR. Global integrated ISR is the linchpin of effects-based operations and enables integration and synchronization of assets, people, processes, and information across all domains, to inform the commander's decision cycle.

The evolution of technology and information enabled a move from the segregation to integration of operations and intelligence. The elements of ISR are interdependent and mutually supporting to compress the find, fix, track, target, engage, and assess (F2T2EA) process from days to minutes.

Global integrated ISR operations enable operations throughout the range of military operations (ROMO) in permissive and non-permissive environments by serving as a theater capability. Global integrated ISR operations also facilitate the integration and synchronization of joint, Departmental, national, and coalition ISR capabilities. Other Services may focus organic elements of ISR efforts towards the tactical level of war, specifically in support of organic component operations (i.e., supporting a specific mission or unit). These forces are typically organic to a service echelon. Coalition members or allies will tend to focus their ISR efforts to meet their own informational needs.

The Air Force currently uses the majority of its ISR assets to directly support national objectives and the joint force commander's (JFC's) strategic and operational goals and component-level requirements. One of the most valuable attributes of airpower is its flexibility, the inherent ability to project power dynamically across large swaths of an operational area. Flexibility of ISR operations is exponentially enhanced with distributed ops. Global integrated ISR monitors both friendly and adversary movements and capabilities in a dynamic environment, and drives the F2T2EA process. The Air Force may designate some assets as organic assets to satisfy Service-specific collection requirements. An example is the use of unmanned aerial systems (UAS) to support base defense or special operations or cyberspace sensors to protect the AF network.

The Air Force conducts global integrated ISR operations through a five-phase process: planning and direction; collection; processing and exploitation; analysis and production; and dissemination (PCPAD). The process is not linear or cyclical, but rather represents a network of interrelated, simultaneous functions that can, at any given time, feed and be fed by other functions.

See facing page for an overview and further discussion of PCPAD.

II. Planning and Direction; Collection; Processing and Exploitation; Analysis and Production; and Dissemination (PCPAD)

Ref: Annex 2-0, Global Integrated ISR Operations (29 Jan '15), pp. 4 to 5.

The Air Force conducts global integrated ISR operations through a five-phase process: planning and direction; collection; processing and exploitation; analysis and production; and dissemination (PCPAD). The process is not linear or cyclical, but rather represents a network of interrelated, simultaneous functions that can, at any given time, feed and be fed by other functions.

Planning and Direction

The planning and direction phase begins the process by shaping decision-making with an integrated and synchronized ISR strategy and collection plan that links global integrated ISR operations to the JFC's intelligence requirements and integrates them into the air tasking order (ATO) and its reconnaissance, surveillance, and target acquisition (RSTA) annex.

Collection

The collection phase occurs when the mission is executed and the sensors actually gather raw data on the target set. The collected data in its raw form has relatively limited intelligence utility.

Processing and Exploitation

The processing and exploitation phase increases the utility of the collected data by converting it into useable information. During the analysis and production phase analysts apply critical thinking and advanced analytical skills by fusing disparate pieces of information and draw conclusions resulting in finished intelligence.

Dissemination

Finished intelligence is crucial to facilitating informed decision-making, but only if it is received in a timely manner. Dissemination, the final phase of PCPAD, ensures the commander, planners, and operational forces receive the derived intelligence in time to make effective decisions and conduct effective operations. The Air Force's distributed operations capability enables it to conduct global integrated ISR operations and provide timely and tailored intelligence on a global level to multiple end users. The analyzed intelligence can be disseminated or stored for future use. Properly formatted and archived data makes previously collected and exploited information readily available to correlate and provide context to data.

III. Global Integrated ISR Enduring Capabilities

Ref: Annex 2-0, Global Integrated ISR Operations (29 Jan '15), pp. 6 to 7, and p. 13.

Global integrated intelligence, surveillance, and reconnaissance (ISR) enables decision advantage for the joint and coalition warfighter through five integrated capabilities: battlespace characterization; collection operations; targeting; production of intelligence mission data (IMD) for information based weapons and platforms; and, intelligence support to weapon system design and acquisition (see figure).

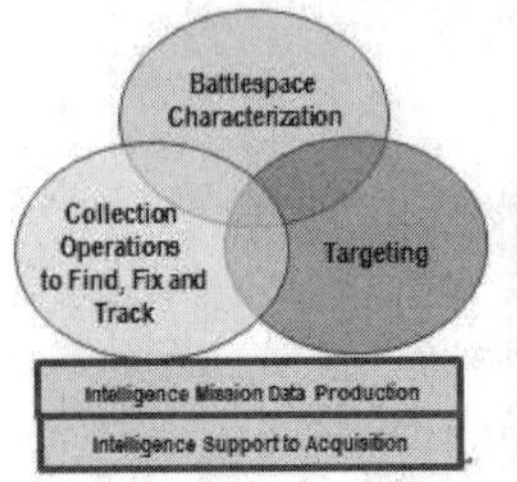

Global Integrated ISR Enduring Capabilities

Battlespace Characterization

Battlespace characterization is the ability to understand and predict adversary capabilities, tactics, techniques and procedures (TTPs), threat dispositions, centers of gravity, and courses of actions within the context of the operating environment in order to provide indications and warning, identify potential vulnerabilities to our forces and identify opportunities to achieve our combat objectives. In short, through battlespace characterization, global integrated ISR captures what is known, what is not known, and what is believed and continuously updates and tests those conclusions to prepare for and execute joint operations across the range of military operations.

Collection Operations to Find, Fix, Track and Characterize

Collection operations is the command and control and synchronization of ISR sensors, platforms and exploitation resources to find, fix, track and characterize adversary activities and infrastructure as well as the operating environment. The aim is to test beliefs, confirm knowledge and discover intelligence gaps in order to enhance our battlespace characterization, targeting, IMD and acquisition support requirements.

Targeting

Targeting is the process for selecting and prioritizing targets and matching appropriate actions to those targets to create specific desired effects that achieve objectives, taking account of operational requirements and capabilities. Interactions with information and intelligence gathered during find, fix and track activities and battlespace characterization are used to conduct deliberate (preplanned) and dynamic (time-sensitive) targeting. The targeting cycle spans development of commander's objectives, guidance and intent; through target development, vetting, validation, nomination and prioritization; to commander decision and force assignment, planning and execution, and finally assessment.

Intelligence Mission Data Production

IMD production is the ability to derive, produce, and rapidly update the intelligence used for programming platform mission systems in development, testing, operations and sustainment including, but not limited to, the following functional areas: signatures, electronic warfare integrated reprogramming, order of battle, characteristics and performance, and GEOINT. The proliferation of information-based weapon systems—weapons and platforms that require detailed intelligence information to operate as designed—has significantly increased the need for the production and rapid integration of IMD into Air Force operations.

Intelligence Support to Acquisition

Intelligence support to acquisition and its associated analytical processes enable the acquisition community to impact weapon system design through future threat projections and to capture the intelligence sensitivity of a particular development program. Threat assessments, extending from the current risk out beyond 20 years, include production of System Threat Assessments and System Threat Assessment Reports. Programs also must be analyzed for intelligence sensitivity to determine if they require intelligence data during development or to perform their mission, or require the direct support of intelligence personnel or influence intelligence data at any point in the planning and direction; collection; processing and exploitation; analysis and production; and dissemination cycle. An Intelligence Supportability Analysis will be developed for intelligence sensitive systems.

IV. Cross-Domain Integration and Global Integrated ISR

Global integrated intelligence, surveillance, and reconnaissance (ISR) operations are conducted in, from, and through all domains (air, land, maritime, space and cyberspace), across all phases of operations, in permissive and non-permissive environments. These operations focus on meeting the joint force commander's intelligence requirements within complex operational environments. Integrated planning and direction, collection, processing and exploitation, analysis and production, and dissemination (PCPAD) capabilities include integration of cross-domain collection activities using the full-spectrum of sensors (e.g., signals intelligence, radar, electro-optical, infra-red, human, and ground-based); integrated processing and exploitation and analysis and production activities in air operation centers (AOCs), Air Force Distributed Common Ground System, and national production centers; and integrated intelligence products disseminated to tactical, operational and strategic users. Ultimately, cross-domain integrated capabilities enable global integrated ISR forces to quickly analyze collected data, and feed the resulting intelligence—real-time in many instances—to warfighters.

Net-Centric Operations

Global integrated ISR systems use networks, satellite communications, and datalinks to execute global integrated ISR missions. This net-centric structure is known as distributed operations and requires that global integrated ISR operations be cross-domain integrated. For example, a single global integrated ISR mission may collect on maritime target sets using an airborne platform and transmit collected data over space-based satellite communications to analysts in another part of the world who then create and disseminate intelligence products through cyberspace.

For this reason, an open and secure net-enabled architecture is essential to cross-domain integrated analysis and dissemination. The processed data from collection platforms must move on global networks to multiple analysis sites for exploitation and further dissemination. The results should be stored in such a way that they are readily discoverable and retrievable to improve the timeliness, depth and accuracy demanded by multiple customers.

V. Joint Intelligence Preparation of the Operational Environment (JIPOE)

Joint Intelligence Preparation of the Operational Environment (JIPOE) is a valuable methodology focusing intelligence, surveillance, and reconnaissance (ISR) for the commander and the commander's supporting command and control (C2) elements by getting "inside" the enemy's decision-making cycle. Specifically, JIPOE focuses analysis on the adversary, the operational environment (OE), and the effect of the OE on both friendly and enemy COAs. JIPOE and intelligence preparation of the battlespace (IPB) are key tools for conducting analysis and production that directly support C2 planning and direction processes. The key distinction between JIPOE and IPB is the supported commander. Combatant commands and joint forces conduct JIPOE while Service components provide service-level IPB. JIPOE/IPB results in the production of an intelligence estimate, potential adversary courses of action (COAs), named areas of interest, and high-value targets, which are inputs to the joint operation planning process (JOPP) and the joint force commander (JFC) and commander, Air Force forces (COMAFFOR) planning and targeting processes. The JIPOE process includes integrating analysis, production, collection management, and targeting processes to shape decision making and enable operations. Finally, JIPOE is a significant enabler for commanders to leverage or support the full range of instruments of national power (diplomatic, information, military and economic) to ensure shaping ops deter adversary actions or if deterrence fails, set conditions for successful operations should conflict occur.

Detailed threat analysis is critical for friendly force mission planning and defense suppression across all domains. JIPOE assesses how the enemy doctrinally organizes, trains, equips and employs their forces against friendly force vulnerabilities. JIPOE also assesses the cultural, social, religious, economic, and government elements of the country/region and determines the possible effects of enemy and friendly COAs on them.

JIPOE alerts decision-makers at all echelons to potential emerging situations and threats. JFC guidance provided during planning shapes the overall concept of operations, which in turn drives planning requirements for air, space, and cyberspace employment. The challenge of the JFACC/commander, Air Force forces (COMAFFOR) is determining where and when to focus attention in order to influence events early, ready forces, and begin setting conditions for future operations. Therefore, preparation of the operational environment is essential to supporting the commander's visualization process, determining (component-level) intelligence requirements, anticipating critical decision points during operations, and prescribing rules of engagement (ROE).

JIPOE contributes to those enabling functions to plan and prepare for potential follow-on military operations. For example, global integrated ISR provides the intelligence needed to understand how an adversary's networks can be affected by non-kinetic (cyberspace and information operations) capabilities. The global integrated ISR contribution includes but is not limited to identifying data, system/network configurations, or physical structures connected to or associated with the network or system, determining system vulnerabilities, and suggesting actions warfighters can take to assure future access and/or control of the system, network, or data during anticipated hostilities.

JIPOE, target system analysis, and target development processes highlight an adversary's centers of gravity (COGs), key capabilities and vulnerabilities, possible intentions, and potential COAs. By identifying known adversary capabilities, JIPOE provides the conceptual basis for the JFC and COMAFFOR to visualize how the adversary might threaten friendly forces or influence mission accomplishment. JIPOE is the process in which critical thinking skills are applied to effectively counter an adversary's denial and deception strategy and anticipate surprise. Ultimately, JIPOE shapes the JOPP and by extension, the air component's air operations plan, operations order, and air operations directives.

VI. ISR in the Air Operations Center

Ref: Annex 2-0, Global Integrated ISR Operations (29 Jan '15), pp. 43 to 45.

The intelligence, surveillance, and reconnaissance division (ISRD) is responsible for effectively orienting the commander, Air Force forces (COMAFFOR) to current and emerging enemy capabilities, threats, courses of action (COAs), centers of gravity (COGs), global integrated ISR operations management and targeting intelligence support. The ISRD accomplishes this task by integrating the global integrated ISR and air tasking order (ATO) processes. The ISRD provides intelligence crucial to the air mobility, strategy, combat plans, and combat operations divisions planning and executing operations. This intelligence helps achieve the commander's objectives as well as provides the means by which the effects of the operations are measured.

The ISRD has primary responsibility to support the planning, tasking, and execution of theater air, space, and cyberspace global integrated ISR operations. The ISRD serves as the senior intelligence element of the theater air control system (TACS), and integrates global integrated ISR platforms and capabilities (internal and external to the AOC) in support of the joint force. Additionally, the ISRD ensures that global integrated ISR within its responsibility is optimally managed to operate within the context of a complex national and joint intelligence architecture.

The ISRD chief is the SIO for the AOC and reports to the AOC director. The ISRD Chief works in close coordination with other division chiefs and senior AOC staff to determine the best utilization of ISR personnel throughout the AOC to support AOC processes and requirements. The ISRD chief works closely with the COMAFFOR A2 to ensure the ISRD and A2 staffs are working together effectively.

- Provides analyses of the enemy and a common threat picture to the JFACC, staff planners, AOC divisions and other Air Force elements in theater.
- Provides combat ISR support assessment activities for air component operations planning and execution in conjunction with the strategy, combat plans, and combat operations divisions.
- Directs and manages the air component's global integrated ISR operations, to include reachback, distributed, and federated operations.
- Provides direct targeting support to the ATO cycle in response to COMAFFOR guidance.
- Provides all-source intelligence support to other AOC divisions to enhance the execution of their core processes.

ISRD and PCPAD

The ISRD is key in all elements of the PCPAD process. Two central functions of the ISRD within the PCPAD process are ISR operations management and analysis.

The Combat Operations Division (COD)

The COD is responsible for executing "today's war." ISR personnel within the Combat Ops Division form the Senior Intelligence Duty Officer (SIDO) Team which is responsible for providing up-to-date intelligence inputs in order to provide maximum situational awareness for the Chief of Combat Operations. SIDO responsibilities include leading a team responsible for current global integrated ISR operations. This team maintains an accurate threat picture, supports dynamic operations (i.e., personnel recovery and prosecution of dynamic targets), and monitors execution of the ATO and RSTA annex governing global integrated ISR operations.

The SIDO Team is also responsible, through close coordination with platform and processing, exploitation and dissemination (PED) liaison officers (LNOs), for dynamic retasking of theater air, space, and cyberspace global integrated ISR assets and requisite PED support for JFC objectives.

VI. Intelligence Disciplines

Ref: Annex 2-0, Global Integrated ISR Operations (29 Jan '15), pp. 43 to 45.

ISR resources collect data that become finished intelligence when processed, analyzed, and integrated. These data can be collected through a wide variety of means. The following is a list of intelligence collection disciplines relevant to Air Force ISR operations:

Geospatial Intelligence (GEOINT)

Geospatial intelligence (GEOINT) is defined as exploitation and analysis of imagery and geospatial information to describe, assess, and visually depict physical features and geographically referenced activities on the Earth. Imagery intelligence (IMINT) involves the collection of images that are recorded and stored (on film, digitally, on tape, etc.). These images can be used to help identify and locate adversary military forces and facilities and give the commander insight into the adversary's capabilities.

Signals Intelligence (SIGINT)

SIGINT is defined in JP 1-02 as a category of intelligence comprising either individually or in combination all communications intelligence (COMINT), ELINT, and foreign instrumentation signals intelligence (FISINT), however transmitted. More specifically, SIGINT uses intercepted electromagnetic emissions to provide information on the capabilities, intentions, formations, and locations of adversary forces. **COMINT** consists of information derived from intercepting, monitoring, and locating the adversary's communications systems. COMINT exploits the adversary's communications, revealing the adversary's intentions. **ELINT** consists of information derived from intercepting, monitoring, and locating the adversary's noncommunication emitters. It exploits the adversary's radar, beacon, and other noncommunication signals, allowing friendly forces to locate adversary radars and air defense systems over a wide area. **FISINT** consists of technical information derived from the intercept of electromagnetic emissions, such as telemetry, associated with the testing and operational deployment of foreign air and space, surface, and subsurface systems.

Measurement and Signature Intelligence (MASINT)

Measurement and signature intelligence (MASINT) is scientific and technical intelligence obtained by quantitative and qualitative analysis of data (metric, angle, spatial, wavelength, time dependence, modulation, plasma, and hydromagnetic) derived from specific technical sensors for the purpose of identifying any distinctive features associated with the target. The detected feature may be either reflected or emitted. Examples of MASINT might include distinctive infrared signatures, electronic signals, or unique sound characteristics. MASINT can be collected by ground, airborne, sea, and space-based systems.

Human Resources Intelligence (HUMINT)

HUMINT is the intelligence collection discipline that uses people in the area of interest to identify or provide insight into adversary plans and intentions, research and development, strategy, doctrine, and capabilities. Some examples of HUMINT collection include clandestine acquisition of photography, documents, and other material; overt collection by air attaches in diplomatic and consular posts; debriefing of foreign nationals and US citizens who travel abroad; official contacts with foreign governments, aircrew and ground personnel debriefing, and SOF intelligence collection missions.

Open-Source Intelligence (OSINT)

Open-source intelligence (OSINT) is the intelligence collection discipline that uses information of intelligence value that is available to the general public. Particular sources are newspapers, other publications, radio and television media, and the internet. OSINT processing transforms text, graphics, sound, and video into usable information in response to user requirements.

VII. Strategic Attack (SA)

Ref: Annex 3-70, Strategic Attack (25 May '17).

Department of Defense Directive (DODD) 5100.1, Functions of the Department of Defense and Its Major Components, states that one of the functions of the Air Force is to "organize, train, equip, and provide forces to...conduct global precision attack, to include strategic attack... and prompt global strike." Formerly, strategic attack (SA) was defined in terms of nuclear delivery systems or weapons. This is no longer true. SA is not defined in terms of weapons or delivery systems used—their type, range, speed, or destructiveness—but by its effective contribution to achieving strategic objectives.

Strategic Attack (SA) is offensive action specifically selected to achieve national strategic objectives. These attacks seek to weaken the adversary's ability or will to engage in conflict, and may achieve strategic objectives without necessarily having to achieve operational objectives as a precondition.

Strategic Attack

SA includes analysis, planning, targeting, command and control (C2), execution, and assessment in combination to support achievement of strategic objectives. An analysis of the definition clarifies SA:

"**Strategic**" refers to the highest level of an enemy system that, if affected, will contribute most directly to the achievement of our national security objectives. It is not limited to the use of nuclear weapons, although in some instances the weapon most appropriate for a particular set of circumstances may be nuclear. (System: "A functionally, physically, and/or behaviorally related group of regularly interacting or interdependent elements forming a unified whole.")

"**Attack**" entails offensive action. It implies proactive and aggressive operations against an enemy (whether a state, a non-state actor, or other organization) and may be used preemptively and without regard to enemy military force. Attacks may employ kinetic or non-kinetic means, from nuclear and conventional destructive weapons, to forms of cyberspace power like offensive cyberspace operations, in order to create lethal and non-lethal effects.

SA is an approach to war focused on the adversary's overall system and the most effective way to target or influence that system. SA planners should examine the full spectrum of that system: political, military, economic, social, infrastructure, and information in the context of stated national security objectives. SA involves the combination of effects that most effectively and efficiently achieves those objectives at the strategic level. In the Air Force context, SA is a discrete set of military operations aimed at achieving those strategic objectives. Airpower offers the quickest and most direct means to conduct those operations.

SA involves the systematic application of lethal and non-lethal capabilities against an enemy's strategic centers of gravity (COG), to undermine the enemy's will and ability to threaten our national security interests.

The aim of SA is to help directly achieve national security objectives by generating effects that significantly influence adversary COGs. SA operations are essentially effects-based and should be planned, executed, and assessed as a unified, adaptive

I. Strategic Attack and Warfighting Strategy

Ref: Annex 3-70, Strategic Attack (25 May '17), pp. 10 to 16.

Strategic attack (SA) represents one key element of a unified national approach to handling a conflict and should not be employed in isolation. A sound, unified approach will comprise diplomatic, informational, military, and economic activities orchestrated carefully to achieve national security objectives. It is most effectively used in a manner that complements and is complemented by other operations. For example, action against an enemy's forces may expose critical targets and increase their consumption of war-sustaining resources. Such operations may also be necessary to enable SA, as the defeat of the Luftwaffe through offensive counterair operations did during World War II. Certain coercive applications of SA simply may not work in the absence of complementary diplomatic, political, or economic actions.

Regardless of these considerations, the United States can pursue a comprehensive strategy designed to place maximum stress on the enemy system (nation or organization). The process of developing this strategy should start with the desired end state and then be worked backwards from big to little, strategic to tactical. The enemy should be analyzed as a system and an effects-based approach should be used to determine required effects and actions. Striking an enemy's centers of gravity (COG) should be accomplished as quickly and from as many directions and sources as possible, in order to place overwhelming strain on the system.

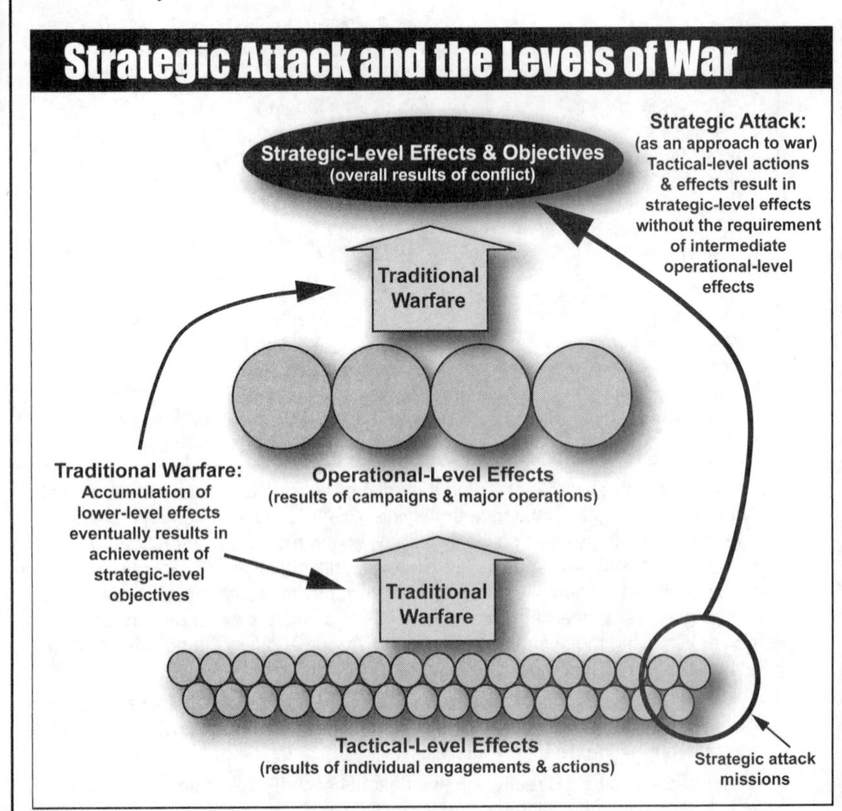

Victory in any conflict requires some mechanism for changing the enemy's behavior. Behavior can be influenced by affecting the enemy's capability to fight or by influencing his will to fight (by creating effects on enemy systems); most situations will involve aspects of both. There are several mechanisms that can be used to implement a coercive strategy.

Centuries of traditional surface warfare have conditioned leaders of world powers to raise armies and navies, the primary attributes of which are mobility, armor, firepower, depth, and sustained presence in foreign lands. These attributes help to withstand force-on-force engagements until strategic breakthrough can be attained. Military force is one instrument of national power. Bypassing surface forces or simultaneously attacking other instruments of national power and centers of gravity (COGs) may result in a change of an adversary's ability or will to fight.

Strategic Objectives

Ends, not means, drive the strategic attack (SA) effort. Successful SA requires clear and attainable objectives. Objectives and desired end states should be clearly understood by planners and commanders orchestrating the SA effort and should be tied to the SAs themselves by a clear, logical mechanism of cause and effect. SA operations are designed to produce political, military, economic, social, infrastructure, cyberspace, and information effects that contribute directly to achieving the strategic objectives of the joint force commander (JFC) and higher authorities. The senior commander and national leaders should also weigh SA operations against potential unintended effects, since attacking certain COGs could have undesired impacts on populations and neighboring countries. Strategic objectives, like those at all levels, should be measurable. Commanders and national leaders should know when those objectives are achieved.

Strategic Effects

SA seeks to achieve the greatest effect for the least cost in lives and resources by systematically applying force to COGs within the pertinent systems. Systematic application of force should not be confused with sequential application, but instead refers to a systematic approach to planning and executing attacks to achieve desired effects. System change that drives enemy compliance is the goal of SA. This system change will most effectively be achieved by applying force through parallel operations where the targeted systems are struck in a compressed timeframe. This type of attack has the highest probability of pushing a system beyond its ability to react or adapt. Attempting to change the system through attacks on its periphery will not be as effective as overwhelming system-wide parallel attack.

- SA achieves objectives through **indirect effects**. SA, even more than other forms of attack, is concerned with higher-level indirect effects. Direct effects are the results of actions with no intervening causal mechanism between act and outcome. Direct effects trigger additional outcomes—intermediate effects or mechanisms that produce higher-order outcomes or results.
- Sequential operations generally yield **sequential effects.** Also, the type of system being attacked, the action taken against it, the number of nodes struck and the amount of time used to carry out the attacks will affect whether effects are cascading—sudden, catastrophic changes in system states that often affect surrounding related systems—or cumulative—building sequentially in small amounts toward system change.
- **Systemic Effects.** Every party to a conflict, whether a modern nation-state or terrorist organization, is a complex, adaptive system. Every system has elements critical to its functioning: key strengths and sources of power. Some aspects or elements of every system are vulnerable to attack or influence. The key to understanding systemic effects is understanding how these two are related: what the system's critical vulnerabilities are.
- **Decisive Effects.** SA offers commanders many options for winning conflicts outright or for shaping them in decisive ways. It supports or underpins a variety of potential strategies. For example, attacks on leadership can often provide significant strategic leverage. SA can deny an enemy the means and resources it requires to continue a conflict. SA can deny an enemy strategic options or choices.

course of action (COA), starting with the desired outcome and working backwards to determine the required effects and actions. It is focused on the objectives achieved rather than the platforms, weapons, or methods used.

SA is oriented on the adversary's system, changing it to conform to our national objectives. SA accomplishes this change by affecting (positively or negatively) the COGs in the enemy (not just military) system that will force the overall system to change as desired in the shortest possible period of time. COGs are the leverage points in the system that, when affected, create significantly more change than would be achieved by affecting parts of the system that are not centers of gravity. COGs can be physical things like leaders, key production, structures, people, or organizations. COGs may also encompass intangible things, such as an enemy's moral strength. Affecting COGs will yield results disproportionate to the effort expended, that is, they will provide the highest payoff (enemy system change) for the least cost (lives, resources, time, etc.).

A center of gravity is defined in joint doctrine as a source of power that provides moral or physical strength, freedom of action, or will to act. In the context of SA against enemy systems, COGs are focal points that hold a system or structure together, draw power from a variety of sources, and provide purpose and direction to that system. In practical terms, COGs have critical requirements, some of which may be vulnerable to attack—critical vulnerabilities. These critical vulnerabilities may yield decisive points: geographic places, specific key events, critical factors, or functions that, when acted upon, allow commanders to gain a marked advantage over an adversary or contribute materially to creating a desired effect. Affecting these decisive points should exploit a COG's critical vulnerabilities in a manner that creates desired effects against the COG itself. SA may often be the function of choice for exploiting adversary decisive points.

SA affects conflict-sustaining resources. While it may often be difficult to directly target an adversary's will, we can often target the means the adversary employs to conduct or continue the conflict. Modern high-technology warfare is resource intensive; the support necessary to sustain it provides many lucrative targets which, when attacked, may help to speed enemy collapse and remove options. This is true across the range of military operations and not just for modern, high-technology combat. The target sets may change, but the principles remain the same.

SA affects the enemy's strategy. Sun Tzu said the best policy in war is to defeat the enemy's strategy; this requires we hold at risk what the enemy holds dear or deny them the ability to obtain what they seek. While other forms of military or national power can also deny the enemy strategic choices, SA can often do so more effectively and efficiently.

II. Basic Characteristics

Strategic attack (SA) seizes upon the unique capability of air, space, and cyberspace to achieve objectives by striking at the heart of the enemy, disrupting critical leadership functions, infrastructure, and strategy, while at the same time avoiding a sequential fight through layers of forces.

Unless the enemy's military forces are deemed to be a strategic centers of gravity (COG), they are not useful as SA targets. In fact, the goal of SA operations is to bypass the fielded forces to the maximum extent possible. A way to illustrate this concept is to think of the military as a tool being used by a person, nation, or organization to enforce its will. It very often makes more sense to attack the person, nation, or organization using the tool rather than the tool itself. SA's goal is to exert influence on the decision-maker rather than the tool being used by the decision-maker.

Next, SA conducted against an enemy system in a deliberate, systematic way generates strategic-level effects without first having to fight the enemy's fielded forces. SA seeks to prevent an enemy from achieving goals (reactive) or enabling us to achieve our goals (proactive). By affecting strategic-level COGs, the results should be greater than those generated by a similar effort against peripheral systems or targets.

SA can also act on the psychology of the enemy leadership by changing the political climate or denying options or choices. These attacks could indirectly affect the adversary's will to fight.

III. The Role of Airpower in Strategic Attack

Strategic attack (SA) capabilities are founded on the characteristics of the air, space, and cyberspace domains and the resources used in them. These characteristics (range, speed, precision, flexibility and lethality) enable a joint force commander (JFC) to access to the depths of adversary's centers of gravity (COGs) where land and maritime forces cannot easily reach. Added to traditional domains is cyberspace. Employing the Air Force's cyber capabilities throughout this domain, in combination with the traditional capabilities of air and space power, allows synergy and flexibility across a range of lethal and non-lethal options. Airpower used in SA comprises the JFC's maneuver elements that can operate in three dimensions and time. Through cyberspace, the commander has access to the cognitive and information dimensions that can give an asymmetric advantage and unprecedented access to an adversary's decision-making cycle. Through the employment of Air Force forces and capabilities, the JFC is not limited to two-dimensional warfare, using his surface forces in a series of tactical battles to position maneuver elements that threaten an adversary's COGs. The application of airpower, integrated with information operations (IO), creates effects more rapidly than surface forces alone, thereby accessing the psychological and cognitive aspects of warfare directly. Properly employed, SA can be the Air Force's most decisive warfighting capability.

IV. Command and Control of Strategic Attack

Effective command and control (C2) arrangements and relationships are crucial to the success of strategic attack (SA). Unity of effort is key to the success of SA operations and can only be achieved through command and control arrangements that ensure unity of command. The commander, Air Force forces (COMAFFOR), who is normally the joint force air component commander (JFACC) should be the supported commander for SA operations who, in turn, supports the achievement of the joint force commander's (JFC) objectives. The concept of centralized control and decentralized execution of airpower is vital to effective SA because the synergy of all applied force elements is needed to debilitate the adversary's willingness and capability to wage war. The fragmented air command structure used during the Vietnam War proved that piecemeal application of force by the various Air Force and other Service force elements dilutes the effectiveness of an operation and often serves to extend an operation without achieving US national or military objectives.

Strategic attack (SA) is employed in a joint construct in a unified command structure under the authority of a combatant commander (CCDR) tasked at the direction of the President and Secretary of Defense (SecDef). In this context, air forces organize, train, equip, and plan as an integral element of a joint or multinational force. However, the air component can be employed independently of the surface components in a joint force to help a joint force commander (JFC) achieve objectives. This is particularly true for operations with strategic objectives that require direct attack. The criteria to attack using airpower independent of surface components of the joint force depend on the expected effectiveness and availability of capabilities appropriate to creating the desired effects. In most instances, deep-ranging Air Force forces would be employed in conjunction with other component air elements of the joint force.

V. Elements of Effective Employment

Ref: Annex 3-70, Strategic Attack (25 May '17), pp. 43 to 54.

Parallel versus Sequential Operations

Strategic attack (SA) is normally most effective when employed using parallel operations. Strikes on centers of gravity (COG) are almost always necessary, but a parallel approach—simultaneously striking a wide array of targets chosen to cause maximum shock effects across an enemy system—limits an adversary's ability to adapt and react and thus places the most stress on the system as a whole. This may offer the best opportunity to trigger system-wide shock, thus inducing paralysis or collapse. The object is to effectively control the opponent's strategic activity through rapid decisive operations. Even when this is not fully realized, parallel attack should work synergistically with other actions to cause favorable changes in enemy behavior.

Coercion

Coercion is a concerted effort to modify an adversary's behavior by manipulating the actual or perceived costs and benefits of continuing or refusing to pursue a certain course of action. A coercive strategy may involve one or more of several potentially overlapping mechanisms to include denial, leadership attack, power base erosion, unrest and weakening.

The mechanism by which SA can most effectively coerce the enemy is through denial, whereby it threatens the enemy with outright defeat or otherwise prevents them from achieving their military objectives. In this way, denial seeks to change enemy behavior by hindering or destroying his capability to fight. Denial can be implemented in two ways; counterforce or counter-strategy. Counterforce reduces the enemy's capability to carry out its intended actions by affecting their ability to fight while counter-strategy seeks to convince the enemy that their actions will not succeed, instilling a sense of hopelessness. Denial convinces the enemy that defeat is inevitable and that it would be more prudent to capitulate sooner rather than later. In other cases, denial induces strategic paralysis within entire enemy systems, thus rendering effective resistance impossible, i.e., denying the enemy the ability to act, at least temporarily.

Complementary Operations and Synergy

While SA offers commanders independent, potentially decisive options, it is usually most effective when employed in conjunction with surface forces and other instruments of national power. SA contributes to and benefits from the synergistic effects of other operations. Space control and information operations (IO) separate an adversary from indigenous or third party support, preventing enemy space or information systems from interfering with SA. Surface maneuver benefits from and supports SA by creating a dynamic environment that the enemy must confront with degraded capabilities. Land offensives create high demands upon both enemy infrastructure and fielded forces by speeding consumption of vital war materiel, thus potentially creating enemy critical vulnerabilities.

Pitfalls and Limitations

Strategic attack (SA) has a proven record of success, but it has also failed in application in a number of cases. Failure was generally due to poor understanding of the enemy or of the pitfalls inherent in a conceptually difficult form of force application. Success requires careful planning; thorough, sophisticated understanding of the enemy; complete knowledge of one's own capabilities, requirements, and vulnerabilities; and anticipation of the effects that problems like friction, incrementalism, misprioritization, and restraints/constraints can have on operations.

VIII. Nuclear Operations

Ref: Annex 3-72, Nuclear Operations (19 May '15).

The Air Force's responsibilities in nuclear operations are to organize, train, equip, and sustain forces with the capability to support the national security goal of deterring nuclear attack on the United States, our allies, and partners. The primary purpose of US nuclear operations is to promote stability which results in:

- Deterring adversaries from attacking the United States and its interests with their nuclear arsenals or other weapons of mass destruction (WMD)
- Dissuading adversaries from developing WMD
- Assuring allies and partners of the US' ability and determination to protect them, thus obviating the need to develop or acquire their own nuclear arsenals
- Holding at risk a specific range of targets

"Deterrence in the twenty-first century demands credible, flexible nuclear capabilities, linked to comprehensive strategies and matched to the modern strategic environment. That environment will continue to include nation-states with nuclear arsenals that could pose an existential threat to the United States. It will also include: multiple near-peer states with increasingly modernized nuclear capabilities that challenge regional stability; various nuclear aspirant states who resist global non-proliferation norms and whose emerging capabilities threaten U.S. allies; and non-state entities seeking nuclear capabilities. In the future, the flexibility and resilience of our triad of nuclear deterrent forces will continue to play an important role in strategic stability and underpin other tools of statecraft."

-- Flight Plan for the Air Force Nuclear Enterprise

Nuclear weapons are as important in 21st century global environment as they ever have been. Our nuclear deterrent is the ultimate protection against a nuclear attack on the United States, and through extended deterrence, it also serves to reassure our distant allies of their security against regional aggression. It also supports our ability to project power by communicating to potential nuclear-armed adversaries that they cannot escalate their way out of failed conventional aggression.

Paradoxically, while the number of nuclear powers has increased since the end of the Cold War, the total number of nuclear weapons has decreased. Yet, the number of nuclear-capable nations continues to grow. Fewer US nuclear weapons have forced a transformation in Air Force thinking and analysis, especially in a military environment that has grown more complex due to conventional capabilities, missile defense, and the proliferation of antiaccess/area denial capabilities. Maintaining strategic stability will be an important challenge in the years ahead as both state- and non-state actors seek to acquire new capabilities or to modernize and recapitalize existing nuclear systems. Each nuclear actor brings their own decision calculus. Some actors may possess a limited ability, if any, to correctly discern US operations, detect changes in US posture, or recognize US intent. Likewise, US decision makers, including combatant commanders, subordinate joint force commanders, and commanders and staffs of Air Force components require understanding of both adversary and ally decision-making processes and behaviors. Nuclear operations in a proliferated, multipolar world is no longer reducible to a bipolar, Cold War paradigm.

Finally, a special note about nuclear operations doctrine. Normally, doctrine provides guidance to commanders for their consideration in campaign design as well as during the course of executing an operation and they adjust their forces to seize opportunities and respond to adversary initiatives. However, since nuclear operations have the potential to achieve effects at the strategic, operational, and tactical levels simultaneously, the conduct of nuclear operations is strictly controlled to ensure a unified effort across all instruments of national power. As such, subordinate nuclear commanders have very little flexibility in adjusting the execution of a nuclear plan. Also, detailed force planning is performed at the joint, not Service, level; hence, there is little Service doctrinal guidance herein on such normally expected topics as planning considerations at the Service component level. Airmen may be called upon to provide weapons system expertise, or regional expertise within a regional planning context.

I. Fundamentals of Nuclear Operations

The end of the Cold War has had a major impact on the perceived utility and role of nuclear weapons in the United States. Reduced tensions between former Cold War adversaries had reduced the specter of a large-scale, Cold War-type nuclear exchange enabling force reductions; however, as long as nuclear weapons exist, the possibility of their use remains. This risk is aggravated as some state- and non-state actors seek to acquire new capabilities while others modernize and recapitalize existing nuclear systems. Thus, while the prospect of a massive nuclear exchange seems remote, the potential for a limited nuclear attack has actually grown. For this reason, nuclear weapons are as important as they have ever been.

US nuclear policy is not static and is shaped by numerous considerations. As the civilian leadership changes US policy due to new threats or technologies, the Air Force will need to develop new concepts, systems, and procedures. For instance, the concepts of "mutual assured destruction" and "flexible response" required different types of weapons, different plans, and different degrees of survivability for command and control systems. Stated policies also affect the ability to deter an enemy. As an example, US policy on using nuclear weapons to respond to an adversary's battlefield use of weapons of mass destruction (WMD) is purposely vague. The ambiguous nature of US policy makes it impossible for an enemy to assume such a response would not be forthcoming. Even though there is no guarantee nuclear force would be used to respond to a WMD attack, planners are responsible for making alternative options available for civilian policymakers.

Physical employment of nuclear weapons should remain an option for the United States. To maintain credibility, actual employment should be a plausible consideration in certain circumstances. Without that possibility, the value of deterrence and assurance will likely be undermined.

The employment of nuclear weapons is normally considered a form of strategic attack. Strategic attack is defined as "offensive action specifically selected to achieve national strategic objectives. These attacks seek to weaken the adversary's ability or will to engage in conflict, and may achieve strategic objectives without necessarily having to achieve operational objectives as a precondition." Strategic attack is intended to accomplish national, multinational, or theater strategic-level objectives without necessarily engaging an enemy's fielded military forces. However, this does not preclude operations to destroy the enemy's fielded forces if required to accomplish strategic national objectives.

The employment of nuclear weapons at any level requires explicit orders from the President. The nature of nuclear weapons -- overwhelmingly more significant than conventional weapons -- is such that their use can produce political and psychological effects well beyond their actual physical effects. The employment of nuclear weapons may lead to such unintended consequences as escalation of the current conflict or long-term deterioration of relations with other countries. For this reason above all others, the decision whether or not to use nuclear weapons will always be a political decision and not a military one.

II. Strategic Effects: Deterrence, Assurance, Dissuasion, and Defeat

Air Force nuclear forces consist of delivery systems; nuclear command, control, and communications (NC3) capabilities; personnel; and the physical infrastructure for sustainment. Intercontinental ballistic missiles (ICBMs) and dual-capable bombers and fighters are the Air Force's delivery platforms. Combined with the Navy's submarine-launched ballistic missiles (SLBMs) and other assets, these forces form the nuclear triad. Each nuclear-capable system offers distinct advantages. SLBMs offer survivability whereas ICBMs are the most responsive, offering prompt, on-alert capability combined with dispersed fielding; also, attacks on ICBMs are unambiguous attacks against the United States. Dual-capable bomber and fighter aircraft offer mission flexibility and a capability to provide distinct signaling in a crisis through posturing to alert and through shows of force.

Deterrence, extended deterrence, assurance, dissuasion, and defeat stem from the credibility of our nuclear capabilities in the minds of those we seek to deter, assure, or dissuade. The objectives of deterring adversaries and assuring allies require visible and credible nuclear capabilities. This credibility is attained through focused day-to-day training, periodic exercises, and regular inspections which underpin the credibility of US nuclear capability.

Show of Force

Show of force is defined as "an operation designed to demonstrate US resolve that involves increased visibility of US deployed forces in an attempt to defuse a specific situation that, if allowed to continue, may be detrimental to US interests or national objectives."

Shows of force are frequently used to deter adversaries and assure allies, frequently in the same stroke. The deployment of an additional number of bombers or fighters to a tense region is one very familiar example using Air Force capabilities. Another is the deployment of additional intelligence, surveillance, and reconnaissance assets, such as Predator remotely piloted aircraft, to signal increased US presence.

Deterrence, assurance, and dissuasion apply across the range of military operations and during all phases of planning and execution, most normally as part of global and theater shaping (see following chart). Although deterrence activities are more typically envisioned as occurring mainly in the "shape" and "deter" phases within the joint operational planning construct, deterrence may actually occur in any phase. Influencing an adversary's risk/benefit calculus to reduce their available options -- a form of escalation control -- can take place while other operations (including other nuclear operations) are ongoing.

Although joint doctrine nominally labels deterrence as a Phase 1 activity within the plan phasing construct, deterring adversaries (especially in weapon of mass destruction-related actions) and assuring allies continues even after escalation has increased to the point of nuclear or conventional weapons employment. The objective of stability does not cease once other military operations begin. Indeed, deterrence can occur before, during, or after military operations.

See following pages (pp. 4-74 to 4-75) for an overview and further discussion of "Strategic Effects: Deterrence, Assurance, Dissuasion, and Defeat."

Strategic Effects: Deterrence, Assurance, Dissuasion, and Defeat

Ref: Annex 3-72, Nuclear Operations (19 May '15), pp. 5 to 17.

Deterrence

Deterrence is defined as "the prevention of action by the existence of a credible threat of unacceptable counteraction and/or belief that the cost of action outweighs the perceived benefits." Deterrence is critical to US national security efforts. Both nuclear and conventional operations contribute to the effect. Although nuclear forces are not the only factor in the deterrence equation, our nuclear capability underpins all other elements of deterrence.

Deterrence requires US nuclear operations to be visible to the target audience. To have credibility, an adversary must believe that the Air Force has the capability to act quickly, decisively and successfully. The cumulative effects of deterrence and assurance stem from the credibility of nuclear capabilities in the minds of those we seek to deter, assure, or dissuade. This credibility is attained through activities such as day-to-day training, periodic exercises, and regular inspections which demonstrate Air Force nuclear force capability and readiness.

Nuclear delivery system testing and treaty inspections are distinct messaging opportunities. Both are highly visible examples of strategic messaging. Successful capability testing and treaty inspections provide the world evidence of the credibility of the US' safe, secure, and effective nuclear deterrent.

Nuclear operations can also be used to deter conventional threats. Nuclear operations in the 21st century may be tied to more complex situations, combining both conventional and nuclear operations. Today's Air Force recognizes that many adversaries are willing to employ nuclear operations under many different circumstances.

Extended Deterrence

Historically, the United States provides for the security of its allies by threatening a nuclear response in the event of an enemy attack. This threat of retaliation serves as the foundation for what is defined as extended deterrence.

Extended deterrence is sometimes described as providing a nuclear umbrella over allies and partners. The United States pledges use of its own nuclear arsenal to allies in order to provide for their security and serves as a nonproliferation tool by obviating the need for allies and partners to develop or acquire and field their own nuclear arsenals.

In the case of the North Atlantic Treaty Organization (NATO), the continued deployment of US nuclear weapons in Europe is a strategic alliance issue. This on-going forward basing of US nuclear capabilities not only extends deterrence of adversaries on behalf of European allies, but also assures NATO partners that the Air Force is capable of helping ensure their collective national security.

According to the NATO Deterrence and Defense Posture Review, "Nuclear weapons are a core component of NATO's overall capabilities for deterrence and defense alongside conventional and missile defense forces. As long as nuclear weapons exist, NATO will remain a nuclear alliance. Allies agree … to develop concepts for how to ensure broadest possible participation of Allies concerned in their nuclear sharing arrangements." For the United States' Pacific partners, the Air Force provides a nuclear umbrella over Japan and South Korea, as well as Australia and New Zealand.

Extended deterrence and assurance of allies are two sides of the same coin. Shows of force, which are "operations designed to demonstrate US resolve that involves increased visibility of US deployed forces in an attempt to defuse a specific situation that, if allowed to continue, may be detrimental to US interests or national objectives," shape both allied and adversary beliefs.

Assurance

Complementing extended deterrence, where the objective is to influence the decision-making of an adversary, assurance involves the easing of the fears and sensitivities of allies and partners.

US assurance of allies and partners has been conveyed through various alliances, treaties, and bilateral and multilateral agreements. For example:

- The Treaty of Mutual Cooperation and Security Between the US and Japan specifies a commitment to defense cooperation, regular consultations, and peace and security in the Far East
- The Mutual Defense Treaty Between the United States and the Republic of Korea declares the countries' shared determination to defend themselves and preserve peace and security in the Pacific area
- The North Atlantic Treaty reaffirms the goal of promoting stability, uniting efforts for collective defense, and for the preservation of peace and security among NATO partners

A key Air Force contribution to assurance is through shows of force.

If proliferation increases, it can be expected that allies and partners will demand tangible assurance from the US. This, in turn, will continue to drive demands on the force structure and capability requirements.

Dissuasion

Dissuasion, also closely related to deterrence, consists of actions taken to demonstrate to an adversary that a particular course of action is too costly, or that the benefits are too meager. The intent is thus to dissuade potential adversaries from embarking on programs or activities that could threaten our vital interests, such as developing or acquiring nuclear capabilities. Dissuasion differs from deterrence in that it is a concept aimed at precluding the adversary from developing or acquiring nuclear capabilities. Dissuasion is most often conducted using instruments of national power in concert, such as a combination of diplomatic, economic, and military measures. Air Force nuclear forces may play an important role in the latter, often by providing a credible deterrent.

Defeat

To convince an adversary to surrender or to end a war on terms favorable to the United States, the President may authorize defeat of an enemy using nuclear weapons. Defeat is an objective (and thus technically an effect) that may be achieved using nuclear weapons, by themselves or in conjunction with other forces, should the decisive and culminating nature of their effects be required to resolve a conflict. Operations seeking outright defeat of an enemy using nuclear weapons will likely use other effects of nuclear operations (any or all of the other nuclear operations effects) simultaneously to influence the decision making process of all parties involved.

Defeat may entail prevailing over the enemy's armed forces, destroying their war-making capacity, seizing territory, thwarting their strategies, or other measures in order to force a change in the enemy's behavior, policies, or government. Escalation control is a major consideration for this effect. Escalation control entails the ability to increase the enemy's cost of defiance, while denying them the opportunity to neutralize those costs. In addition, the high level of commitment required for the use of nuclear weapons by the United States is a tangible demonstration of our resolve and likely to affect our ability to defeat the will of an enemy.

Nuclear weapons have been used in combat only twice, of course: at Hiroshima and Nagasaki, culminating World War II in the Pacific.

Refer to "Practical Design: The Coercion Continuum" in Annex 3-0, Operations and Planning, for additional discussion on effects.

III. Nuclear Surety

Ref: Annex 3-72, Nuclear Operations (19 May '15), pp. 27 to 30.

The Air Force implements a stringent nuclear surety program to assure nuclear weapons and their components do not become vulnerable to loss, theft, sabotage, damage, or unauthorized use. All individuals involved with nuclear weapons and nuclear weapon components are responsible for the safety and security of those devices at all times.

A. Nuclear Surety

"The goal of the Air Force Nuclear Weapons Surety Program is to incorporate maximum nuclear surety, consistent with operational requirements, from weapon system development to dismantlement." (AFI 91-101, Air Force Nuclear Weapons Surety Program). This program applies to materiel, personnel, and procedures that contribute to the safety, security, and control of nuclear weapons, thus assuring no nuclear accidents, incidents, loss, or unauthorized or accidental use. The Air Force continues to pursue safer, more secure, and more reliable nuclear weapons consistent with operational requirements.

Adversaries and allies should be highly confident of the Air Force's ability to secure nuclear weapons from accidents, theft, loss, and accidental or unauthorized use. This day-to-day commitment to precise and reliable nuclear operations is the cornerstone to the credibility of deterrence.

Per Department of Defense (DOD) Directive 3150.02, DOD Nuclear Weapons Surety Program, "[f]our DoD nuclear weapon system surety standards provide positive measures to:

- Prevent nuclear weapons involved in accidents or incidents, or jettisoned weapons, from producing a nuclear yield.
- Prevent deliberate prearming, arming, launching, or releasing of nuclear weapons, except upon execution of emergency war orders or when directed by competent authority.
- Prevent inadvertent prearming, arming, launching, or releasing of nuclear weapons in all normal and credible abnormal environments.
- Ensure adequate security of nuclear weapons." [bold in original]

Whether working with continental US (CONUS)-based nuclear forces or conducting theater nuclear operations, commanders must ensure the safety, security, and reliability of their weapons and associated components. While the appropriate infrastructure already exists at CONUS bases with nuclear forces, geographic combatant commanders should consider the additional needs incurred if they are going to have nuclear weapons deployed into their area of responsibility.

Nuclear surety is the capstone construct that contains nuclear safety, security, and reliability programs, each of which is summarized below.

B. Safety

All individuals involved with nuclear weapons are responsible for the safety of those devices. Because of the destructive potential of these weapons, and the possibility that their unauthorized or accidental use might lead to war, safety is paramount.

The four previously mentioned standards include inherent warhead design features that prevent accidental or unauthorized nuclear yields, delivery platform design features, and operational procedures that prevent accidental or unauthorized use. The positive measures may take the form of mechanical systems, such as permissive action links that do not allow the arming or firing of a weapon until an authorized code has been entered. They may also involve personnel monitoring systems, such as the Personnel Reliability Program (PRP), the Arming and Use of Force (AUoF) by Air Force Personnel, or the Two-Person Concept. Commanders are responsible for ensuring that appropriate systems are in place, as described by appropriate Air Force policies. To track the implementation of these positive measures, the Air Force certifies its nuclear weapons systems. The Air Force's Nuclear Certification Program includes safety design, weapon compatibility, personnel reliability, technical guidance, specific job qualifications, inspections, and Weapons System Safety Rules (WSSR).

Refer to AFI 63-125, Nuclear Certification Program, AFI 91-101, Air Force Nuclear Weapons Surety Program, and AFI 31-117, Arming and Use of Force by Air Force Personnel, for more specific guidance.

Weapon System Safety Rules (WSSR)

Weapon system safety rules (WSSR) ensure that nuclear weapons are not detonated, intentionally or otherwise, unless authorized. Safety rules apply even in wartime. While commanders may deviate from a specific rule in an emergency, they may not expend a nuclear weapon until an authentic execution order has been received. This has led to the so-called "usability paradox." Nuclear weapons must be "usable enough" so an enemy is convinced they may be rapidly employed in the event of an attack. They must not be so "usable," however, as to allow for the unauthorized use due to individual action or mechanical error.

C. Security

Nuclear weapons and their components must not be allowed to become vulnerable to loss, theft, sabotage, damage, or unauthorized use. Nuclear units must ensure measures are in place to provide the greatest possible deterrent against hostile acts. Should this fail, security should ensure detection, interception, and defeat of the hostile force before it is able to seize, damage, or destroy a nuclear weapon, delivery system, or critical components.

Commanders are accountable for the safety, training, security, and maintenance of nuclear weapons and delivery systems, and reliability of personnel at all times. Whether on a logistics movement or during an airlift mission, commanders should limit the exposure of nuclear weapons outside dedicated protection facilities consistent with operational requirements. Commanders must ensure that nuclear weapons and nuclear delivery systems are maintained according to approved procedures. Commanders are responsible for considering the additional needs incurred if nuclear capabilities are deployed into their operational area.

A security infrastructure exists at bases that routinely handle nuclear weapons. However, weapons and their delivery systems may be moved to other bases to enhance survivability or may be deployed into a theater. Commanders at such locations must ensure appropriate storage facilities are established and proper security measures are in place. The storage of nuclear weapons on a base not only requires a secure location and additional security personnel, but also impacts other areas such as driving routes, local flying area restrictions, aircraft parking areas, the use of host-nation or contract personnel, and other aspects of day-to-day operations. Note, too, that weapons are most vulnerable in transit or when deployed for use, so special care must be taken at those times. Commanders and, in fact, all Airmen have a responsibility for force protection, and the security of nuclear weapons is a key component of that concept.

D. Reliability

The Air Force employs positive measures to ensure the reliability of its nuclear weapons systems and personnel to accomplish the mission. Reliability is also a product of the system's safety features, including safety design, weapon compatibility, personnel reliability, technical guidance, specific job qualifications, and nuclear technical inspections. Independent inspections and staff assistance visits are also an integral part of maintaining nuclear surety.

IV. Nuclear Command and Control System

The nuclear command, control, and communications (NC3) system refers to the "collection of activities, processes, and procedures performed by appropriate commanders and support personnel who, through the chain of command, allow for decisions to be made based on relevant information, and allow those decisions to be communicated to forces for execution" (AFI 13-550, Air Force Nuclear Command, Control, and Communications [NC3]).

The President's ability to exercise nuclear authority is through the Nuclear Command and Control System (NCCS).

"The NCCS supports the Presidential nuclear C2 of the combatant commands in the areas of integrated tactical warning and attack assessment, decision making, decision dissemination, and force management and report back. To accomplish this, the NCCS comprises those critical communications system components of the DOD information networks that provide connectivity from the President and Secretary of Defense through the National Military Command System to the nuclear combatant commanders and nuclear execution forces. It includes the emergency action message dissemination systems and those systems used for tactical warning/attack assessment, conferencing, force report back, reconnaissance, retargeting, force management, and requests for permission to use nuclear weapons. The NCCS is integral to and ensures performance of critical strategic functions of the Global Command and Control System. The Minimum Essential Emergency Communications Network provides assured communications connectivity between the President and the strategic deterrent forces in stressed environments." (Joint Publication 1, "Doctrine for the Armed Forces of the United States")

Because only the President of the United States can authorize the employment of US nuclear weapons, nuclear operations require NC3 systems to provide national leaders with situational awareness, advance warning, and command and control capabilities. Deterrence, stability, and escalation control require that these capabilities endure nuclear attack so that no adversary can contemplate a disarming first strike.

Positive Control

The President may direct the use of nuclear weapons through an execute order via the Chairman of the Joint Chiefs of Staff to the combatant commanders and, ultimately, to the forces in the field exercising direct control of the weapons.

To allow for the timely execution of these orders, emergency action procedures allow for a timely response to an execution message and ensure an execution order is valid and authentic. Air Force personnel involved in the actual employment of nuclear weapons are intensively and continuously trained and certified in these procedures so they can quickly and accurately respond to the order.

Positive Release Orders

To prevent unauthorized employment of nuclear weapons, cryptologic systems are used to validate the authenticity of nuclear orders. Access to these systems and codes are tightly controlled to ensure unauthorized individuals are not permitted to gain access to the means to order or terminate nuclear weapons employment. Conversely, once appropriate orders have been sent, weapon system operators must respond in a timely manner using standard procedures. Knowledge of these procedures could allow an adversary to determine the time required to conduct operations and the methods crew members will use to accomplish them, allowing that adversary to take more effective measures to counter or limit a nuclear strike.

As with all components of force protection, information security and operations security are critical to mission success.

IX. Personnel Recovery

Ref: Annex 3-50, Personnel Recovery (23 Oct '17).

Our adversaries clearly understand there is great intelligence and propaganda value to be leveraged from captured Americans that can influence our national and political will and negatively impact our strategic objectives. For these reasons, the Air Force maintains a robust and well trained force to locate and recover personnel who have become "isolated" from friendly forces. Personnel recovery (PR) is an overarching term that describes this process, and the capability it represents.

Personnel Recovery (PR)

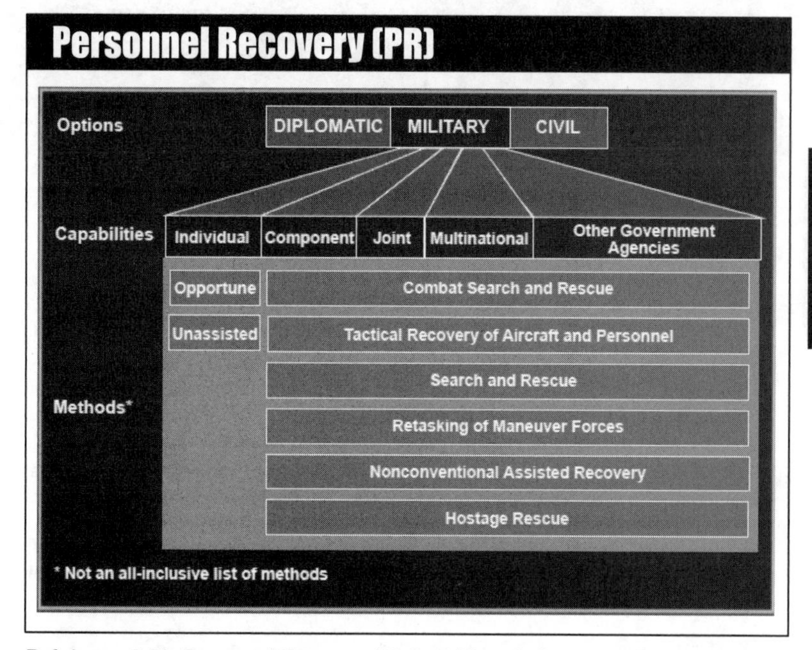

Ref: Annex 3-50, Personnel Recovery (23 Oct '17), p. 8. Personnel Recovery Options, Capabilities, and Methods.

PR is defined as "the sum of military, diplomatic, and civil efforts to prepare for and execute the recovery and reintegration of isolated personnel." (Joint Publication [JP] 3-50, Personnel Recovery). Chairman of the Joint Chiefs of Staff Instruction 3270.01B, Personnel Recovery and Presidential Policy Directive 30, Hostage Recovery Activities, and Executive Order 13698, Hostage Recovery Activities, expand PR responsibilities to: prevent, plan for, and coordinate a response to isolating events to include all US Government (USG) departments and agencies.

The Air Force conducts PR using the fastest and most effective means to recover IP. Air Force PR forces deploy to recover personnel or equipment with specially outfitted aircraft/vehicles, specially trained aircrews and ground recovery teams with PR support personnel and capabilities in response to geographic combatant commander (CCDR) taskings. Traditionally the Air Force focused on the recovery of downed aircrews; however, recent experience has proven that Air Force PR forces are responsible for the recovery of many types of IP.

I. Personnel Recovery Functions

Ref: Annex 3-50, Personnel Recovery (23 Oct '17), pp. 4 to 16.

The DOD mandates each Service to plan and conduct PR in support of its own operations. PR has four functions: preparation, planning, execution, and adaptation:

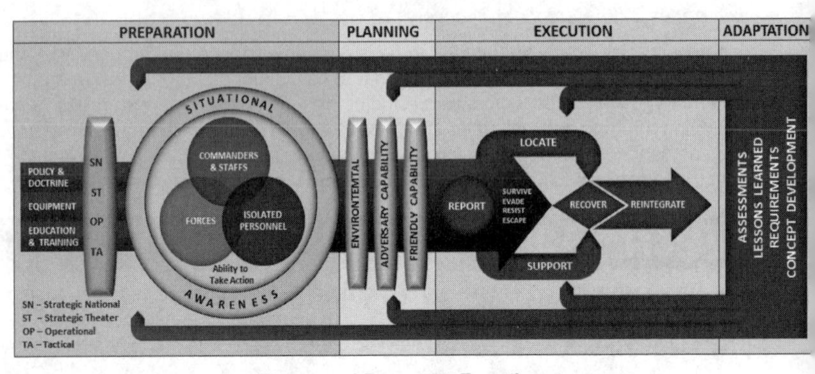

Personnel Recovery Functions

Preparation

PR is, by nature, an event fraught with variables and complexities that are difficult to predict prior to their occurrence. It is also one that should be executed quickly to increase the likelihood of success. Historically, the successful recovery of an individual who has been on the ground behind enemy lines for longer than four hours falls below 20 percent. While PR events don't lend themselves to a great deal of prior planning, there is much that can be done long before an event is declared to shorten the decision and execution processes—this is preparation.

Preparation involves the development of policy, doctrine, equipment, education and training in a standardized fashion as seen through the tactical, operational, theater strategic and national strategic lenses. All of this is directed at commanders and staffs, forces that could participate in a PR event, to provide greater situational awareness that enhances their abilities to take expedient, decisive action, including the rescue of relief of isolated personnel (IP).

Planning

Planning occurs during recurring deliberate or contingency planning processes. Air Force planning is conducted using the joint operations planning process for air with specific outcomes based on commander's intent. It entails detailed PR mission analysis, course of action development, and wargaming based on the plan's mission, goals and tasks developed for the PR appendix (Appendix 5 to Annex C) of a joint force commander's (JFC's) basic operation plan, or the joint air operations plan at the component level. A completed PR mission analysis will be the foundation for the PR operational concept and be used in the development of the PR appendix of the basic plan. Planning options available are found in the definition of PR: Diplomatic, military and civil.

Execution (PR Essential Tasks)

Report. Awareness and notification initiate the PR process. Rapid and accurate notification is essential for a successful recovery. Threat conditions permitting, IP should attempt to establish contact with friendly forces in accordance with (IAW) notification procedures as outlined in the PR special instructions (SPINS) portion of the ATO.

Initial Response. Once an actual or potential PR incident or potential isolating event is reported, the PRCC initially assumes the duties of PR mission coordinator, initiates PR planning, and provides search and rescue incident reports and search and rescue situation reports to inform the JPRC. As the PR mission coordinator, the PRCC tasks and coordinates mission requirements with subordinate PR-capable units.

Locate. Methods used to locate IP may include: theater electronic surveillance, reconnaissance, C2 aircraft, global satellites, wingman reports, and visual and electronic search by dedicated PR forces. Even with precise coordinates that can pinpoint the isolated person's location, PR forces still have to authenticate the isolated person's identity prior to facilitating successful support and recovery operations.

Support. Support is the planned effort necessary to ensure the physical and psychological sustainment of IP. The five objectives in supporting an IP are: Situational awareness, protection, establishing two-way communications, providing morale-building support, and aerial resupply (including aerial escort to a supply cache or more secure area). Protection may also encompass the suppression of enemy threats to the IP. This may preclude capture for the isolated person and disruption of the adversary's response to rescue efforts. When possible, combat rescue officers/pararescuemen and/or equipment may be pre-positioned to support the IP until the recovery phase. Besides support to the IP, this task also includes physical and psychological assistance to the IP's family.

Recover. This task reflects activities by commanders, staff, recovery force, and IP to physically recover the IP. CSAR is the Air Force's preferred recovery mechanism. As information of a potential PR incident becomes available, the PRCC should assess the situation quickly, determine mission feasibility, and disseminate data to units that may participate in the rescue mission. Once mission execution appears feasible, units may be tasked to initiate or continue planning, or launch from alert. If they launch, the recovery force should include all the necessary supporting forces required to execute a recovery operation. The JFC or the designated PR supported commander can issue the "execution order." The JFC's PR concept of operations or SPINS will direct specific launch and execution authority as determined by the JFC.

Alert. Immediate response missions commence from a dedicated ground or airborne alert posture. In order to decrease flight time to the anticipated recovery area and reduce air refueling requirements, rescue forces may be located on the ground at a forward location or loitering in anticipation of an execution order. Additionally, these forces may be embedded in existing airborne missions to further reduce response time.

Deliberate. Commanders choose this method when an immediate response may not be possible due to environmental, political, or threat considerations. Deliberately planned missions give planners the flexibility to utilize all necessary assets to complete the recovery.

Hold. A PR mission is never closed because of risk or inability to locate the IP; however, a mission may be placed on hold for these reasons. Generally a "hold" status on a mission means the information required to execute does not meet commander's execute criteria (e.g., location, intelligence, etc.).

Restructure Reintegration. Reintegrate is the task that allows Department of Defense to provide medical care and protect the well-being of recovered personnel through decompression, while conducting debriefings to gather necessary intelligence and SERE information.

Adaptation

Adaptation is dependent on the collection of PR information and data from after-action reports, PR mission logs, debriefings, and oral interviews. This information enables a process that includes continuous analysis of everything that is going on in PR as it happens, the recognition of what is working correctly and what is not, and implementing change when and where needed. Adaptation can re-enter the PR functional chart where needed, through updated policy, doctrine, equipment, or training in the preparation function, to different tactics used in the recovery task in the execution function. The purpose of adaptation is to make changes that promote more effective and safer PR and achieve higher rates of success.

II. Personnel Recovery Missions

The Air Force organizes, trains, and equips its rescue forces to provide unique capabilities to CCDRs. The primary mission of Air Force Rescue is to use a combination of specially trained Airmen and systems to recover IP. Diverse skill sets allow dedicated PR forces to accomplish many collateral missions. These collateral missions may include: casualty evacuation (CASEVAC), civil search and rescue (SAR), counter-drug activities, emergency and/or traditional aeromedical evacuation homeland security, humanitarian assistance, international aid, non-combatant evacuation operations, support for National Aeronautics and Space Administration (NASA) flight operations, mass rescue operations, theater security cooperation, specialized air and ground mobility, PR command and control and the complex reintegration, infiltration and exfiltration of personnel in support of air component commander missions, special operations missions, and rescue of special operations forces.

Combat Search and Rescue (CSAR)

Combat search and rescue (CSAR) is the Air Force's preferred mechanism for PR in uncertain or hostile environments and denied areas. CSAR is often the only feasible means the Air Force has to execute PR. While PR is not limited to combat operations, CSAR, by definition, is a combat task and not conducted in humanitarian assistance, Civil SAR or CASEVAC.

Airpower

III. Personnel Recovery System

While Air Force PR missions can collaterally recover IP from any Service, each Service is primarily responsible for PR for their own operations. Until recently, the Air Forces' focus was to provide for the recovery of Airmen. Currently, each Service is now committed to the recovery of any captured, missing, or IP from uncertain or hostile environments and denied areas.

DOD personnel recovery systems ensure a complete and coordinated effort to recover US military, DOD civilians and DOD contractor personnel, and other personne directed by the President or SecDef. Air Force capabilities, tactics, techniques, and procedures represent an integral part of the joint PR system. This system consists o the preparation, planning, execution and adaptation functions. Although the activities within these functions can happen consecutively, they generally occur concurrently or, at a minimum, they overlap in execution. There are three primary PR responsibilities: prevent, plan for, and coordinate/respond to isolating events.

IV. Joint Personnel Recovery Center (JPRC)

Geographic CCDRs should establish a standing joint personnel recovery center (JPRC) or functional equivalent. Joint force commanders (JFCs), who may be CCDRs, subordinate unified commanders, or joint task force commanders, normally designate the responsibility for the joint PR mission area to the joint operations directorate (J-3) or to a component commander. The JPRC is integrated into the appropriate operations center.

If the JFC chooses to coordinate joint PR through a component commander, the JFC should also designate them as supported commander for the joint PR mission area. The JFC delegates to the supported commander the necessary authority to successfully accomplish the five execution tasks. This relationship should be evaluated as operations progress through the different campaign phases. At the same time, component commanders have primary authority and responsibility to plan and conduct PR in support of their own operations. In other words, whether the JFC elects to coordinate PR through the J-3 or a component commander, all Service components should maintain a PR coordination cell (PRCC) capability in order to execute component PR responsibilities.

X. Special Operations

Ref: Annex 3-05, Special Operations (9 Feb '17).

Special operations are operations requiring unique modes of employment, tactical techniques, equipment and training often conducted in hostile, denied, or politically sensitive environments and characterized by one or more of the following: time sensitive, clandestine, low visibility, conducted with and/or through indigenous forces, requiring regional expertise, and/or a high degree of risk.

Air Force Special operations are an integral part of special operations and provide combatant commanders and ambassadors discreet, precise, and scalable operations that integrate military operations with other activities. They are designed to assess, shape, and influence in foreign political and military environments unilaterally or by working with host nations, regional partners, and indigenous populations in a culturally attuned manner that is both immediate and enduring in order to enable the nation to prevent and deter conflict or prevail in war. Special operations can be conducted independently; however, most require a networked approach in conjunction with operations of conventional forces (CF) other government agencies, or host nations, and may include operations with indigenous, insurgent, or irregular forces. Special operations may differ from conventional operations in degree of strategic, physical, and political risk; operational techniques; modes of employment; and dependence on intelligence and indigenous assets.

Additionally, Air Force special operations forces (AFSOF) are relatively small forces that may operate independently from other friendly forces. Air Force special operations are often conducted at great distances from major bases in a distributed manner with relatively small footprints. They employ sophisticated communications systems and special means of infiltration, support, and exfiltration to penetrate and return from hostile, denied, or politically sensitive areas. AFSOF should complement and not compete with, nor be a substitute for CF. As an example, an AC-130 gunship should not be employed when a conventional aircraft would be more appropriate for the target and the operational conditions. The need to attack or engage strategic or operational targets with small units drives the formation of special units with specialized, highly focused capabilities. Although not always decisive on their own, when properly employed, special operations (SO) can be designed and conducted to create conditions favorable to US strategic goals and objectives. Often, these operations may require clandestine or low visibility capabilities.

The most important element of the Air Force's SO capabilities resides in its aircrews, Special Tactics units, combat aviation advisory teams, and support personnel specially trained to conduct and support a wide array of missions. Certain AFSOF units are regionally-oriented, culturally astute, and include personnel experienced and conversant in cultures and languages found in specific operational areas. When required, special operations forces (SOF) elements should provide liaisons to facilitate conventional, multinational, and interagency interoperability, which highlights the complementary nature of conventional and SOF.

AFSOF are composed of SO aviation units (including unmanned aircraft systems), Special Tactics personnel (including combat control teams, pararescue personnel, special operations weather teams, and select tactical air control party), dedicated SOF intelligence, surveillance, reconnaissance units, aviation foreign internal defense units, and support capabilities such as command and control, information operations, and combat support functions.

I. USSOCOM Core Activities

Ref: Annex 3-05, Special Operations (9 Feb '17), pp. 12 to 14.

Special operations (SO) core activity definitions and descriptions are primarily derived from JP 3-05, USSOCOM Pub 1 and other supporting SO doctrine publications. It is important to note, core activities are mutually supporting and interoperable in most cases. Rarely, if ever, will a special operation occur that does not support, include or impact multiple core activities. The execution of one core activity may have operational or strategic influence on other core activities being performed or planned. As an example, an unconventional warfare campaign may include elements of direct action and special reconnaissance. The SOCOM core activities are:

Direct Action (DA). Short-duration strikes and other small-scale offensive actions conducted as a special operation in hostile, denied, or diplomatically sensitive environments and which employ specialized military capabilities to seize, destroy, capture, exploit, recover, or damage designated targets. Air Force special operations forces (AFSOF) support DA by employing specialized air mobility (SAM), precision strike, and Special Tactics core mission areas.

Special Reconnaissance (SR). Reconnaissance and surveillance actions conducted as a special operation in hostile, denied, or politically sensitive environments to collect or verify information of strategic or operational significance, employing military capabilities not normally found in conventional forces. AFSOF support SR by employing SAM, intelligence, surveillance, reconnaissance (ISR), and Special Tactics core mission areas.

Countering Weapons of Mass Destruction (CWMD). SOF support US government efforts to curtail the development, possession, proliferation, use, and effects of weapons of mass destruction (WMD), related expertise, materials, technologies, and means of delivery by state and non-state actors. WMD are chemical, biological, radiological, and nuclear weapons capable of a high order of destruction or causing mass casualties and exclude the means of transporting or propelling the weapon where such means is a separable and divisible part from the weapon. The strategic objectives of CWMD operations are to reduce incentives to obtain and employ WMD; increase barriers to acquisition and use of WMD; manage WMD risks emanating from hostile, fragile, failed states, and/or havens; and deny the effects of current and emerging WMD threats. USSOCOM supports geographic combatant commanders through technical expertise, materiel, and special teams to complement other CCMD teams that locate, tag, and track WMD; DA in limited access areas; helping build partnership capacity to conduct CWMD activities; military information support operations (MISO) to dissuade adversaries from reliance on WMD; and other specialized capabilities. AFSOF supports CWMD through its Specialize Air Mobility, Precision Strike, and Special Tactics core missions.

Counterterrorism. Actions taken directly against terrorist networks and indirectly to influence and render global and regional environments inhospitable to terrorist networks. AFSOF primarily support these actions with SAM, precision strike, and Special Tactics core mission areas.

Unconventional Warfare (UW). Activities conducted to enable a resistance movement or insurgency to coerce, disrupt, or overthrow a government or occupying power by operating through or with an underground, auxiliary, and guerrilla force in a denied area. AFSOF primarily support UW activities with SAM and Special Tactics core missions.

Foreign Internal Defense (FID). Participation by civilian and military agencies of a government in any of the action programs taken by another government or other designated organization to free and protect its society from subversion, lawlessness, insurgency, terrorism, and other threats to its security. AFSOF support this core activity through the AFSOF aviation foreign internal defense (AvFID) and Special Tactics core missions.

Security Force Assistance (SFA). The DoD activities that contribute to unified action by the US Government to support the development of the capacity and capability of foreign security forces and their supporting institutions. AFSOF support SFA activities through the AFSOC AvFID and Special Tactics core missions.

Note: FID and SFA are similar at the tactical level where advisory skills are applicable to both. At the operational and strategic levels, both FID and SFA focus on preparing foreign security forces (FSF) to combat lawlessness, subversion, insurgency, terrorism, and other internal threats to their security; however, SFA also prepares FSF to defend against external threats and to perform as part of an international force. Although FID and SFA are both subsets of security cooperation, neither is considered a subset of the other and can both be executed simultaneously.

Hostage Rescue and Recovery. Hostage rescue and recovery operations are sensitive crisis response missions in response to terrorist threats and incidents. Offensive operations in support of hostage rescue and recovery can include recapture of US facilities, installations, and sensitive material outside the continental US. AFSOF support these activities through the full range of their core missions.

Counterinsurgency. Comprehensive civilian and military efforts designed to simultaneously defeat and contain insurgency and address its root causes. SOF are particularly suited for counterinsurgency (COIN) operations because of their regional expertise, language, and combat skills, and ability to work among populations and with or through indigenous partners. AFSOF support COIN activities employing the full range of their core missions.

Foreign Humanitarian Assistance (FHA). DoD activities conducted outside the United States and its territories to directly relieve or reduce human suffering, disease, hunger or privation. AFSOF supports HA/DR by employing Command and Control, SAM, ISR, and Special Tactics core mission areas.

Military Information Support Operations (MISO). The planned operations to convey selected information and indicators to foreign audiences to influence their emotions, motives, objective reasoning, and ultimately the behavior of foreign governments, organizations, groups, and individuals. The purpose of MISO is to induce or reinforce foreign attitudes and behavior favorable to the originator's objectives. AFSOF supports MISO by employing the Information Operation core mission area.

Civil Affairs Operations. Those military operations conducted by civil affairs forces that (1) enhance the relationship between military forces and civil authorities in localities where military forces are present; (2) require coordination with other interagency organizations, intergovernmental organizations, nongovernmental organizations, indigenous populations and institutions, and the private sector; and (3) involve application of functional specialty skills that normally are the responsibility of civil government to enhance the conduct of civil-military operations. AFSOF can support these activities by establishing measures to help the host nation (HN) gain support of the local populace and the international community, and reduce support or resources to those destabilizing forces threatening legitimate processes of the HN government. AFSOF support Civil Affairs Operation activities by employing the full range of their core missions.

Refer to TAA2: Military Engagement, Security Cooperation & Stability SMARTbook (Foreign Train, Advise, & Assist) for further discussion. Topics include the Range of Military Operations (JP 3-0), Security Cooperation & Security Assistance (Train, Advise, & Assist), Stability Operations (ADRP 3-07), Peace Operations (JP 3-07.3), Counterinsurgency Operations (JP & FM 3-24), Civil-Military Operations (JP 3-57), Multinational Operations (JP 3-16), Interorganizational Cooperation (JP 3-08), and more.

Airpower

II. AFSOC Core Activities

Ref: Annex 3-05, Special Operations (9 Feb '17), pp. 11 to 14.

As an Air Force major command, and the Air Force component to US Special Operations Command (USSOCOM), AFSOC is responsible for providing specially tailored aviation related capabilities to conduct or support special operations core activities and other SecDef directed taskings. AFSOC refers to these capabilities as core missions. AFSOC core missions include:

Agile Combat Support (ACS)

Enables all AFSOC core missions and capabilities across the range of military operations. Protects, fields, prepares, deploys, maintains, sustains, and reconstitutes Air Force special operations personnel, weapons systems, infrastructure, and information in support of USSOCOM core activities.

Aviation Foreign Internal Defense (AvFID)

AFSOC combat aviation advisors (CAA) assess, train, advise, assist, and equip (ATAAE) partnered forces aviation assets in airpower employment, sustainment, and integration. CAA conduct special operations activities by, with, and through foreign aviation forces. CAA mission priorities are focused on mobility, ISR, and precision strike missions, with associated surface integration tasks that enable the air-to-ground integration of partnered forces.

Command and Control (C2)

C2 is the exercise of the commander's authority and direction over assigned and attached forces. Operational C2 elements consist of personnel and equipment with specialized capability to plan, direct, coordinate, and control forces to conduct joint / combined special operations.

Information Operations (IO)

IO is an integrated approach utilizing information-related capabilities during all phases of operations to influence, disrupt, corrupt, or usurp adversarial human and automated decision making while protecting our own by producing effects across the entire battlespace. The resulting information superiority allows friendly forces the ability to collect, control, exploit, and defend information without effective opposition. IO is successful by identifying and using any combination of information-related capabilities necessary to achieve the desired effects.

Intelligence, Surveillance, and Reconnaissance (ISR)

ISR synchronizes and integrates sensors, assets, and processing, exploitation and dissemination in direct support of current and future SOF operations. It consists of manned and remotely piloted aircraft and Distributed Common Ground Systems that deliver actionable intelligence to the special operations forces (SOF) operator. ISR produces detailed, specialized products tailored to mission, customer, and pace of operation that gives SOF a decisive advantage against our adversaries.

Precision Strike (PS)

PS provides the joint force commander and the SOF operator with specialized capabilities to find, fix, track, target, engage and assess (F2T2EA) applicable targets. F2T2EA can use a single weapon system or a combination of systems to fulfill elements of the kill chain. PS missions include close air support, air interdiction, and armed reconnaissance. Attributes associated with PS include persistence, robust communications, high situational awareness, precise target identification, lethality, and survivability (as required for the environment).

Specialized Air Mobility (SAM)

SAM missions include both specialized mobility and refueling. Specialized mobility is the rapid global infiltration, exfiltration, and resupply of personnel, equipment, and material using specialized systems and tactics. Specialized refueling is the rapid, global refueling using specialized systems and tactics, thereby increasing mission flexibility and aircraft range. This is done via in-flight refueling either as a tanker or receiver and can additionally be conducted on the ground through a forward arming and refueling point (FARP). These missions may be clandestine, covert, low visibility, or overt and through hostile, denied, or politically sensitive airspace using manned or unmanned platforms with a single aircraft or part of a larger force package. SAM aircraft operate across the range of military operations in all environmental regions (e.g. arctic, desert, littoral, mountainous, sea, tropical, etc.), day and night, and during adverse weather conditions to include transient exposure to chemical, biological, radiological, and nuclear effects.

Special Tactics (ST)

ST uses highly specialized, combat proven capabilities to integrate, synchronize, and control air assets to achieve tactical, operational, and strategic objectives. ST is comprised of the total force consisting of combat controllers (CCT), pararescue (PJ), special operations weather teams (SOWT), tactical air control party (TACP), special operations surgical team (SOST), and specialized combat mission support. ST capabilities consist of air traffic control; assault zone assessment, establishment and control; terminal attack control; fire support; operational preparation of the environment; special reconnaissance; command and control communications; full spectrum personnel and equipment recovery; humanitarian relief; and battlefield trauma care. ST supports and optimizes airpower effects. Agile ST forces enable projection and integration of SOF power across domains, geographic boundaries, and operational environments in support of the ST core capabilities Global Access, Precision Strike, and Personnel Recovery requirements. Through an integrated warfighting approach, ST is uniquely capable of delivering airpower against hard problem sets that are not otherwise within operational reach of the joint force.

Special Operations Relationship to Irregular Warfare

Adaptive adversaries such as terrorists, insurgents, criminal networks, and rogue states, resort to irregular forms of warfare as effective ways to challenge US forces. Irregular warfare (IW) conflicts are a violent struggle among state and non-state actors for legitimacy and influence over the relevant population(s).

Since many irregular threats are not purely military problems, many of the responses required are not purely military either. Due to the complex nature of these threats, such conflicts may not end with decisive military victory. They are more likely to require long-term involvement to remedy, reduce, manage, or mitigate the conflict. To prevent, deter, disrupt, and defeat irregular threats, US forces should seek to work in concert with other government agencies and multinational partners, and, where appropriate, the partnered actors to understand the situation in depth, plan and act in concert, and continually assess and adapt their approach in response to the dynamic and complex nature of the problem. Because of inherent capabilities, characteristics, and specialized training, SOF are ideally suited to participate in US efforts to counter IW adversaries and threats.

Refer to Annex 3-2, Irregular Warfare, for more information on IW.

III. AFSOF Command Relationships

AFSOF are under the combatant command (COCOM) authority of the US Special Operations Command (USSOCOM) and under administrative control (ADCON) of the Commander, Air Force Special Operations Command (AFSOC). USSOCOM is a functional combatant command, with Service-like responsibilities in areas unique to special operations (SO), and when established as a supported command, plans and conducts certain global SO missions.

USSOCOM exercises COCOM authority over theater special operations commands (TSOC) for organize, train, and equip responsibilities. A geographic combatant commander (GCC), exercises operational control of special operations forces (SOF) through the commander, TSOC (CDRTSOC). The CDRTSOC may also be designated as the joint force special operations component commander (see Figure titled Notional Theater Command Structure).

When a GCC establishes and employs multiple JTFs and independent task forces, CDRTSOC may establish and employ a Special Operations Joint Task Force (SO-JTF) and/or multiple joint special operations task forces (JSOTF) to command and control SOF assets. The GCC normally establishes support relationships between JSOTF commanders and JTF/task force commanders.

AFSOC retains Service ADCON of all assigned active component and exercises specified elements of ADCON over reserve component AFSOF personnel. AFSOC may share selected elements of Service ADCON with other Air Force component commands in order to obtain regional beddown support.

Regardless of the arranged command relationship, commanders should provide for a clear, unambiguous chain of command (unity of command).

Chap 4

XI. Cyberspace Operations

Ref: Annex 3-12, Cyberspace Operations (30 Nov '11).

Cyberspace is a global domain within the information environment consisting of the interdependent network of information technology infrastructures, including the Internet, telecommunications networks, computer systems, and embedded processors and controllers.

The employment of cyberspace capabilities where the primary purpose is to achieve military objectives or effects in or through cyberspace.

Cyberspace Superiority

Cyberspace superiority is the operational advantage in, through, and from cyberspace to conduct operations at a given time and in a given domain without prohibitive interference.

Cyberspace superiority may be localized in time and space, or it may be broad and enduring. The concept of cyberspace superiority hinges on the idea of preventing prohibitive interference to joint forces from opposing forces, which would prevent joint forces from creating their desired effects. "Supremacy" prevents effective interference, which does not mean that no interference exists, but that any attempted interference can be countered or should be so negligible as to have little or no effect on operations. While "supremacy" is most desirable, it may not be operationally feasible. Cyberspace superiority, even local or mission-specific cyberspace superiority, may provide sufficient freedom of action to create desired effects. Therefore, commanders should determine the minimum level of control required to accomplish their mission and assign the appropriate level of effort.

Airpower

I. The Cyberspace Domain

Cyberspace is a domain. Cyberspace operations are not synonymous with information operations (IO). IO is a set of operations that can be performed in cyberspace and other domains. Operations in cyberspace can directly support IO and non-cyber based IO can affect cyberspace operations.

Cyberspace is a man-made domain, and is therefore unlike the natural domains of air, land, maritime, and space. It requires continued attention from humans to persist and encompass the features of specificity, global scope, and emphasis on the electromagnetic spectrum. Cyberspace nodes physically reside in all domains. Activities in cyberspace can enable freedom of action for activities in the other domains, and activities in the other domains can create effects in and through cyberspace.

Even though networks in cyberspace are interdependent, parts of these networks are isolated. Isolation in cyberspace exists via protocols, firewalls, encryption, and physical separation from other networks. For instance, classified networks such as the US

Refer to CYBER: The Cyberspace Operations SMARTbook (in development). U.S. armed forces operate in an increasingly network-based world. The proliferation of information technologies is changing the way humans interact with each other and their environment, including interactions during military operations. This broad and rapidly changing operational environment requires that today's armed forces must operate in cyberspace and leverage an electromagnetic spectrum that is increasingly competitive, congested, and contested.

Armed Forces Secure Internet Protocol Router network (SIPRnet) are not hardwired to the Internet at all times, but connect to it via secure portals. Additionally, the construction of some hard-wired networks isolates them from most forms of radio frequency (RF) interference. These factors enable these networks to be isolated within cyberspace, yet still allow controlled connectivity to global networks.

Cyberspace segments are connected and supported by physical infrastructure, electronic systems, and portions of the electromagnetic spectrum (EMS). As new systems and infrastructures are developed, they may use increasing portions of the EMS, have higher data processing capacity and speed, and leverage greater bandwidth. Systems may also be designed to change frequencies (the places where they operate within the EMS) as they manipulate data. Thus, physical maneuver space exists in cyberspace.

Logical maneuverability in cyberspace is often a function of the security protocols used by host systems. Systems seeking connectivity with a secure host will have more difficulty gaining access than systems seeking connectivity with unsecured hosts. Additionally, defense against entry by undesired systems resides in the code or logic of the host system. Once a connection between systems is established, a potential intruder must exploit a fault in logic to enter the system. Code writing can thus be a form of logical maneuver in cyberspace. The potential intruder writes malicious code to gain maneuverability against targeted systems. As a defender becomes aware of unwanted presence within the system, the defender will alter the system's code to deny entry. The intruder, wishing to remain "on target," adapts the malicious code accordingly. This process is the equivalent of forces maneuvering to gain positions of advantage in the traditional air, land, space, and maritime domains. Both logical and physical maneuver space is required — one is often useless without the other.

II. The Operational Environment

The cyberspace domain is now a primary conduit for transactions vital to every facet of modern life. Our society and military are increasingly dependent on cyberspace. Cyberspace is a source of both strength and vulnerability for modern society. While cyberspace operations enable a modern society, they also create critical vulnerabilities for our adversaries to attack or exploit. Manufacturing controls, public utilities distribution, banking, communications, and the distribution of information for national security have shifted to networked systems. While this 30-year evolution has significantly benefited society, it has also created serious vulnerabilities. Increased wireless dependence and expanded interconnectivity has exposed previously isolated critical infrastructures vital to national security, public health, and economic well-being. Adversaries may attempt to deny, degrade, manipulate, disrupt, or destroy critical infrastructures through cyberspace attack, thus affecting warfighting systems and the nation as a whole. Recent incursions into Department of Defense (DOD) and Air Force networks underscore today's cyberspace challenge.

Adversaries in cyberspace are exploiting low-entry costs, widely available resources, and minimal required technological investment to inflict serious harm, resulting in an increasingly complex and distributed environment. The expanded availability of commercial off-the-shelf (COTS) technology provides adversaries with increasingly flexible and affordable technology to adapt to military purposes. Low barriers to entry significantly decrease the traditional capability gap between the US and our adversaries. Adversaries are fielding sophisticated cyberspace systems and experimenting with advanced warfighting concepts.

Cyberspace Infrastructure Relationships

The Air Force depends upon the US' critical infrastructure and key resources for many of its activities, including force deployment, training, transportation, and normal operations. Physical protection of these is no longer sufficient as most critical infrastructure is under the control of networked and interdependent supervisory control and data acquisition (SCADA) or distributed control systems (DCS).

Since private industry is the primary catalyst for technological advancements, the military may become increasingly reliant on COTS technology. This reliance may present three primary vulnerabilities:

U.S. National Cyberspace Policy

Ref: Annex 3-12, Cyberspace Operations (30 Nov '11), p. 8.

There are many policy documents pertaining to cyberspace operations policy.

National Strategy to Secure Cyberspace

The National Strategy to Secure Cyberspace is the comprehensive strategy for the US to secure cyberspace. It spells out three strategic priorities:

- Prevent cyber attacks against America's critical infrastructure
- Reduce national vulnerability to cyber attacks
- Minimize damage and recovery time from cyber attacks

The National Strategy to Secure Cyberspace outlines the framework for organizing and prioritizing US Government efforts in cyberspace. This strategy guides federal government departments and agencies that secure cyberspace. It identifies the steps every individual can take to improve our collective cyberspace security.

National Military Strategy for Cyberspace Operations (NMS-CO)

The National Military Strategy for Cyberspace Operations (NMS-CO) is the comprehensive strategy for US Armed Forces to ensure US superiority in cyberspace. There are four strategic priorities of the NMS-CO:

- Gain and maintain initiative to operate within adversary decision cycles
- Integrate cyberspace capabilities across the range of military operations (ROMO)
- Build capacity for cyberspace operations
- Manage risk for operations in cyberspace

The NMS-CO describes the cyberspace domain, articulates cyberspace threats and vulnerabilities, and provides a strategic framework for action. The NMS-CO is the US Armed Forces' comprehensive strategic approach for using cyberspace operations to assure US military strategic superiority in the domain. The integration of offensive and defensive cyberspace operations, coupled with the skill and knowledge of our people, is fundamental to this approach.

- **Foreign ownership, control, and influence of vendors**. Many of the COTS technologies (hardware and software) the Air Force purchases are developed, manufactured, or have components manufactured by foreign countries. These manufacturers, vendors, service providers, and developers can be influenced by adversaries to provide altered products that have built-in vulnerabilities, such as modified chips.
- **Supply chain**. The global supply chain has vulnerabilities that can potentially lead to the interception and alteration of products. These vulnerabilities are present throughout the product life cycle, from the inception of the design concept, to product delivery, and to product updates and support.
- **COTS and government off-the-shelf (GOTS) balance**. The vast majority of the Air Force's cyberspace operations components and capabilities are from COTS and to a much smaller degree, GOTS technologies.

III. Challenges of Cyberspace Operations

Cyberspace operations offer unique military challenges. The paragraphs below address some of the known challenges: mission assurance, a compressed decision cycle, anonymity and the attribution challenge, and various threats inherent to cyberspace itself.

There is a requirement to balance defensive cyberspace actions within cyberspace with their impact on ongoing air, space, and cyberspace operations. The lack of situational awareness among domains can cause serious disconnects in one, significantly hindering operations in others.

Mission Assurance

Mission assurance consists of measures required to accomplish essential objectives of missions in a contested environment. Mission assurance entails prioritizing mission essential functions (MEFs), mapping mission dependence on cyberspace, identifying vulnerabilities, and mitigating risk of known vulnerabilities.

Mission assurance ensures the availability of a secured network to support military operations by assuring and defending the portion of the network directly supporting the operation. Cyberspace mission assurance begins by mapping the operation to the supporting architecture. Then, deliberate actions are taken to assure the availability of that portion of the network. These may include adding backups to single points of failure in the network or delaying certain maintenance actions to ensure the network will meet mission requirements. Second, the proactive actions are taken to ensure the network is secure and defended. These actions may include focusing the attention of defensive cyberspace operations assets on the slice of the network supporting the operations and conducting operations to ensure no threats are resident on the network.

A "contested cyber environment" involves circumstances in which one or more adversaries attempt to change the outcome of a mission by denying, degrading, disrupting, or destroying our cyber capabilities, or by altering the usage, product, or our confidence in those capabilities.

Warfighters should realize risks and vulnerabilities are often created by the interdependencies inherent in the networking and integration of systems through cyberspace. Integration of cyberspace operations involves actions taken to enable decision superiority through command and control (C2), innovation, integration, and standardization of systems across air, space, and cyberspace domains. Integration means are tested via operational experiments like the Joint Expeditionary Force Experiment. Identifying vulnerabilities is difficult within a contested cyber environment. Our systems are open to assault and are difficult to defend. Some known examples of vulnerabilities in cyberspace operations are listed in the National Military Strategy for Cyberspace Operations (NMS-CO).

Assuring missions via cyberspace operations involves risk. Since the nature of cyberspace is interconnectivity, all cyberspace operations have inherent risk requiring constant attention and mitigation. Cyberspace is a domain with its own set of risks.

Anonymity and the Inherent Inherent Attribution Challenge

Ref: Annex 3-12, Cyberspace Operations (30 Nov '11), p. 12.

Perhaps the most challenging aspect of attribution of actions in cyberspace is connecting a cyberspace actor or action to an actual, real-world agent (be it individual or state actor) with sufficient confidence and verifiability to inform decision- and policy-makers. Often this involves significant analysis and collaboration with other, non-cyberspace agencies or organizations. While cyberspace attribution (e.g., identifying a particular IP address) may be enough for some actions, such as establishing access lists (e.g., "white" or "black" lists of allowed or blocked IP addresses), attribution equating to positive identification of the IP address holder may be required for others, such as offensive actions targeting identified IP addresses.

The nature of cyberspace, government policies, and international laws and treaties make it very difficult to determine the origin of a cyberspace attack. The ability to hide the source of an attack makes it difficult to connect an attack with an attacker within the cyberspace domain. The design of the Internet lends itself to anonymity.

Anonymity is maintained both by the massive volume of information flowing through the networks, and by features that allow users to cloak their identity and activities. Nations can do little to combat the anonymity their adversaries exploit in cyberspace; however, the same features used by terrorists, hackers, and criminals, strengthen state surveillance and law enforcement capability, in modified form. Actions of anonymous or unidentified actors are akin to an arms race. Illicit actors continually amaze those in global law enforcement with the speed at which they stay one step ahead in the technology race. Nevertheless, nations have the advantage of law and the ability to modify the technological environment by regulation.

Anonymity is a feature of the Internet because of the way information moves through it and the way it is governed. The underlying architecture was intended to be robust, distributed, and survivable. The anonymous nature of the Internet is literally written into the structure of the Internet itself and cannot be dislodged without physically destroying many networks. The Internet was also designed where the intelligence was placed at the ends of the network, not in the network itself. Routing tools, software applications, and information requests come from the ends, in contrast to a traditional telephone network in which the switches, routing protocols, etc., are in the network itself. The difference makes it much harder to trace individual bits of information once they are in the network. The Internet's governance structure reflects its design. This makes attribution a challenge.

In this domain, a risk assumed by one is potentially assumed by all. Mitigation of risk can result in a decreased risk level considered acceptable to continue conducting operations. Examples of this kind of approach toward handling risk can be seen in many aspects. The implementation of firewalls, training, education, and intrusion detection and prevention systems represent types of risk mitigation.

Just as in the air domain, we do not defend the entire Cyberspace domain; we defend what is relevant to our operations. In cyberspace, this means protecting pathways and components, since action against critical systems could seriously degrade our ability to fight and win. Whether used offensively or defensively, however, conducting particular cyberspace operations may require access to only a very small "slice" of the domain. This does not mean "localized" in the sense of a limited geographical area (although that too may sometimes be required), but perhaps just a string of internet protocol (IP) addresses, which may span the globe but represent only a miniscule portion of data flow bandwidth. Similarly, it may involve the ability to hack through one particular firewall that may physically reside upon several servers, but which is never engaged physically only through virtual means. Finally, many operations may span only seconds from inception to conclusion, given the speed at which the Internet operates. Successfully operating in cyberspace may require abandoning common assumptions concerning time and space.

Freedom of action in cyberspace is a basic requirement for mission assurance. However, having the cyberspace capacity to achieve this freedom of action should not be taken for granted. Just as operating in the air domain requires having the capacity to do so (airborne platforms, runways, etc.), the Air Force should ensure it acquires sufficient capacity (bandwidth, components, etc.) to operate within cyberspace. Since access to cyberspace permeates daily activities, it is easy to overlook this requirement and assume that sufficient capacity will simply exist.

Cyberspace operations seek to ensure freedom of action across all domains for US forces and allies, and deny that same freedom to adversaries. Specifically, cyberspace operations overcome the limitations of distance, time, and physical barriers present in other domains. Exploiting improved technologies makes it possible to enhance the Air Force's global operations by delivering larger information payloads and increasingly sophisticated effects. Cyberspace links operations in other domains thus facilitating interdependent defensive, exploitative, and offensive operations to achieve situational advantage.

Potential adversaries wish to undermine mission assurance actions via cyberspace operations. The Air Force ensures it can establish and maintain cyberspace superiority and fight through cyberspace attacks at any time regardless if the US requires the use of military forces. Our adversaries have also demonstrated that they can create civil instability through cyber attacks. The Air Force maintains a capability to provide defense support to civil authorities in cyberspace when called upon by national leadership. Potential adversaries have declared and demonstrated their intent; Russia's relatively crude ground offensive into Georgia in 2008 was preceded by a widespread and well-coordinated cyberspace attack.

One last point to highlight concerning mission assurance is homeland infrastructure protection from threats or natural disaster. The Air Force should prepare to respond rapidly to mitigate effects of such threats or events and reconstitute lost critical infrastructure capabilities while also providing support to civil authorities as directed by competent authority. The Air Force should establish policies and guidance to ensure the execution of mission essential functions for critical infrastructure protection, in the event that an emergency threatens or incapacitates operations.

Compressed Decision Cycle of Cyberspace Operations

The fact that operations can take place nearly instantaneously requires the formulation of appropriate responses to potential cyberspace attacks within legal and policy constraints. The compressed decision cycle may require predetermined rules for intelligence, surveillance, and reconnaissance (ISR) actions.

XII. Information Operations (IO)

Ref: Annex 3-13, Information Operations (28 Apr '16).

The purpose of information operations (IO) is to affect adversary and potential adversary decision making with the intent to ultimately affect their behavior in ways that help achieve friendly objectives. Information operations is defined as "the integrated employment, during military operations, of information-related capabilities [IRCs] in concert with other lines of operation to influence, disrupt, corrupt, or usurp the decision making of adversaries and potential adversaries while protecting our own." Deliberate targeting of an adversary's decision making process is enabled by understanding the cognitive factors related to that process, the information that they use, and how they receive and send information. IO is an integrating function, which means that it incorporates capabilities to plan, execute, and assess the information used by adversary decision makers, with the intent of influencing, disrupting, corrupting, or usurping that process. This is not the same as integrating non-lethal capabilities and activities, which may or may not have a behavior-related objective as their primary purpose.

IRCs Employed Across the ROMO

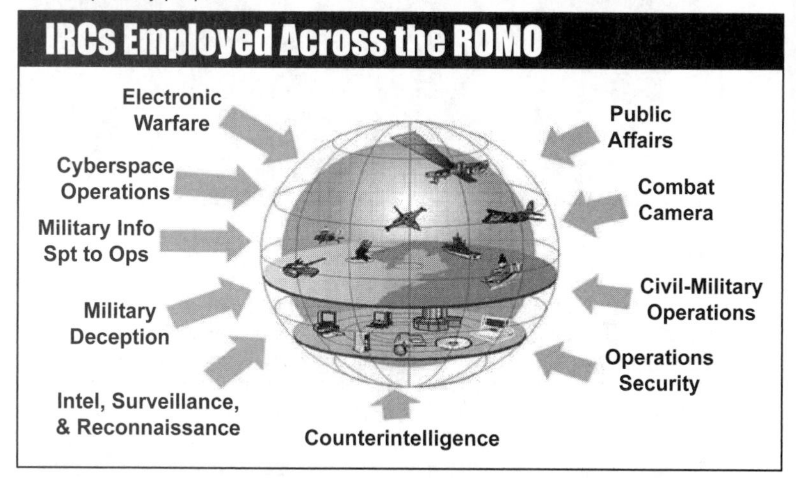

IO is fundamental to the overall military objective of influencing an adversary. IO involves synchronizing effects from all domains during all phases of war through the use of kinetic and non-kinetic actions to produce lethal and non-lethal effects. The planning and execution processes begin with the commander's operational design that guides planners as they coordinate, integrate, and synchronize the IRCs and other lines of operation. IO planning should be integrated into existing planning

Refer to JFODS5: The Joint Forces Operations & Doctrine SMARTbook (Guide to Joint, Multinational & Interorganizational Operations) for further discussion. Topics and chapters include joint doctrine fundamentals (JP 1), joint operations (JP 3-0), joint planning (JP 5-0), joint logistics (JP 4-0), joint task forces (JP 3-33), information operations (JP 3-13), multinational operations (JP 3-16), interorganizational cooperation (JP 3-08), plus more!

processes, such as the joint operation planning process (JOPP). IO planning is not a standalone process. In fact, JP 5-0 clearly identifies IO as a key output resulting from course of action development.

Additionally, IO is complementary to the practices, processes, and end goals of an effects-based approach to operations. IO facilitates targeting development and intelligence requirements, and matches actions with intended messages. Through planning, execution, and assessment processes, IO provides the means to employ the right capabilities (kinetic and non-kinetic) to achieve the desired effects to meet the combatant commander's objectives while supporting the commander's communication synchronization strategy.

The decision-making process can be modeled with a cycle of steps referred to as the observe, orient, decide, act (OODA) loop. The steps of this model occur within the information environment and consist of three targetable dimensions: 1) informational, 2) physical, and 3) cognitive.

Information Dimension

The information dimension represents the content of the information used by the decision maker. Once someone applies meaning to any data element, the data element is transduced into information. This distinction is subtle; but the impact is profound.

Cognitive Dimension

The cognitive dimension is where the decision maker transforms the data from the physical dimension into meaningful information. While we can't directly target the adversary's cognitive processes, we can indirectly target them through the information and physical dimensions. This is accomplished by understanding the adversary's culture, organization, and individual psychology, which enables us to affect the adversary's OODA loop and ultimately their behavior.

I. Role of Information Operations

Information operations (IO) presents viable options to combatant commanders (CCDRs) for conducting operations throughout the range of military operations (ROMO) and all phases of war. IO enables forces to achieve objectives and possibly deter aggression. It enables the use of information-related capabilities (IRCs) in restricted, contested, or politically sensitive areas where traditional air, land, and sea operations may not be permitted. Historically, commanders have employed various IRCs to prevent escalation and enable security.

For example during a humanitarian assistance operation, a commander may influence host nation and even regional cooperation through the integration of public affairs (PA) activities and military information support operations (MISO) messaging designed to facilitate safe and orderly humanitarian assistance among the local populace. During a major operation, the commander may influence region-wide perceptions as well as local behavior through integration of electronic warfare (EW), MISO, and cyberspace operations (CO) with other kinetic or non-kinetic missions against key targets.

See figure on previous page for examples of IRCs employed across the ROMO.

II. Airman's Perspective on IO

Air Force information operations (IO) primarily exists at the air component level as part of the joint IO effort under the joint force commander (JFC) and combatant commander (CCDR). "At the operational level of war, IO ensures synchronized messaging from all IRCs and ensures information-related capabilities (IRCs) comple-

IO Definitions and Descriptions

Ref: Annex 3-13, Information Operations (28 Apr '16), pp. 3 to 4.

Commander's Communication Synchronization

Commander's communication synchronization (CCS) is the Department of Defense's primary approach to implementing United States Government (USG) strategic communication guidance as it applies to military operations. The CCS is the joint force commander's (JFC's) approach for integrating all IRCs, in concert with other lines of effort and operation. It synchronizes themes, messages, images, and actions to support the JFC's objectives. Commander's intent should be reflected in every staff product. Air Force component commanders should similarly conduct their own commander's communication synchronization program. This component level communication synchronization coordinates themes, messages, images, and actions to support the commander, Air Force forces' objectives.

Information Environment

The information environment is defined as "the aggregate of individuals, organizations, and systems that collect, process, disseminate, or act on information." The information environment is comprised of the physical, informational, and cognitive dimensions. IO primarily focuses on affecting the cognitive dimension, where human decision making occurs, through the physical and information dimensions.

Information-Related Capabilities (IRCs)

IRCs are defined as "tools, techniques, or activities using data, information, or knowledge to create effects and operationally desirable conditions within the physical, informational, and cognitive dimensions of the information environment." IRCs create both lethal and nonlethal effects. When IRCs are employed with the primary purpose of affecting the cognitive dimension, it is typically considered IO. IRCs may also include activities such as counterpropaganda, engagements, and shows-of-force, as well as techniques like having the host nation designated as the lead for night raids or not using dogs to search houses. IRCs can be employed individually or in combination to create lethal and non-lethal effects.

Informational Dimension

The informational dimension encompasses where and how information is collected, processed, stored, disseminated, and protected. It is the dimension where the command and control (C2) of military forces is exercised and where the commander's intent is conveyed.

Physical Dimension

The physical dimension is composed of C2 systems, key decision makers, and supporting infrastructure that enable individuals and organizations to create effects. The physical dimension includes, but is not limited to, human beings, C2 facilities, newspapers, books, microwave towers, computer processing units, laptops, smart phones, tablet computers, and any other objects that are subject to empirical measurement. The physical dimension is not confined solely to military or nation-based systems and processes; it is a defused network connected across national, economic, and geographical boundaries."

Cognitive Dimension

The cognitive dimension encompasses the minds of those who transmit, receive and respond to, or act on information. These elements are influenced by many factors, including individual and cultural beliefs, norms, vulnerabilities, motivations, emotions, experiences, morals, education, mental health, identities, and ideologies.

Target Audience

A target audience is defined as "an individual or group selected for influence."

ment each other and do not detract from or interfere with any IO-related/messaging objectives. It includes informing and attempting to affect behavior and decision making as it applies to all relevant non-US audiences. IO should not be confused with integrating non-lethal capabilities. IO planners should be aware of capabilities for creating both lethal and non-lethal effects, as well as plans to ensure any cognitive effects they have will enhance and not detract from IO-related/messaging objectives. IO planners work with all other planners and IRC liaisons, using standard planning and execution steps of the joint operation planning process for air, air tasking cycle, and targeting cycle to accomplish commander's objectives. IO-specific by-products include items such as synchronization matrices, coordinated narratives and themes, and target audience analysis. There is no separate IO plan.

The targeting of a select audience's decision-making process is not new for Airmen. In addition to the requisite understanding of the information content and connectivity used by targeted decision makers, the Air Force has developed an analysis capability called behavioral influence analysis (BIA). BIA provides an understanding of the decision makers' behavior to include culture, organization, and individual psychology (e.g., perceptual patterns, cognitive style, reasoning and judgment, and decision selection processes). It is this knowledge, coupled with an Airman's ability to strike information-related targets that is the essence of Air Force IO. The integrated employment of capabilities to affect information content and connectivity of an adversary provides military advantage to friendly forces.

Air Force IO also includes the integrated planning, employment, monitoring, and assessment of themes, messages, and actions (verbal, visual, and symbolic) as part of the commander's communication synchronization (CCS) at the component level. The CCS will include pertinent portions of the joint force commander's or combatant commander's communication strategy, which may include communication synchronization themes and messages as well as any relevant component commander's themes and messages. At the air component level, Air Force IO planners should ensure these themes, messages, and actions (e.g., IRCs) are integrated across all lines of operation.

III. Information-Related Capabilities (IRCs)

In 2011, the definition of information operations (IO) was revised to eliminate references to specific capabilities and describe those generically as information-related capabilities (IRCs). As a result, the Air Force no longer distinguishes and categorizes IO capabilities with terms like "core capabilities", "influence operations," or "integrated control enablers." The Air Force now references tools, techniques, and activities when used to affect the information environment.

The distinction of IO's role as an integrating function merits emphasis. IO is not a capability in and of itself. IO does not "own" individual capabilities but rather plans and integrates the use of IRCs, tools, techniques, and activities in order to create a desired effect—to affect adversary, neutral, and friendly decision making, which contributes towards a specified set of behaviors. IRCs can be employed by themselves or in combination to conduct or support a wide range of missions. For example, IO planners should help ensure electronic attack (EA), offensive space control, air attacks, and cyberspace operations are coordinated and deconflicted from the perspective of cognitive/behavioral effects. The coordination process should also strive to resolve conflict between actions and messages. Individually, IRCs have wider application than IO employment. What unites capabilities as IRCs is a common IO battlespace—the information environment—whether those capabilities operate in it or affect it. Numerous Air Force capabilities have potential to be employed for IO purposes.

See following pages (pp. 4-100 to 4-102) for an overview and further discussion.

IV. Organization of Information Operations

Ref: Annex 3-13, Information Operations (28 Apr '16), pp. 29-31.

Air Force information-related capability (IRC) planners operating in-theater under the operational control (OPCON) of the commander, Air Force forces (COMAFFOR) are typically assigned or attached to an air expeditionary task force (AETF). Within the AETF, IRC forces are normally attached to an air expeditionary wing, group, or squadron. The COMAFFOR normally exercises command and control (C2) of the AETF through an A-staff and an air operations center (AOC). See Annex 3-30, Command and Control, for further discussion of C2 mechanisms.

The Air Force embeds information operations (IO) and IRC expertise within the AFFOR staff, AOC IO team or the joint force commander's (JFC's) IO staff or cell to facilitate IRC integration and operations. Component staffs address component objectives and the desired effects required to achieve them. Also, the Air Force may augment other staffs with IO and IRC expertise to assist with tasking IRCs in-theater and integrating global IRCs and effects. IO planning should be coordinated during planning at the JFC and air component level, by AFFOR and AOC staffs. Planners at both levels should coordinate adaptive planning processes with supporting commands for IO.

Presentation of IO Planners and IRC Forces

When directed, the Air Force presents information operations (IO) planners and information-related capability (IRC) forces to combatant commanders (CCDRs) to meet national-level and theater-level taskings.

The IO staff and planning function for a theater component is typically presented as a function within an air operations center (AOC) and on the commander, Air Force forces' (COMAFFOR's) staff. The AFFOR and AOC IO planners typically serve as a focal point for coordinating requirements for reachback support from IRCs outside of theater and should ensure their plans and support are in line with joint IO across the joint operations area.

An AOC normally includes an IO team that coordinates with all of the AOC divisions and with counterpart IO elements at other commands and task forces. The IO team may be attached to the AOC's strategy division and coordinate with the other AOC divisions, or the IO team may report direct to the AOC commander as a cross-cutting specialty team. Also within the AOC, an IO duty officer is typically assigned to work alongside other specialty duty officers for the senior duty officer or directly for the chief of combat operations.

Service and Functional IO Responsibilities

IO planners and IRC specialists on the Service and functional component staffs fill critical roles needed to successfully integrate IRC tasks and effects into theater operations. AFFOR and AOC specialists share a common effort in support of the commander's objectives and complement each other's responsibilities. The two staffs coordinate regularly to ensure consistency in focus and that their respective responsibilities and external relationships are appropriately deconflicted.

Reachback and Federated Support for IO

Commanders and their staffs should consider leveraging other resources and capabilities available through reachback and federation to support theater IO and IRC activities. There are many Service, joint, Department of Defense, interagency, and national organizations referenced in this publication that can provide additional support to theater IO efforts. For instance, the National Air and Space Intelligence Center and the 363d Intelligence Surveillance Reconnaissance Group may be able to provide behavioral analysis and targeting products to meet IRC operational requirements.

Information-Related Capabilities (IRCs)

Ref: Annex 3-13, Information Operations (28 Apr '16), pp. 9 to

Electronic Warfare (EW)

Electronic warfare (EW) is defined as "military action involving the use of electromagnetic and directed energy to control the electromagnetic spectrum [EMS] or to attack the enemy." EW consists of three divisions: electronic attack (EA), electronic protection (EP), and electronic warfare support (ES). EW contributes to the success of information operations by using offensive and defensive tactics and techniques in a variety of combinations to shape, disrupt, and exploit adversarial use of the EMS while protecting friendly freedom of action in that spectrum. During combat operations, the commander, Air Force forces (COMAFFOR)/joint force air component commander (JFACC) is usually designated as EW control authority (EWCA) and jamming control authority for the employment of EW assets, associated policy, and processes in the joint operations area. The COMAFFOR/JFACC typically stands up an EW coordination cell to employ EA to negate an adversary's effective use of the EMS by degrading, neutralizing, or destroying combat capability. To deconflict intended effects, the following activities should be closely coordinated: EA, EP, ES, offensive cyberspace operations, offensive space control, military deception, operations security, and intelligence.

See pp. 4-103 to 4-108 for further discussion.

Military Information Support Operations (MISO)

Military information support operations (MISO) are defined as "planned operations to convey selected information and indicators to foreign audiences to influence their emotions, motives, objective reasoning, and ultimately the behavior of foreign governments, organizations, groups, and individuals in a manner favorable to the originator's objectives."1 MISO may attempt to either induce change in foreign attitudes and behavior or reinforce existing attitudes and behavior. MISO at the combatant command level usually resides in the combatant commander's (CCDR) J39 directorate or in a military information support task force (MISTF), which includes a MISO planner as a member of the joint IO cell or joint IO staff.

Refer to JP 3-13.2, Military Information Support Operations, for more information.

Military Deception (MILDEC)

Military deception (MILDEC) is defined as "actions executed to deliberately mislead adversary military, paramilitary, or violent extremist organization decision makers, thereby causing the adversary to take specific actions (or fail to take actions) that will contribute to the accomplishment of the friendly mission."1 Deception operations can span all levels of war and can include, at the same time, both offensive and defensive components. During planning, MILDEC can be integrated into the early phases of an operation. The MILDEC role during the early phases of an operation will be based on the specific situation of the operation or campaign to help set conditions that will facilitate phases that follow. Deception can distract the adversary from legitimate friendly military operations and can confuse and dissipate adversary forces. MILDEC affects the adversary's information systems, processes, and capabilities to create desired behavior. MILDEC planners require adversary and potential adversary decision maker analysis for a sufficiently detailed understanding of how the information environment supports the adversary's decision-making process.

Each information-related capability (IRC) has a part to play in successful MILDEC credibility over time, so information operations (IO) facilitates close coordination with military information support operations (MISO), operations security (OPSEC), public affairs (PA), and commander's communication synchronization (CCS) personnel within the joint IO cell or staff. Whereas MISO, PA, and CCS activities may share a common specific audience with MILDEC, only MILDEC actions are designed to mislead. There is a delicate balance between successful deception efforts and media access to ongoing operations. Inappropriate media access may compromise deception efforts. Conversely, MILDEC must not intentionally target or mislead the news media, the US public, or Congress. Deception activities potentially visible to the US public should be closely coordinated with PA operations so as to not compromise operational considerations or diminish the credibility of PA operations in the national media. Due to the sensitive nature of MILDEC plans and objectives, a strict need-to-know policy should be enforced. Additionally, approval authorities for conducting MILDEC actions are typically at the joint force commander-level or above, so the approval action may require sufficient lead time for staffing.

Refer to JP 3-13.4, Military Deception, for more information.

Continued on next page

Operations Security (OPSEC)

Operations security (OPSEC) is defined as "a process of identifying critical information and subsequently analyzing friendly actions attendant to military operations and other activities." OPSEC denies adversaries critical information and observable indicators about friendly forces and intentions. OPSEC identifies any unclassified activity or information that, when analyzed with other activities and information, can reveal protected and important friendly operations, information, or activities. A critical information list should be developed and continuously updated in peacetime as well as conflict. The critical information list helps ensure military personnel and media are aware of non-releasable information.

The information operations (IO) team enables the OPSEC planner to maintain situational awareness of friendly information and actions and to assist other air operations center planners in incorporating OPSEC considerations during the planning process. Once the OPSEC process identifies vulnerabilities, other information-related capabilities (e.g., military deception, military information support operations, electronic warfare, cyberspace operations) can be used to ensure OPSEC requirements are satisfied.

Public Affairs (PA)

Public affairs (PA) provides information operations (IO) with an open and credible means to reach key public audiences. PA consists of public information, command information, and civic engagement activities that are directed toward both the external and internal publics with interest in the DOD. The external public may include allied, neutral, and adversary audiences. Truth is foundational to the credibility of all public affairs operations. Timely and agile dissemination is essential to help achieve desired information effects. PA plays a significant role throughout the range of military operations, with PA being one of the most prominent information-related capabilities (IRCs) used prior to the outset of hostilities and during stability operations. While PA cannot provide false or misleading information, it must be aware of the intent of other IRCs such as military deception, military information support operations (MISO) and operations security to lessen the chance of compromise. PA integration with other IRCs is vital to ensure the capabilities complement rather than conflict with each other.

See pp. 4-109 to 4-112 for further discussion.

Audience Engagements

Audience engagements are an important contributor to information operations (IO) because of their ability to convey key messages where they are needed to assist in accomplishing military objectives. Engagements permit interface directly with a specific audience through traditional methods of information exchange. Engagements are broadly described as interactions that take place between military personnel and audiences. Audiences may be key leaders or mass populations, and those audiences may be military or civilian. Engagements may be in person or virtual (e.g., a teleconference), impromptu encounters or planned events, such as during civil-military operations (CMO).

Continued on next page

Counterintelligence (CI)

Counterintelligence (CI) is defined as "information gathered and activities conducted to identify, deceive, exploit, disrupt, or protect against espionage, other intelligence activities, sabotage, or assassinations conducted for or on behalf of foreign powers, organizations or persons or their agents, or international terrorist organizations or activities. CI support to information operations (IO) includes identifying threats within the information environment through CI collections and analysis and assessing those threats through reactive and proactive means.

Space Operations

Two mission areas of space operations concern the information environment—global space mission operations and space control. **Global space mission operations** capitalize on the information environment to provide force-enhancing capabilities, which include: intelligence, surveillance, and reconnaissance; launch detection; missile tracking; environmental monitoring; satellite communications; and positioning, navigation, and timing. **Space control** is defined as "operations to ensure freedom of action in space for the United States and its allies and, when directed, deny an adversary freedom of action in space."

See pp. 4-37 to 4-40 for further discussion.

Cyberspace Operations (CO)

Cyberspace operations (CO) are defined as "the employment of cyberspace capabilities where the primary purpose is to achieve objectives in or through cyberspace."1 CO use specific cyberspace capabilities to create effects that support operations across all domains. In contrast, information operations (IO) integrates information-related capabilities (IRCs) with its focus on the decision making of adversaries and allies alike. When employed in support of IO, CO include offensive and defensive capabilities exercised through cyberspace, as an integrated aspect of a larger effort to affect the information environment. CO may be employed independently or in conjunction with other IRCs to create effects in the adversary's battle space and ensure US forces' freedom of maneuver in the information environment.

See pp. 4-89 to 4-94 for further discussion.

Signature Management (SM)

Signature management (SM) is a process used to profile day-to-day observable activities and operational trends at wings/installations and at each of their resident or associate units. SM incorporates the analytical methods of OPSEC creating synergies and resource efficiencies for the wing/installation OPSEC program. These result in identifying details that can be used in efforts to defend or exploit operational advantages at a given military installation and inherent to a unit's operational mission.

Other Information Operations Capabilities

Information operations (IO) planners should consider all available options and/or combinations of lethal and non-lethal, kinetic and non-kinetic means in order to achieve the desired lethal and/or non-lethal effects.

Modern military operations require the ability to engage a target audience with a combination of lethal and non-lethal means, to produce both lethal and non-lethal effects. Non-kinetic and non-lethal means are not reserved only for friendly or neutral audiences. The ability to influence and affect an adversary through non-lethal means may prove to be a better option.

IO planners should maintain close coordination with the special technical operations element to integrate, synchronize, and deconflict operations, as appropriate.

Refer to JP 3-13, Information Operations, for more information.

Continued from previous page

Continued from previous page

XIII. Electronic Warfare (EW)

Ref: Annex 3-51, Electronic Warfare (10 Oct '14).

Electronic Warfare (EW) is waged to secure and maintain freedom of action in the electromagnetic spectrum (EMS). Military forces rely heavily on the EMS to sense, communicate, strike, and dominate offensively and defensively across all warfighting domains. EW is essential for protecting friendly operations and denying adversary operations within the EMS.

The term EW refers to military action involving the use of electromagnetic (EM) and directed energy (DE) to control the EMS or to attack the enemy. This is not limited to radio or radar frequencies but includes infrared (IR), visible, ultraviolet, and any other free-space electromagnetic radiation. EW is critical to air, space, and cyberspace forces gaining freedom of action within contested environments.

EW consist of three divisions: electronic attack (EA), electronic warfare support (ES), and electronic protection (EP). All three contribute to the success of air, space, and cyberspace operations. Proper employment of EW capabilities produces the effects of detection, denial, deception, disruption, degradation, exploitation, destruction, and protection. Capabilities inherent to the EW divisions can be used for both offensive and defensive purposes and are coordinated through electromagnetic battle management (EMBM).

See following pages for an overview of the three EW divisions (pp. 4-104 to 4-105) and EW effects (pp. 4-106 to 4-108).

EW Principal Activities

EW operations have developed over time to exploit the opportunities and vulnerabilities inherent in the physics of EM energy. The principal activities used in EW include the following: countermeasures, EMBM, EM compatibility; EM deception; EM hardening, EM interference resolution, EM intrusion, EM jamming, electromagnetic pulse (EMP), EMS control, electronic intelligence collection, electronic masking, electronic probing, electronic reconnaissance, electronics security, EW reprogramming, emission control, joint electromagnetic spectrum operations (JEMSO), joint electromagnetic spectrum management operations (JEMSMO), low-observability/stealth, meaconing, navigation warfare (NAVWAR), precision geolocation, and wartime reserve modes.

EW Across the Range of Military Operations (ROMO)

Employed across the range of military operations (ROMO), EW can enhance the ability of operational commanders to achieve an advantage over adversaries. Commanders rely on the EMS for intelligence; communication; positioning, navigation, and timing (PNT); sensing; command and control (C2); attack; ranging; data transmission; and information and storage. Therefore, control of the EMS is essential to the success of military operations and is applicable at all levels of conflict. EW considerations must be fully integrated into operations in order to be effective. Additionally, the scope of these operations is global and extends from the earth's surface into space. Unfettered access to selected portions of the EMS is critical for weapon system effectiveness and protection of critical assets. EW is a force multiplier that can create effects throughout ROMO. When EW actions are properly integrated with other military capabilities, synergistic effects may be achieved, losses are minimized, and effectiveness is enhanced.

I. Electronic Warfare Divisions

Ref: AFDD3-13.1, Electronic Warfare (Sept '10), pp. 7 to 9.

EW consist of three divisions: electronic attack (EA), electronic warfare support (ES), and electronic protection (EP). All three contribute to the success of air, space, and cyberspace operations. Capabilities inherent to the EW divisions can be used for both offensive and defensive purposes and are coordinated through electromagnetic battle management (EMBM).

Electronic Warfare Divsions

Electronic Attack (EA)

Electronic Warfare Support (ES)

Electronic Protection (EP)

Airpower

A. Electronic Attack (EA)

EA is the division of EW involving the use of electromagnetic (EM), directed energy (DE), or antiradiation weapons to attack personnel, facilities, or equipment with the intent of degrading, neutralizing, or destroying enemy operational capability. EA prevents or reduces an enemy's use of the electromagnetic spectrum (EMS). It can be accomplished through detection, denial, disruption, deception, and destruction. EA includes lethal attack with assets like high-speed antiradiation missiles (HARMs); active applications such as decoys (flares or chaff), EM jamming, and expendable miniature jamming decoys; and employs EM or DE weapons (lasers, radio frequency weapons, particle beams, etc.).

EM jamming and the suppression of enemy air defenses (SEAD) are applications of EA:

Electromagnetic Jamming

EM jamming is the deliberate radiation, reradiation, or reflection of EM energy for the purpose of preventing or reducing an enemy's effective use of the EMS, with the intent of degrading or neutralizing the enemy's combat capability. Early Air Force EW efforts were primarily directed toward electronically jamming hostile radars to hide the number and location of friendly aircraft and to degrade the accuracy of radar-controlled weapons. Currently, jamming enemy sensor systems can limit enemy access to information on friendly force movements and composition and cause confusion. Jamming can degrade the enemy's decision making and implementation process when applied against command and control systems. An adversary heavily dependent on centralized control and execution for force employment presents an opportunity for EA.

Suppression of Enemy Air Defenses (SEAD)

SEAD is that activity which neutralizes, destroys, or temporarily degrades surface-based enemy air defenses by destructive and/or disruptive means. The goal of SEAD operations is to provide a favorable situation in which friendly tactical forces can perform their missions effectively without interference from enemy air defenses. In Air Force doctrine, SEAD is not part of EW, but it is a broad term that may include the use of EW. In Air Force doctrine, SEAD is part of the counterair framework and directly contributes to offensive counterair and obtaining air superiority. This may involve using EM radiation to neutralize, degrade, disrupt, delay, or destroy elements of an enemy's integrated air defense system (IADS). During hostilities, enemy IADS will probably challenge friendly air operations. EW systems tasked to perform SEAD may be employed to locate and degrade, disrupt, neutralize, or destroy airborne and ground-based emitters. Typically, SEAD targets include radars for early warning/ground-controlled intercept (EW/GCI), acquisition (ACQ), surface-to-air missiles (SAMs), and antiaircraft artillery (AAA). Many Air Force functions can be enhanced with the employment of SEAD operations.

B. Electronic Warfare Support (ES)

ES responds to taskings to search for, intercept, identify, and locate sources of intentional and unintentional radiated electromagnetic energy for the purpose of threat recognition. Commanders, aircrews, and operators use ES to provide near-real-time information to supplement information from other intelligence sources. Additionally, ES information can be correlated with other intelligence, surveillance, and reconnaissance (ISR) information to provide a more accurate picture of the electromagnetic operational environment and therefore a better understanding of the battlespace. This information can be developed into an electronic order of battle for situational awareness and may be used to develop new countermeasures. The relationship between ES and signals intelligence (SIGINT), which includes electronic intelligence (ELINT) and communications intelligence (COMINT), is closely related because they share common functions of search, interception, identification, location, and exploitation of electromagnetic radiation. The distinction lies in the type and use of information, and who has tasking authority. ES resources are tasked by or under direct control of operational commanders. The operational commander may have authority to task national SIGINT assets to provide ES or may have direct operational control over tactical resources capable of providing ES. In either case, ES is distinguished by the fact that the operational commander determines aspects of resource configuration required to provide ES that meets immediate operational requirements. SIGINT is tasked by national authorities. The passive nature of ES allows it to be effectively employed during peacetime.

Refer to Joint Publication 3-13.1, Electronic Warfare, and Chairman of the Joint Chiefs of Staff Instruction (CJCSI) 3210.03C, Joint Electronic Warfare Policy, for a more in-depth discussion of the relationship and distinctions between ES and SIGINT.

C. Electronic Protection (EP)

EP includes the actions taken to protect personnel, facilities, and equipment from any effects of friendly, neutral, or enemy use of the EMS, as well as naturally occurring phenomena that degrade, neutralize, or destroy friendly combat capability. Examples of EP include frequency agility, changing pulse repetition frequency, emission control (EMCON), and low observable technologies. Integration of EP and other security measures can prevent enemy detection, denial, disruption, deception, or destruction. Friendly force reliance on advanced technology demands comprehensive EP safeguards and considerations. Proper frequency management is a key element in preventing adverse effects (i.e., jamming friendly forces) by friendly forces. Much of the success of EP occurs during the design and acquisition of equipment. EMCON and low observable technologies are passive applications of EP.

II. Electronic Warfare (EW) Effects

Ref: AFDD3-13.1, Electronic Warfare (Sept '10), pp. 10 to 14.

Electronic Warfare (EW) involves the use of electromagnetic energy (EM) and directed energy (DE) to control the electromagnetic spectrum (EMS) or the attack the enemy. Military forces depend on the EMS for applications that include: intelligence; communication; positioning, navigation, and timing (PNT); sensing; command and control (C2); attack; ranging; data transmission; and information and storage. Control of the EMS, while denying the adversary the same, is critical to the success of military operations.

Control

To control is to dominate the EMS, directly or indirectly, so that friendly forces may exploit or attack the adversary and protect themselves from exploitation or attack. Control is accomplished through applications of electronic attack (EA), electronic warfare support (ES), and electronic protection (EP). EA limits adversary use of the EMS; EP secures use of the EMS for friendly forces; and ES enables commanders' ability to identify and monitor actions in the EMS throughout the operational environment.

While control of the EMS through the proper application of EW is advantageous, when improperly used without coordination may result in EM interference or EM fratricide, and consequently unintended effects like disruption of friendly cyberspace/information networks. Additionally, an ill-timed jamming package may highlight an otherwise unseen force or deny the use of a frequency by friendly forces. An incorrect or wrongly interpreted radar warning receiver (RWR) indication may cause an inappropriate action to be taken. Electromagnetic battle management (EMBM) ensures effective control of the electromagnetic operational environment (EMOE). EMBM is the dynamic monitoring, assessing, planning and directing of joint electromagnetic spectrum operations (JEMSO) in support of the commander's scheme of maneuver. EMBM will proactively harness multiple platforms and diverse capabilities into a networked and cohesive sensor/decision/target/engagement system, as well as protect friendly use of the EMS while strategically denying benefits to the adversary.

Refer to JP 3-13.1, Electronic Warfare, and JP 6-01, Joint Electromagnetic Spectrum Operations, for additional information JEMSO and joint electromagnetic spectrum management operations (JEMSMO) .

EW has offensive and defensive aspects that work in a "move- countermove" fashion. Often, these aspects are used simultaneously and synergistically. In the same way that air superiority allows friendly forces the freedom from attack, freedom to maneuver, and freedom to attack, the proper coordinated use of EW allows friendly forces to use the EMS. As examples, the offensive denial of a C2 network by EM jamming disrupts the adversary's ability to control forces that would otherwise engage a friendly strike force. The proper use of EP allows friendly radar and communications to continue operating in the presence of enemy jamming.

EW is not limited to manned airborne application; it is also applied from land, sea, space, and cyberspace. The proper employment of EW involves various applications of control to achieve detection, denial, deception, disruption, degradation, exploitation, protection and destruction.

Detection

Detection is identification of potential enemy EM emissions through use of ES measures. It involves assessing the electromagnetic environment (EME) to include radar/radio frequency, electro-optics/laser, and the infrared (IR) spectrums using active and passive means. It is the first step in EW because effective mapping of the EME is essential to develop an accurate electronic order of battle (EOB). The EOB is critical for EW decision making and for using the EMS to meet mission objectives. The various means of detection include on-board receivers, space-based systems, unmanned aircraft (UA), human intelligence (HUMINT), and other intelligence, surveillance, and reconnaissance (ISR) systems. Detection supports all divisions of EW and enables the avoidance of known hostile systems. When avoidance is not possible, it may become necessary to deny, deceive, disrupt, or destroy the enemy's electronic systems.

Denial

Denial is defined as the prevention of access to or use of systems or services. In an EW context, it is the prevention of an adversary from using EMS-dependent systems (e.g., communications equipment, radar) by affecting a particular portion of the EMS in a specific geographical area for a specific period of time. Denial involves controlling the information an enemy or adversary receives, preventing the acquisition of accurate information about friendly forces. Denial is accomplished through EA techniques (degradation, disruption, or deception); expendable countermeasures; destructive measures; network applications; tactics, techniques, and procedures (TTP); and/or emission control (EMCON).

Continued on next page

Deception

Deception is measures designed to mislead the adversary by manipulation, distortion, or falsification of evidence to induce the adversary to react in a manner prejudicial to the adversary's interests. Through the use of the EMS, EW manipulates the decision-making loop of the opposition, making it difficult to distinguish between reality and the perception of reality. If an adversary relies on EM sensors to gather intelligence, deceptive information can be channeled into these systems to mislead and confuse. Deception efforts must stimulate as many adversary information sources as possible to achieve the desired objective. Multisensor deception can increase the adversary's confidence about the "plausibility" of the deception story. Deception efforts are coordinated with the military deception officer and considered during development of an overall deception plan, IO plan, and the overall operations or campaign plans. Operational security is critical to an effective deception plan.

EM deception as it applies to EW is the deliberate radiation, reradiation, alteration, suppression, absorption, denial, enhancement, or reflection of EM energy in a manner intended to convey misleading information to an enemy or to enemy EM-dependent weapons, thereby degrading or neutralizing the enemy's combat capability. Deception jammers/transmitters can place false targets on the enemy radar's scope, or cause the enemy radar to assess incorrect target speed, range, or azimuth. Such jammers/transmitters operate by receiving the pulse of energy from the radar, amplifying it, delaying or multiplying it, and reradiating the altered signal back to the enemy's transmitting radar.

There are three types of EM deception: manipulative, simulative, and imitative.

- **Manipulative EM** deception involves an action to eliminate revealing or to convey misleading EM telltale indicators that may be used by hostile forces. An example of this is to mislead the enemy by transmitting a simulated unique system signature from a nonlethal platform, thereby allowing the enemy sensors to receive and catalog those systems as actual threats in the area.
- **Simulative EM** deception is action to simulate friendly, notional, or actual capabilities to mislead hostile forces. Examples of simulative EM detection include the use of chaff to simulate false targets so that the enemy has the impression of a larger strike package or the use of a jammer to transmit a deceptive technique that misleads an adversary's target tracking radar so that it cannot find the true location of its target.
- **Imitative EM** deception introduces EM energy into enemy systems that imitate enemy emissions. Any enemy receiver can be the target of imitative electromagnetic deception. This might be used to screen friendly operations. An example is the use of a repeater jamming technique that imitates enemy radar pulses. These pulses, when received by the tracking radar, input incorrect target information into the enemy's system.

Continued on next page

II. Electronic Warfare (EW) Effects (Cont.)

Ref: AFDD3-13.1, Electronic Warfare (Sept '10), pp. 10 to 14.

Continued from previous page

Disruption

Disruption is to interrupt the operation of adversary EMS dependent systems. Effective disruption limits adversary capabilities by degrading or interfering with the adversary's use of the EMS to limit the enemy's combat capabilities. Disruption is achieved by using EM jamming, EM deception, EM intrusion, and physical destruction. These will enhance attacks against hostile forces and act as a force multiplier.

Degradation

Degradation is to reduce the effectiveness or efficiency of adversary EMS-dependent systems. Employing EM jamming, EM deception, and/or EM intrusion is intended to degrade adversary systems thus confusing or delaying actions of adversary operators.

Exploitation

Exploitation is using adversary EM radiation for friendly advantage. EM energy may provide tactical, operational, and strategic situational awareness of the EMOE, and is used to develop an EOB. Additionally, EM energy is used to identify, recognize, characterize, locate, and track EM radiation sources to support current and future operations. Data transmissions produce EM energy for exploitation by signals intelligence (SIGINT), provide targeting for EM or destructive attacks, and develop awareness of operational trends.

Protection

Protection is the preservation of the effectiveness and survivability of mission-related military and nonmilitary personnel, equipment, facilities, information, and infrastructure deployed or located within or outside the boundaries of a given operational area. This includes ensuring that EW activities do not electromagnetically destroy or degrade friendly intelligence sensors; communications systems; positioning, navigation, and timing capabilities; and other EMS-dependent systems and capabilities. Protection is achieved by component hardening, EMCON, EMS management and deconfliction, and other means to counterattack and defeat adversary attempts to control the EMS. Spectrum management and EW work collaboratively to accomplish active EMS deconfliction, which includes the capabilities to detect, characterize, geolocate, and mitigate EMI that affects operations. Additionally, structures such as a joint force commander's electronic warfare staff (JCEWS) or the Commander, Air Force Forces, (COMAFFOR) electronic warfare coordination cell (EWCC) enhance operational-level EP through coordination and integration of EW into the overall scheme of maneuver.

Continued from previous page

Destruction

When used in the EW context, destruction is the use of EA to eliminate targeted adversary personnel, facilities, or equipment. Target tracking radars and C2 nodes may be high value targets because their destruction seriously hampers an adversary's effectiveness. Destruction requires determining the exact location of the target. This location may be determined through the effective application of ES measures. Adversary EM systems can be destroyed by a variety of weapons and techniques, ranging from bombardment with conventional munitions to intense radiation and high energy particle beam overloading. Destruction of EM capabilities has the most sustained effects and may be the best means of denying adversary use of the EMS. The duration of the destructive effects depends on an adversary's' capability to reconstitute. An example of EW application of destruction would be the use of a high-speed antiradiation missile (HARM) against enemy radars.

Chap 4

XIV. Public Affairs (PA) Operations

Ref: Annex 3-61, Public Affairs Operations (28 Jul '17).

Public affairs (PA) is defined as "communication activities with external and internal audiences" (Joint Publication 3-61, Public Affairs)

Air Force PA "advances Air Force priorities and achieve mission objectives through integrated planning, execution, and assessment of communication capabilities. Through strategic and responsive release of accurate and useful information, imagery, and musical products to Air Force, domestic, and international audiences, PA puts operational actions into context; facilitates the development of informed perceptions about Air Force operations; helps undermine adversarial propaganda efforts; and contributes to the achievement of national, strategic, and operational objectives". Truth is the foundation of all PA operations, both in terms of credibility and capability.

Department of Defense Guidance

It is Department of Defense (DOD) policy to make available timely and accurate information so the public, the Congress, and the news media may assess and understand the facts about national security and defense strategy. DOD Directive (DODD) 5122.05, Enclosure 2, Principles of Information, delineates principles of information that apply in supporting the DOD policy.

Refer to Annex 3-61: Appendix A for the complete listing of those principles.

Airpower

I. A Commander's Responsibility

The public affairs (PA) program is a commander's responsibility and commanders are ultimately responsible for successful integration of PA capabilities into the organization's operations; as such, commanders require a clear understanding of PA's role in operations to help achieve desired effects.

PA operations give commanders awareness of the information environment (IE) and the means put joint operations in context, facilitate informed perceptions about military operations, undermine adversarial propaganda directed at domestic and international audiences, and help achieve national, strategic, and operational objectives. Additionally, PA operations are a force multiplier by analyzing and influencing the IE's effect on military operations and delivering increased battlespace awareness to the commander through analysis of the IE. PA capabilities are most effective when planned and executed as an integral part of an overall operation.

PA operations also facilitate open and honest two-way communication within the Air Force and between the Air Force and the public. Commanders instill trust and enhance morale by personally communicating to the Airmen in their commands. As spokespersons for the Air Force, the Department of Defense (DOD), and the US Government, commanders and their representatives play a vital role in building public support for military operations and communicating US resolve to international audiences.

The PA officer (PAO) is the commander's principal spokesperson, senior PA adviser, and a member of the personal staff. The PAO must have the knowledge, skills, resources, access, and authority to provide timely, truthful, and accurate information, visual information, and context to the commander, the staff, and subordinate and supporting commanders, and to rapidly release information in accordance with DOD policy and guidance to the news media and the public.

Providing the maximum disclosure of timely and accurate information as rapidly as possible enables the commander to seize the information initiative.

Refer to Air Force Instruction (AFI) 35-101, Public Affairs Responsibilities and Management, for a list of specific responsibilities for commanders regarding PA.

II. PA Core Competencies

Public affairs' (PA) four core competencies, as outlined in Air Force Instruction 35-101, Public Affairs Responsibilities and Management, describe the primary ways PA contributes to overall mission accomplishment. The following are the core competencies of PA operations:

- Trusted counsel to leaders
- Airman morale and readiness
- Public trust and support
- Global influence and deterrence

These synergistic competencies are core contributions of PA operations to the Air Force and are conducted across the range of military operations. PA operations are most effective when they are integrated into strategic, operational, and tactical plans and executed with direct support from commanders at all levels to achieve desired effects.

A. Trusted Counsel to Leaders

PA operations provide commanders and other Air Force leaders with candid, timely, trusted and accurate counsel and guidance concerning the effects of the information environment (IE) on the Air Force's ability to meet operational objectives. This PA competency includes providing predictive awareness of the global public IE through the observation, analysis, and interpretation of domestic and foreign media reporting, public opinion trends, lessons learned from the past, and preparing leaders to engage the media. This counsel helps commanders make well-informed decisions regarding the IE's effect on operations and forecast possible results. PA support to commanders is integral to operational success as media and public interest increase during operations and can affect the outcome.

B. Airman Morale and Readiness

Airman morale and readiness directly translates into combat capability for the Air Force. PA operations enable Airmen to understand their roles in the mission, explaining how policies, programs, and operations affect them and their families.

PA operations convey truthful, credible, and useful information to support Airman morale and readiness, and provide the Air Force with capabilities to counter misinformation and propaganda directed at our forces. PA tools such as articles, commanders' calls, band performances, and social media are some of the components of this PA competency. PA operations counter adversary propaganda efforts and help to minimize the loneliness, confusion, boredom, uncertainty, fear, rumors, and other factors that cause stress and undermine efficient operations.

PA operations also contribute to readiness by helping to increase Airmen's understanding of the law of armed conflict , rules of engagement and respect for the protections provided to noncombatants and detainees through the presentation of such information in radio or TV broadcasts, base websites, etc.

Informed and knowledgeable Airmen have higher morale and can be relied upon to effectively deliver Air Force themes and messages as they explain their mission to media representatives, public groups, or individuals. With Airmen as credible, reliable spokespeople, PA operations can more effectively deliver global influence, deterrence, enhanced public trust and support as Airmen convey Air Force themes and messages in the public IE.

III. Public Affairs (PA) Activities

Ref: Annex 3-61, Public Affairs Operations (28 Jul '17), pp. 11-14.

Public Affairs (PA) operations begin at home, before the first Airman deploys, and continue long after the last Airman is redeployed. PA operations focus on 10 synergistic activities to achieve the desired effects of its core competencies:

PA Functional Management

PA functional management ensures the PA office and assigned personnel are resourced, trained, equipped, and ready to accomplish the mission in garrison or deployed.

Communication Planning

Communication planning is important to the creation of strategic, operational, and tactical effects in PA operations. PA operators must gain awareness of the aspects of the total information environment (IE) affecting their location or operation, and should have the means to evaluate and analyze aspects of the IE.

Security and Policy Review

While adhering to the policy of "maximum disclosure, minimum delay," PA ensures information intended for public release neither adversely affects national security nor threatens the safety, security, or privacy of Air Force personnel.

Media Operations

Working proactively with the media increases trust and two-way communication, and is often one of the most rapid and credible means of delivering the commander's message.

Community Engagement

Community engagement encompasses activities of interest to the general public, businesses, academia, veterans, service organizations, military-related associations, think tanks, and other community entities.

Environmental

PA supports environmental program objectives and requirements by facilitating public notification and involvement and communicating the Air Force's commitment to environmental excellence.

Visual Information

Visual products, such as photo, video, and graphics, are essential to effective communication and document the Air Force's visual history, through the accessioning process, for future generations.

Band Operations

Air Force bands provide a wide spectrum of musical support for events that enhance the morale, motivation, and esprit de corps of our Airmen, foster public trust and support, aid recruiting initiatives, and promote our national interests at home and abroad.

Contingency Operations and Wartime Readiness

PA forces are foremost a deployable combat capability, fully trained and prepared to meet the needs of the joint warfighter inside and outside the wire.

Command Information

PA provides effective and efficient communication tools to link Airmen with their leaders. Command information helps Airmen and their families understand their purpose, role, and value to the Air Force. A free flow of information to Airmen and families creates awareness of and support for the mission, increases their effectiveness as Air Force ambassadors, reduces the spread of rumors and misinformation, and provides avenues for feedback.

C. Public Trust and Support

PA operations support a strong national defense—preparing the nation for conflict and war—by building and sustaining public trust and understanding of Air Force contributions to national security. These operations build public understanding and support of expenditures for readiness, advanced weapons, training, personnel, and the associated costs of maintaining a premier air, space, and cyberspace force. With public support, Air Force leaders are able to successfully recruit, train, and equip Airmen to meet Air Force operational requirements.

PA operations give commanders the means to gain and maintain support for the Air Force among diverse public audiences. These operations strengthen the bonds between the Air Force and the public through open, honest dialogue. Data and imagery, continuously available in near-real time in the IE, can have an immediate effect on public support. Likewise, distorted information and imagery distributed by an adversary can have an adverse effect on national will and the support of Air Force operations. PA capabilities, integrated with other operational capabilities and employed effectively, can ensure the Air Force story is told while also preempting and degrading an enemy's effectiveness in misleading the public.

To fight and win in the information age, commanders should employ PA capabilities in a fashion that fosters ongoing public understanding and support of operational requirements. Commanders can also leverage those capabilities including bands, visual information, social media, and other products to transcend traditional media and audience boundaries. Use of the full range of PA capabilities expands the reach and impact of PA operations and its ability to build and enhance public trust and support for Air Force operations.

D. Global Influence and Deterrence

PA contributes to global influence and deterrence by making foreign leaders and audiences aware of US capabilities and resolve. Commanders should employ PA operations to develop and implement communication strategies to inform national and international audiences about how airpower affects global events. Building the awareness of national and international audiences about US resolve to employ its strength can enhance support from friendly countries. The same information may deter potential adversaries, driving a crisis back to peace before the use of kinetic force becomes necessary.

Information and power projection demonstrating US or friendly force capabilities and resolve to adversary and international public audiences can be effective in causing adversary decision makers to seek other options short of conflict when they may otherwise not be deterred from conflict. In addition to integrating PA operations during the strategy development and planning phases of an operation, commanders strengthen the effectiveness of PA capabilities when PA operations at all levels are unified by common messages and themes. Exclusion of PA in the early stages of strategy and operational planning limits the effectiveness of PA operations to seize the information initiative from the beginning of an operation and consequently degrades the commander's ability to gain information superiority.

PA operations should be planned for and integrated at multiple levels for employment of offensive PA strategies to help ensure operational success. PA operations can also employ defensive strategies to preempt adversary propaganda and misinformation attempts that otherwise could weaken Air Force global influence and deterrence, Airman morale and readiness, and public trust and support.

Chap 5

Operations & Planning (Overview)

Ref: Annex 3-0, Operations & Planning (4 Nov 16), pp. 40 to 41.

Editor's Note: For the purposes of this publication (AFOPS2), the material from Annex 3-0 Operations and Planning is presented in two separate chapters, with chapter four focusing on airpower and chapter five (this chapter) focusing on strategy, effects-based approach to operations, and the common operations framework (operational design, planning, execution, and assessment).

Air Force Doctrine Annex 3-0 is the Air Force's foundational doctrine publication on strategy and operational design, planning, employment, and assessment of airpower. It presents the Air Force's most extensive explanation of the effects-based approach to operations (EBAO) and contains the Air Force's doctrinal discussion of operational design and some practical considerations for designing operations to coerce or influence adversaries. It presents doctrine on cross-domain integration and steady-state operations–emerging, but validated concepts that are integral to and fully complement EBAO. It establishes the framework for Air Force components to function and fight as part of a larger joint and multinational team. Specific guidance on particular types of Air Force operations can be found in other operational-level doctrine as well as Air Force tactics, techniques, and procedures documents. This publication conveys basic understanding of key design and planning processes and how they are interrelated.

The US' national security and national military strategies establish the ends, goals, and conditions the armed forces are tasked to attain in concert with non-military instruments of national power. Joint force commanders (JFCs), in turn, employ strategy to determine and assign military objectives, and associated tasks and effects, to obtain the ends, goals, and conditions stipulated by higher guidance in an effort to produce enduring advantage for the US, its allies, and its interests. Strategy is a prudent idea or set of ideas for employing the instruments of national power in a synchronized and integrated fashion to achieve theater, national, and multinational objectives. Airmen should follow a disciplined, repeatable approach to strategy development in order to maximize airpower's contribution to overarching national aims.

Today, the United States faces many security challenges including an ongoing conflict against implacable extremists, engagement with regimes that support terrorism, and the need to support international partners. Against this backdrop, US military forces may be called upon to conduct a full range of operations in a variety of conflicts and security situations, including major operations and campaigns, irregular warfare , information operation, homeland defense, humanitarian assistance/disaster relief efforts, building partnerships with other nations, and others.

The operational environments in which airpower is employed may be characterized by simultaneous action by Air Force forces against more than one adversary at a time–including the potential for near-peer and peer competitors–who may attempt to achieve objectives against US interests by using asymmetric advantages across all instruments of power: diplomatic, informational, military, and economic. Conflicts may occur with little or no warning and they may stretch the Air Force as it works with JFCs to provide support for the joint force while simultaneously addressing Air Force-unique missions.

I. Strategy

Ref: Annex 3-0, Operations & Planning (4 Nov 16), pp. 4 to 12.

Strategy is a major focus of Annex 3-0. The very broad joint definition of strategy suffices for the most expansive military meanings (such as described in national-level strategy documents), but in its more commonly understood sense, strategy is a method of arranging and managing ways, means, and risks to achieve an end or set of ends. It produces a coordinated set of options an actor can choose from to achieve continuing advantage. Strategy, in its military sense, is the art of creating military courses of action and encompasses the processes of operational design, planning, execution, and assessment.

Effective Strategy Seeks to Gain *Enduring Advantage*. From a strategic perspective, the methods used to achieve objectives and reach the end state(s) generally carry implications beyond the conclusion of an operation. The purpose of military strategy is not just to "win" or conquer, it is to resolve the conflict on favorable terms for the US, and do so in a way that endures for as long as possible. Such resolution is sought by creating conditions that are at least better for friendly interests, and are often better for all parties involved. Thus, a strategy's ultimate purpose is the attainment and maintenance of a set of future conditions – an end state (or states) – that leads to continuing and *enduring advantage* for friendly interests, for as long as possible and that will often create advantage for neutral and formerly hostile interests as well.

Strategy Encompasses Ends, Ways, Means, and Risks. Strategy should illuminate the reasons an operation is being conducted—its purpose—state the objectives and end state(s) (ends); prescribe the methods by which the ends are achieved—military courses of action (COAs) (ways); determine the tools and resources needed to execute the strategy, such as military forces and supplies (means); and clarify the amount of cost, uncertainty, and vulnerability the commander and national leadership are willing to accept and will need to commit in order to execute the strategy, (costs and risks).

Desired Future Conditions and Commander's Intent Should Drive Strategy. The accomplishment of all military objectives should lead to a desired set of future conditions, the military end state. The attainment of military aims, however, is subordinate to attainment of a set of conditions that must be achieved to resolve the situation or conflict on satisfactory terms and gain enduring advantage, as defined by appropriate civilian authority (such as the President or Secretary of Defense [SecDef] at the national strategic level).

Strategy is Adaptive, Not Static. Strategy should adjust as the adversary reacts to friendly moves an as circumstances change. Therefore, strategy creation should be cyclic and iterative.

Strategy and Planning Are Not the Same and Benefit from Discourse. Strategy formulation begins with the process of operational design, which helps frame the problem the joint force is tasked to solve and design a basic construct for solving it that can be further refined in subsequent planning. Operational design is defined as "the conception and construction of a framework that underpins a campaign or major operation plan and its subsequent execution" Ultimately, design results in mission and intent statements that reflect the commander's vision for the overall operation (including end states that lead to continuing advantage). With this guidance clearly given, strategists and planners can concentrate on discrete problems that can be solved through the military's more formalized planning processes. This is akin to engineers taking the architect's sketches or models and turning them into blueprints and schematics that can then be used by craftsmen (the equivalent of tactical-level planners) to flesh out detail and implement the plan.

Strategy is Art and Science. Executing military strategy depends upon operational art, the creative means through which commanders and staffs develop strategies to organize and employ military forces. As such, there is as much art as science to the military commander's craft. There are many aspects of operations that yield to scientific scrutiny. For instance, direct, immediate weapon effects can be accurately anticipated. The further one gets from immediate effects, however, the harder it becomes to predict indirect outcomes. Science can greatly aid strategy formulation, but the utility of science often does not extend beyond immediate effects—assessment and adaptation require judgment and intuition on the part of commanders and strategists.

Strategy Should Integrate Military Power at All Levels with Other Instruments of National and Multinational Power. Effective military operations require careful integration of the efforts of all appropriate "actors" within the operational environment. All the instruments of power (IOPs) that actors (state or non-state) may wield are interrelated. Political considerations are critical, but so are economic, cultural, informational, and other considerations. Strategy should seek to integrate all relevant IOPs in order to deliver an end state that is, itself, a combination of conditions reflecting all aspects of power.

It is usually beyond the scope of authority for commanders, Air Force forces (COMAFFORs) to direct the integration of elements of national power beyond the military forces for which they are directly responsible. In fact, this is often beyond the authority of the JFC or even the combatant commander (CCDR) in whose area of responsibility (AOR) an operation is taking place. Nonetheless, all commanders are usually constrained to operate with other agencies of the United States government, within international coalitions, and with international nongovernmental organizations (NGOs). Sometimes these relationships can restrain commanders' freedom of action, but just as often they open opportunities for integrating diplomatic, informational, and economic IOPs with military efforts and thus give commanders a wider range of options with which to create intended effects. COMAFFORs, who are normally designated as joint force air component commanders (JFACCs) and may also be joint task force (JTF) commanders, should be prepared to operate as part of a multi-agency and multinational team and, in some cases, to direct personnel from non-Department of Defense (DOD) agencies and multinational partners in support of JFC objectives.

Military strategy at the theater level is normally derived from strategy guidance given by US leadership and multinational partners. At the same time, theater strategy (and all efforts down to tactical tasks) seeks to attain an end state that will enhance national strategic interests, and often those of an alliance, coalition, community of interested states or multinational organizations, embodying the doctrinal concept of unity of effort.

Strategists Should Realize that Tactical and Operational "Victory" Do Not Guarantee Strategic Success. Success at the tactical and operational levels should contribute to strategic success, but this is by no means guaranteed. Many times in history, one side has "won all the battles, but lost the war." This implies that failure at lower levels does not guarantee strategic failure. (If this were so, for instance, the American colonies might never have won their revolutionary war.) It is possible—even easy—for commanders and strategists to become so enamored of success at lower levels that they lose sight of larger strategic trends, exaggerate the influence of lower-level assessment "markers," engage in "wishful thinking" when analyzing the effects of ongoing operations, or incline toward strategic overreach.

Refer to JFODS5: The Joint Forces Operations & Doctrine SMARTbook (Guide to Joint, Multinational & Interorganizational Operations) for further discussion. Topics and chapters include joint doctrine fundamentals (JP 1), joint operations (JP 3-0), joint planning (JP 5-0), joint logistics (JP 4-0), joint task forces (JP 3-33), information operations (JP 3-13), multinational operations (JP 3-16), interorganizational cooperation (JP 3-08), plus more!

II. Effects-Based Approach to Operations (EBAO)

The Air Force designs, plans, conducts, and assesses operations according to an effects-based approach. An effects-based approach is "an approach in which operations are planned, executed, assessed, and adapted to influence or change systems or capabilities in order to achieve desired outcomes." In the most basic sense, this entails determining the effects that the military should create in order to accomplish the military objectives that help achieve the military strategy, as it contributes to overall strategic success—and then applying the best combination of capabilities to create those effects. EBAO is not a planning methodology; it is a way of thinking about operations that provides guidance for design, planning, execution, and assessment as an integral whole. In a more comprehensive sense, EBAO is an approach that emphasizes:

- Operations are driven by desired ends (end states and objectives), and should be defined by the effects required to attain these ends, not just by what available forces or capabilities can do, nor by what the Air Force "customarily" does with a given set of forces.
- Commanders should realize they are dealing with interactively complex problems not solvable by deterministic or "checklist" approaches. Interactive complexity carries implications that are important for commanders to realize.
- The "human element," "friction," and the "fog of war" can never be eliminated.
- There is never a single "right" solution. Commanders seek solutions that are "better" or "worse" and solving one set of problems often causes others to emerge.
- Commanders seek solutions that are most effective first–the solutions to achieve the objectives and end state—and then, given that, strive for efficiency.
- Commanders try to maximize options available and thus consider integrated use of all available military means and other instruments of power (IOPs) to gain continuing advantage within a given strategic context.

The concepts and guidelines described in this section are not wedded to the term "effects-based"—they could have as easily been described as an "objectives-," "outcomes-," "results-," "impact-," or "consequence-based" system of thought.

Nonetheless, "effects-based" is the term that is most widely recognized in Air Force circles. Further, this approach fully complements and helps reinforce the general considerations for military operations and strategy described in the previous sections. The section below presents a more complete explanation of the body of sanctioned ideas that define EBAO, but also presents general considerations that are often ignored in military literature on strategy, and which should help shape the thinking of commanders and strategists. (The order in which the explanatory paragraphs are presented does not necessarily represent their relative importance or priority—these may change from operation to operation.)

EBAO is a comprehensive approach—it cuts across all domains and dimensions, all disciplines and partnerships, all levels, and all IOPs. EBAO provides an overarching way of thinking about action that encompasses operational design, planning, execution, and assessment of operations involving all IOPs across the range of military operations (ROMO). It is not directly tied to any specific strategy or type of operation.

EBAO is about creating effects, not about platforms, weapons, or particular methods. An effects-based approach starts with desired outcomes—the end state(s), objectives, and desired effects—then determines the resources needed to achieve them, while identifying critical resource limitations. It does not start with particular capabilities or resources and then decide what can be accomplished with them. It also assigns missions or tasks according to mission-type orders, leaving decisions concerning the most appropriate mix of weapons, units, and platforms to the lowest appropriate levels within a given organization. Air Force commanders should encourage commanders from other Services, when tasking the Air Force or air component, to request particular effects instead of specific assets. Further, while EBAO is not about technology, there are new platforms, weapons, and methods that can enable new types of effects. These do not become truly useful to the warfighter until they are joined with appropriate employment doctrine and strategy.

EBAO integrates strategy—all design, planning, execution, and assessment efforts—into a unitary whole. These should be inextricably bound together, because effective and efficient execution almost always involves design, planning, and assessment in some form as well, even if not as part of a formal or "official" process. Effective operations should be part of a coherent plan that logically supports and ties all objectives and the end state together; the plan to achieve the objectives should guide execution; and that means of measuring success, gaining feedback, and adapting to changes should be planned for and evaluated throughout execution. Strategy encompasses all the means through which courses of action (COAs) are developed and evaluated, such as the Adaptive Planning and Execution (APEX) system at the national level, the joint operation planning process (JOPP) at the joint force commander (JFC) level, and the joint operation planning process for air (JOPPA), formerly known as the "joint air estimate process," at the component level. These are the collaborative, iterative, and adaptive processes that help integrate strategy from national through joint force component levels. The JOPP and JOPPA are integral and complementary to the APEX process: Adaptive planning describes force and logistical requirements, while the JOPP and JOPPA outline the objectives and tasks military forces are to accomplish.

Operational design and planning set the stage for all subsequent planning activities and thus are where sound effects-based principles may have the greatest impact. Execution encompasses and implements all the various tasking processes and the ongoing operational battle rhythm, as well as all the individual unit actions that comprise implementation of airpower operations; integrating, synchronizing, and deconflicting their accomplishment, as well as disseminating mission-critical information to those needing it. Execution that is not effects-based often devolves into a "checklist mentality," that becomes excessively process-driven and loses sight of the larger context (such as the objectives and end state). This can negate sound planning, as when focusing too narrowly on one or another aspect of the battle rhythm—for example, air tasking order production. Execution that is not effects-based runs the risk of devolving into blindly servicing a list of targets, with little or no anticipation of or adaptation to enemy actions or changes in the operational environment like weather. Assessment encompasses all efforts to evaluate effects and gauge progress toward objective accomplishment. Assessment is used to adapt operations as events unfold and thus feeds the revision of plans. One should always attempt to measure performance of actions and the effectiveness of those actions in terms of creating desired effects and achieving objectives.

See pp. 5-23 to 5-25 for related discussion of EBAO to include an overview of effects (direct, indirect, intended and unintended), objectives and actions.

III. The Common Operations Framework

Although the range of military operations (ROMO) is a continuum that extends from continuous and recurring operations, such as security cooperation during peacetime, to major combat operations in war, there are some significant differences between the focus of strategy during steady-state conditions and the focus during contingencies and major operations. During steady state, strategy focuses on shaping the environment for regional and global stability, deterring aggression, and preventing conflict. Time horizons are thus usually much longer and considerations of readiness, budgeting, and the training and equipping of forces—all of which are outside the scope of doctrine—impact strategy significantly. Contingencies and major operations are the traditional subject of military strategy and doctrine, and thus military decision-making processes described in planning and operations doctrine have focused upon them.

Nonetheless, operations in recent decades have shown that there is significant common ground between steady-state and contingency conditions, and there are considerable advantages to designing coherent and comprehensive strategies for shaping the actual steady-state environment. Potential contingencies and major operations are then considered branches to combatant commanders' overarching theater or global campaign plans. Contingency planning and steady-state planning employ a common logical approach and process.

Common Framework for Operations

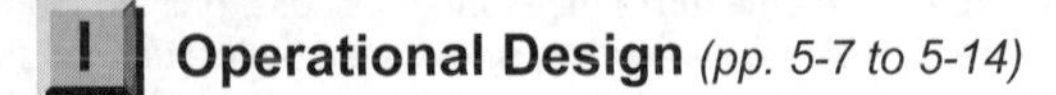

I **Operational Design** *(pp. 5-7 to 5-14)*

II **Planning** *(pp. 5-15 to 5-26)*

III **Execution** *(pp. 5-27 to 5-30)*

IV **Assessment** *(pp. 5-31 to 5-40)*

Ref: Annex 3-0, Operations & Planning (4 Nov 16), p. 40.

A common framework of processes helps to foster coherence in Air Force strategy creation by:

- Creating explicit linkages to national objectives and desired end states.
- Encouraging continuity in thinking used to design and plan operations, regardless of where they occur in the ROMO, whether during steady-state or contingency operations.
- Providing a common method for commanders and staff elements to use in designing and planning contingencies as logical follow-ons to ongoing operations.
- Encouraging logical linkages between resources needed for ongoing operations and those to be flowed in to support emerging contingencies.
- Fundamentals of assessment, including discussions on assessing strategy in general, assessment criteria, assessment measures and indicators, and assessment interpretation.

The common framework for operations is broken into the following general considerations:

- **Operational Design**. Fundamentals of operational design, including discussion of the elements and methods of operational design, the coercion continuum as a practical design construct, and additional considerations specific to airpower. *(See pp. 5-7 to 5-14.)*
- **Planning**. General planning considerations, including discussions on Air Force planning in the context of broader joint planning and the effects-based approach to planning. *(See pp. 5-15 to 5-26.)*
- **Execution**. General execution considerations. *(See pp. 5-27 to 5-30.)*
- **Assessment**. Fundamentals of assessment, including discussions on assessing strategy in general, assessment criteria, assessment measures and indicators, and assessment interpretation. *(See pp. 5-31 to 5-40.)*

Chap 5

I. Operational Design

Ref: Annex 3-0, Operations & Planning (4 Nov 16), pp. 42 to 60.

I. Operational Design Fundamentals

As an element of strategy, operational design is defined as "the conception and construction of the framework that underpins a campaign or major operation plan, and its subsequent execution" (Joint Publication [JP] 5-0, Joint Operation Planning). Operational design helps establish a logically consistent structure from which to understand an operation's aims and, broadly, the methods and means to be used in obtaining them. In other terms, design provides a necessary "front end" to the formal planning processes described in JP 5-0 and elsewhere in Annex 3-0. The "process" of determining the overall focus of an operation—of deciding on the end state, objectives, desired effects, and so on, has been largely a matter of art throughout most of military history. Understanding certain aspects of problem solving can make portions of the commander's art more systematic, although it will never make them "scientific"—in the sense of making them prescriptive and predictable. Approaching operational design deliberately, however, can provide a foundation that facilitates decision-making by creating a structure linking decision analysis to emerging opportunities. Creating such a linkage can substantially reduce the risks associated with an operation and increase the utility of a plan following first contact with an adversary.

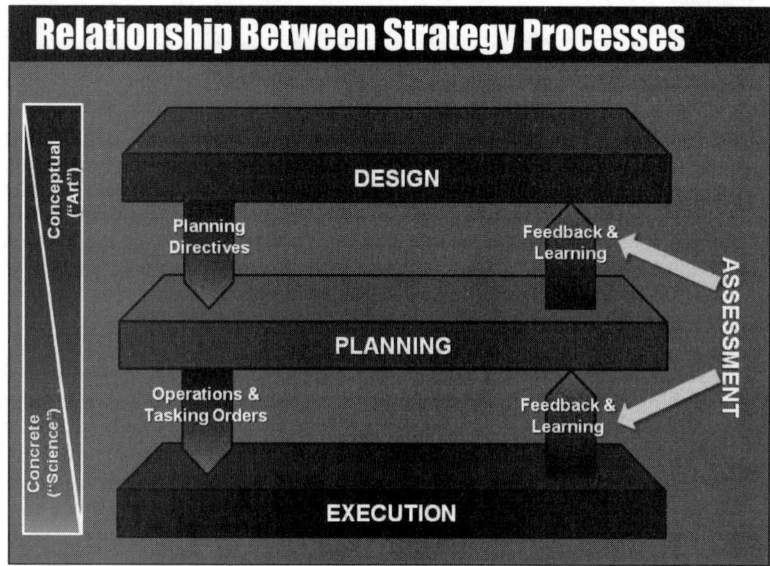

Ref: Annex 3-0, Operations & Planning (4 Nov 16), p. 42. Relationship Between Strategy Processes.

Design consists of three closely interrelated activities, which collectively allow commanders and their staffs to understand and visualize an operation's purpose. These activities are framing the operational environment, framing the problem, and developing the operational approach. Design helps formulate an operational approach and the commander's initial statements of mission and intent, which in turn feed the process of course of action (COA) analysis and selection, which feeds the creation of detailed plans and assessment criteria. Plans are then executed by accomplishing tasks at the tactical level. The results are assessed and operations are adapted based on that assessment, providing input to strategy revision. Design is thus cyclic and iterative, like many other aspects of planning in general.

Operational design is the job of commanders with the support of their strategists and staffs. Planning and design are closely interrelated, since planners take the commander's overarching design concept and intent to create detailed COAs, plans, and orders for operations. Planning and design make it possible to convert broad guidance from national leadership and senior commanders and turn it into discrete tasks at the tactical level.

Design can aid creation of formal planning products as part of deliberate and crisis action planning (CAP). The joint operation planning process (JOPP) activities and products are generally the basis for concurrent joint operation planning process for air (JOPPA) activities, which result in the JFACC's joint air operations plan (JAOP) and the commander, Air Force forces' (COMAFFOR's) component plan. The JAOP and component plans provide operational guidance until the battle rhythm is initiated, at which point strategy guidance is provided through the air operations directive (AOD). The cycle proceeds through execution to feed the reiteration of strategy formulation based on the results of the continuous process of assessment. The first steps of the JOPP and JOPPA reiterate and re-examine the products of operational design, such as the commander's mission and intent statement.

The intermediate planning steps, involving the JOPP, JOPPA, JAOP, and AOD, are discussed in greater detail in chapter six, pp. 6-5 to 6-28.

Design work done by commanders and strategists can be likened to that of an architect in a building project, working directly with the project's "sponsors" (the clients in this illustration; national leadership in a military operation) and the engineers who help realize specific aspects of the architect's design. The engineers are the higher-level planners who accomplish the JOPPA and produce the JAOP and AODs. Tactical planners and controllers (those who produce and execute the air tasking order [ATO]) are like the artisans who create specific details of the plan. Tactical plans tend to solve well-structured problems, where tactics and techniques yield one (or a very few) indisputably correct solutions to objective, empirical problems (like the best ordnance to use on a particular target). Operational plans tend to solve medium-structured problems, where doctrine suggests courses of action that have clear objectives and end state, but may have a number of possible correct solutions (like the best way to win a specific battle). Commanders and strategists, however, usually deal with ill-structured problems, which are far more complex.

The interaction of complex adaptive systems almost always yields ill-structured problems. Warfighters are problem-solvers by nature, but most have been trained to solve either well- or medium-structured problems. With ill-structured problems, however, there is often disagreement even concerning the desired end state or the basic parameters that define the problem to be solved.

Design is a methodology for applying critical and creative thinking to understand, visualize, and define complex, ill-structured problems and develop approaches to solve them. Design requires the right people and the right command climate in order to succeed. Design is not a mechanistic, "checklist," or institutionally-entrenched activity and it cannot be accomplished by any one person, although the commander drives the process and plays a central role. To succeed, the organization practicing design should have a climate that encourages open, honest dialogue and exchange of ideas.

Design requires close interaction among an organization's commander, staff, the commanders and staffs of higher and lower echelons, as well as supporting commanders and their staffs. It is through interchange between different levels that shared understanding and common vision can be achieved. Leaders and staffs at higher echelons may have clear strategic understanding of the problem; those at lower levels may better understand local circumstances. Bridging these perspectives is crucial to achieving a common vision, which enables unity of effort.

Joint functional and Service components need to be involved at various levels in the initial planning stages of joint strategy development. In some cases, however, the joint force air component commander (JFACC) and key air operations center planners may need to volunteer to be included early in the joint force commander's (JFC's) design process. In such cases, joint integration requires that a sufficient number of trained Airmen be included on the JFC planning staff. The air component liaisons, if established, can help make the JFACC aware of pending or ongoing design and planning efforts, but it is also the JFC's responsibility to actively seek airpower expertise. Each theater or joint task force operation will likely be different, and prior coordination is required on how overall joint strategy development may occur and how airpower should be included in that effort. Theater-level design and planning exercises are vital to ensure proper integration when operations commence.

II. Methods of Operational Design

Operational design is the first level of strategy implementation and rests upon operational art, which is defined as the "cognitive approach by commanders and staffs–supported by their skill, knowledge, experience, creativity, and judgment–to develop strategies, campaigns, and operations to organize and employ military forces by integrating ends, ways, and means" (Joint Publication [JP] 5-0, Joint Operation Planning).

Operational art uses the commander's vision and intent to determine broadly what should be accomplished in the operational environment; it is guided by the "why" from the strategic level and implemented with the "how" at the tactical level. In applying operational art, the commander draws on judgment, perception, creativity, experience, education, intelligence, boldness, and character to visualize the conditions necessary for success before committing forces. This visualization is captured in the commander's operational approach, which is a description of the broad actions the force must take to transform current conditions into those desired at the end state (JP 5-0, Joint Operation Planning).

Operational Design Methodology

Design provides an ordered way to create the conceptual framework of a plan. Strategists and planners can then use the joint operation planning process (JOPP) to create detailed subordinate plans and orders. The purpose of design is to create an operational approach that can be "fleshed" into more detailed plans. In order to derive the operational approach, the commander and staff should understand the operational environment and the problems the joint force commander (JFC) has been given to solve. Thus, broadly speaking, operational design consists of framing (or understanding) the operational environment, framing (or defining) the problem, and developing the operational approach.

See following pages (pp. 5-10 to 5-11) for an overview and further discussion.

III. Operational Design Methodology

Ref: Annex 3-0, Operations & Planning (4 Nov 16), pp. 46 to 51.

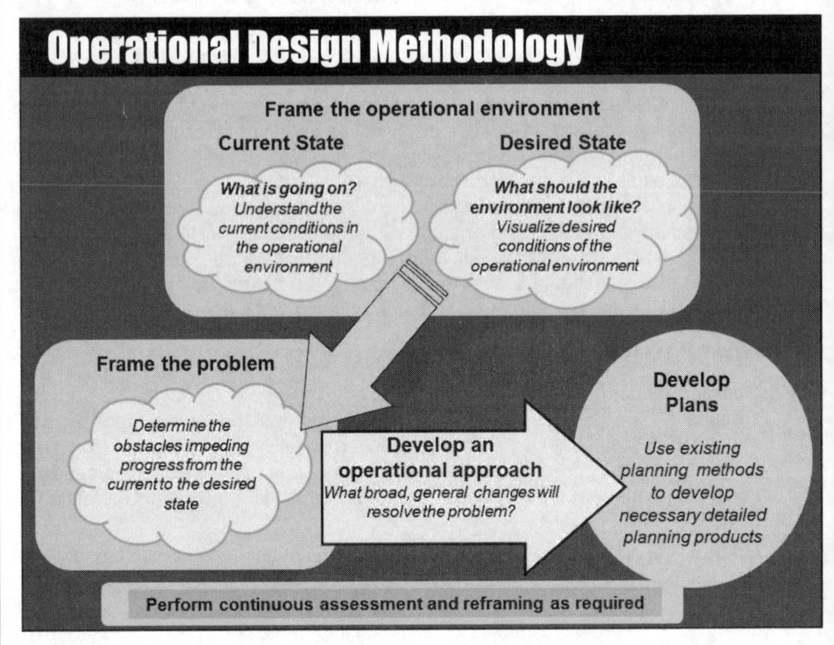

Ref: Annex 3-0, Operations & Planning (4 Nov 16), p. 47. Operational Design Methodology.

A. Framing the Operational Environment

Operational design begins with framing the operational environment (OE)—establishing the larger context of a situation within which the commander should act in order to realize the operation's aims. This entails reviewing all existing guidance from higher authorities (including existing theater campaign and country plans that govern steady-state activities) and examining all actors (opponents, friends, and neutrals) and their relationships within the OE. The aim is to understand existing conditions in order to derive the set of conditions we wish to see at the end of operations (often the restoration of stable steady-state conditions), as well as understanding the competing conditions that other actors would like to see. Based on overarching guidance, the JFC will derive that portion of the end state the military is responsible for delivering (the military end state) and assign the military objectives required to arrive at that end state. These objectives form the basis for the operational approach.

The principal means by which the commander and staff gain understanding of the OE is joint intelligence preparation of the operational environment (JIPOE). Guidance concerning JIPOE can be found in JP 2-01.3, JIPOE and JP 5-0, Chapter IV.

B. Framing the Problem

This part of the process entails reviewing the tendencies and potential actions of all actors within the relevant OE and coming to an understanding of the root causes of the issue at hand. This is not the same as problem solving, which planners do at lower levels to create solutions to medium- and well-structured problems within the conceptual framework created by the commander and strategists. Problem framing entails determining the overall boundaries and aims of the operation, much as an architect does for a building project. In many cases, only the most prominent tendencies and potential actions of all the actors in a situation can be considered in a finite time by a well-informed staff. In-depth understanding may require a lifetime of study and immersion and the military must often go outside its own channels–to the interagency community, regional experts, academics, and local nationals–to leverage such knowledge. When possible, open, collegial dialogue among the commander, "sponsors," other government agencies, and nongovernmental organizations, staff strategists, and planners can be very beneficial during this process. As operational design progresses into planning, the process becomes more formalized and the models strategists and planners work with become more empirical as they engage in course of action (COA) development, analysis, and wargaming.

C. Developing the Operational Approach

The operational approach describes in broad terms how the OE should be changed from existing conditions to desired conditions. It is a commander's means to describe what the joint force must do to achieve objectives that bring about the desired end state. Frame the mission with a clear, concise statement of the purpose to be achieved and the essential tasks to be accomplished—who, what, when, where, and why.

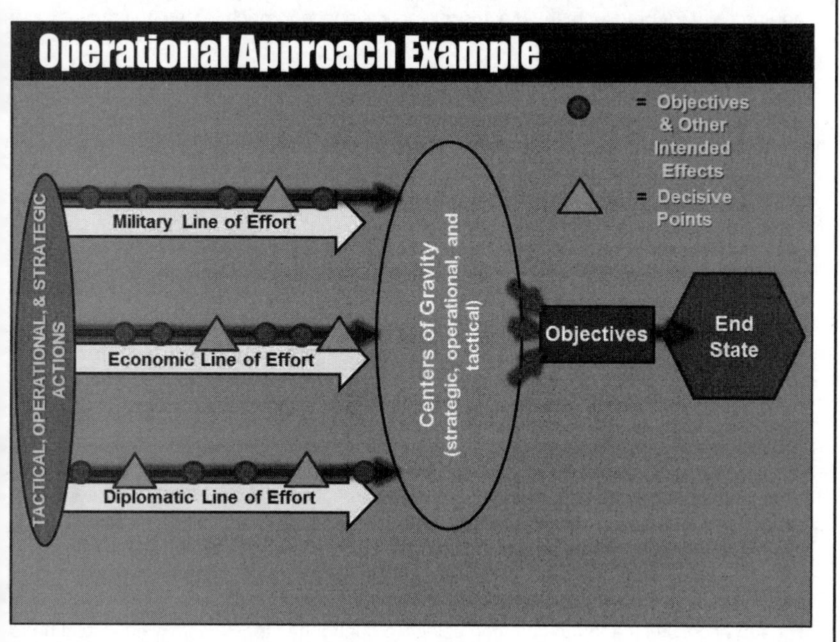

Ref: Annex 3-0, Operations & Planning (4 Nov 16), p. 50. Operational Approach Example.

Refer to JFODS5: The Joint Forces Operations & Doctrine SMARTbook (Guide to Joint, Multinational & Interorganizational Operations) for further discussion. Topics and chapters include joint doctrine fundamentals (JP 1), joint operations (JP 3-0), joint planning (JP 5-0), joint logistics (JP 4-0), joint task forces (JP 3-33), information operations (JP 3-13), multinational operations (JP 3-16), interorganizational cooperation (JP 3-08), plus more!

IV. Steady-State Design: Shaping the Operational Environment

Ref: Annex 3-0, Operations & Planning (4 Nov 16), pp. 96 to 99.

Operational design for the steady state has few differences from operational design for crisis situations. The commander remains the central figure in the entire effort, applying military judgment and experience throughout the process. Commanders should look to steady-state planners to assist in developing the steady-state operational approach. For the commander, Air Force forces (COMAFFOR), steady-state operational design is significantly influenced by the combatant commander, who has likely conducted his or her own operational design effort as part of campaign plan development.

As in any design effort, the commander should define success in the steady state (ends) and allocate forces and resources (means) to achieve the desired ends. The operational approach provides the ways to link steady-state ends and means.

Steady-State Design Objectives

There are five objectives typically associated with the steady state. These provide a starting point for development of a specific operational approach for the commander and situation.

The first objective of steady-state design and planning, and typically the Air Force's highest priority, is to be ready to respond immediately and appropriately to crisis situations. The emphasis here is on force readiness. As an institutional responsibility, force readiness generally falls outside the scope of operational doctrine. Crisis situations are normally unexpected, meaning the readiness of the force at the start of the crisis is the readiness that may apply throughout. Major commands and Headquarters Air Force issue policy and guidance and commit resources to assist operational commanders with force readiness.

A second objective is the need to plan, execute, and assess steady-state operations that contribute to the deterrence and prevention of conflict. It is far preferable to deter or prevent conflict rather than to engage in conflict. The Air Force has many tools available to support this objective, from continuous bomber presence, to theater security packages, to shows of force, to multinational exercises, and more. Another significant deterrence and prevention tool is building partner capacity, which leads to the next steady-state objective.

Building partner capacity is another important objective, especially acknowledging that even the most committed approach to deterring and preventing conflict is not always successful. The steady state should be used to develop international partners with the capability, capacity, and interoperability to respond in crisis with, alongside, or—better yet—instead of the US should deterrence fail. Capable partners can reduce the operational burden on the Air Force in both the short- and long-term.

Theater access is the fourth common objective during the steady state. The ideal time to secure or sustain contingency access is during the steady-state, providing the Air Force with air base access, overflight rights, and host nation and logistics-related agreements vital to the conduct of contingency operations. Strong relationships with partner nation air forces improve the likelihood for theater access exactly where and when Airmen might need it the most.

The final objective is vital to the long-term strength of the Air Force. It is the operational commander's responsibility to participate in Air Force force development activities. Force development ensures the required readiness, capabilities, and capacities to respond appropriately in the mid- to long-term. As with force readiness, major commands and Headquarters Air Force provide policy and guidance that influence how operational commanders (i.e., warfighters) participate in force development.

Security Cooperation Considerations

A basic understanding of security cooperation can also assist the commander in designing an operational approach for the steady state. Three security cooperation considerations are worth highlighting.

First, security cooperation always supports US government interests, and Airmen normally define Department of Defense (DOD) interests in steady-state operational plans such as country plans and campaign support plans. Commanders also ensure Department of State (DOS) buy-in for proposed security cooperation activities, as the DOS is the lead federal agency for diplomacy. The third party to security cooperation is the partner nation itself. The partner nation needs to see the benefit of a relationship or security cooperation event with the US. Therefore, the ideal security cooperation activity simultaneously supports the interests of DOD, DOS, and the partner nation. This ideal is often referred to as the security cooperation "sweet spot." See the figure, "Common Objectives," for a representation of this.

A second security cooperation consideration relates to the establishment of desired partner roles. When considering how partner nations contribute to campaign objectives, early design consideration should be applied to the desired security role for each partner. Commanders and strategists should determine what the US government and US Air Force intend for the partner. In other words, what military role should the partner play to support US interests, such as national sovereignty, regional stability, or global commerce? The establishment of desired roles then leads to an assessment of current capability and the development of specific objectives and activities related to building partner capacity or other security cooperation activities. Just as important, designers may also determine what the US does not want the partner to do. From an Airman's perspective, a partner nation can serve many important security roles; for example:

- Respond to crisis in place of the US Air Force.
- Respond to crisis alongside the US Air Force.
- Lead an air force coalition in responding to crisis.
- Defend its own borders from external air aggression.
- Host a US cooperative security location, forward or main operating base.
- Provide contingency access to US forces.
- Be a supporting partner in regional security framework(s).
- Deny sanctuary to terrorists, insurgents, criminals, or other transnational elements.

The designer should recognize these are roles the US desires the partner to play and may or may not reflect the current desires of the partner. Further design and relationship building efforts may be required to convince a partner to pursue these roles, and then help the partners succeed in developing and performing these roles.

Finally, security cooperation provides an opportunity to mitigate operational risk by strengthening partner capabilities in areas where the US Air Force has its own capability gaps and shortfalls. This consideration relates to the force development discussion above, suggesting operational commanders can support institutional responsibilities by focusing building partner capacity efforts into areas where the Air Force is accepting operational risk.

Refer to TAA2: Military Engagement, Security Cooperation & Stability SMARTbook (Foreign Train, Advise, & Assist) for further discussion. Topics include the Range of Military Operations (JP 3-0), Security Cooperation & Security Assistance (Train, Advise, & Assist), Stability Operations (ADRP 3-07), Peace Operations (JP 3-07.3), Counterinsurgency Operations (JP & FM 3-24), Civil-Military Operations (JP 3-57), Multinational Operations (JP 3-16), Interorganizational Cooperation (JP 3-08), and more.

Operations & Planning

V. Practical Design: The Coercion Continuum

Ref: Annex 3-0, Operations & Planning (4 Nov 16), pp. 52 to 60.

All military strategy seeks to coerce or persuade an adversary or other actor to do one's will. Coercion is convincing an adversary to behave differently than it otherwise would through the threat or use of force. All coercive military action works along a continuum from pure threat (only implied use of force, or using peaceful means to defeat adversary strategies) to pure force (engaging military forces and government control mechanisms), as illustrated in the figure, "The Coercion Continuum."

The Coercion Continuum

Pure Threat		Pure Force
Deterrence Pure Coercion Psychological warfare	"Compellance" Most major operations and campaigns	"Total war" "Brute force" Attrition-based warfare

Ref: Annex 3-0, Operations & Planning (4 Nov 16), p. 52. The Coercion Continuum.

Most combat operations, regardless of size or intensity, reside near the middle of the continuum, however many conflicts may span the entire spectrum. Each conflict has its own character. Many campaigns in World War II (WW II), for example, were close to the "pure force" extreme of the continuum. Operation ALLIED FORCE (OAF), relatively limited in scope and violence, was much closer to the left end of the spectrum. The degree of violence and "brute force" required depends very much upon the national interests at stake, the "target audience," and that audience's determination to resist one's will. It can also be critical to understand that what may be a limited conflict to one side may be viewed as total war by the other—the level of violence and degree of commitment may depend upon the eye of the beholder.

Effective use of airpower can help facilitate conflict resolution closer to the "pure threat" end of the continuum, helping achieve objectives and the end state on more favorable terms, in less time, and more efficiently than might otherwise be possible. However, airpower is capable of creating effects anywhere along the continuum. The destruction of German industry from the air during WW II represented one form of near-pure force strategy, as did the attrition of Iraqi tanks and artillery during Operation DESERT STORM.

II. Planning

Ref: Annex 3-0, Operations & Planning (4 Nov 16), pp. 61 to 77.

I. Air Force Planning in the Context of Joint Planning

Joint operation planning is an integrated process for orderly and coordinated problem solving and decision-making across the spectrum of conflict. In its peacetime application, the process allows the thorough and fully coordinated development of plans for operations during steady-state conditions as well as contingencies. During crises, the process is shortened as needed to support the dynamic requirements of changing events. During execution, the process adapts to accommodate changing factors in the operational environment and maximize the flexibility of operations. For today's commanders, plans are useful as necessary points of departure— planning as a process is still the most important.

Joint operation planning is conducted at every echelon of command, during peacetime as well as conflict, and across the range of military operations. Joint operation planning is accomplished through the adaptive planning and execution (APEX) system, which is "the Department of Defense- (DOD-)level system of joint policies, processes, procedures, and reporting structures, supported by communications and information technology, that is used by the joint planning and execution community to monitor, plan, and execute mobilization, deployment, employment, sustainment, redeployment, and demobilization activities associated with joint operations" (JP 5-0, Joint Operation Planning). The APEX system facilitates iterative dialogue and collaborative planning between the many echelons of command, including between the commander, Air Force forces (COMAFFOR), who usually acts as the joint force air component commander (JFACC), and the joint force commander (JFC) and other components. This helps ensure that the military instrument of national power (IOP) is employed in accordance with national priorities, and that plans are continuously reviewed and adapted to accommodate changes in strategic guidance, resources, the actions of adversaries and other actors, and the operational environment. Joint operation planning also identifies capabilities outside the DOD, and provides the means of integrating military actions with those of other IOPs and multinational partners in time, space, and purpose to create all effects necessary to achieve objectives required to attain the desired end state.

The APEX System formally integrates the activities of the entire joint planning and execution community (JPEC), which facilitates seamless transition from operational design and planning efforts to execution in times of crisis. APEX, and the joint operation planning and execution system (JOPES) technology that underpins it, provides for planning that is integrated from the national level down to theater and component levels.

See chap. six for detailed discussion of planning for joint air operations planning -- including the joint air estimate, joint operation planning process for air (JOPPA), joint targeting, and the joint air tasking cycle.

II. Steady-State Planning

The steady state is a stable condition involving continuous and recurring operations and activities with simultaneous absence of major military, crisis response, and contingency operations (Air Force Instruction [AFI] 10-421, Operations Planning for the Steady State). The steady state is characterized by shaping operations and activities at a relatively low level of intensity, urgency, and commitment of military forces. Steady-state shaping operations are designed to influence the operational environment in order to deter and prevent future conflict, mitigate operational risks, and strengthen United States and partner capabilities to respond to crises and contingencies. Steady-state planning operationalizes combatant commanders' (CCDRs strategies for their geographic theaters or global functional responsibilities. Theater and functional strategies outline a CCDR's vision for integrating and synchronizing military operations with other IOPs, as well as the activities of partner nations and international organizations, in order to achieve strategic objectives.

The DOD's principal steady-state plan is the CCDR theater campaign plan. It is the instrument through which the CCDR militarily executes his or her strategy, by comprehensively and coherently integrating steady-state activities with contingency operations. The CCDR's campaign plan conveys a design for operations that achieve prioritized theater and global campaign objectives, and serves as the integrating framework that informs and synchronizes all subordinate and supporting planning and operations.

CCDR theater campaign plans focus on steady-state activities—including military engagement, security cooperation, and other ongoing operations—considered achievable over a two- to five-year planning horizon. The delineated operations seek to generate and sustain defense posture, deter unwanted adversary behavior, and shape the operational environment so as to proactively defuse strategic problems before they become crises and resolve crises before they reach the stage requiring large-scale military operations.

At the same time, campaign plans should set the conditions for success should contingency operations become necessary. Contingency plans for responding to crises can then be derived from the overarching campaign plan as branch or sequel plans, articulating designs for supporting subsequent operations and campaigns.

COMAFFORs support steady-state planning through their own strategy documents, which outline the COMAFFOR's long-term vision for the Air Force component to the CCDR and provide an Airman's perspective on the CCDR's strategy. Component-specific activities in support or the CCDR's campaign plan are contained in the COMAFFOR's campaign support plan (CSP) and country plans. The COMAFFOR's country plans are theater security cooperation plans at the operational level that align with the CCDR's respective country plans. They focus on achieving country-level objectives related to partner relationships, capacities, and capabilities; as well as access and interoperability.

See facing page for additional discussion. For detailed guidance concerning steady-state planning, refer to AFI 10-421.

Steady-State Planning

Ref: Annex 3-0, Operations & Planning (4 Nov 16), pp. 100 to 101.

Steady-state planning operationalizes the commander, Air Force force's (COMAFFOR's) steady-state strategy. Airmen should employ a process analogous to the joint operation planning process for air to conduct steady-state planning, ensuring they employ a process that at least includes the important steps of mission analysis, design, and plan development. Steady-state planners should consider the availability of resources in developing the plan, so the plan itself should articulate the need for resources required for plan execution.

Just as the campaign plan is the combatant commander's (CCDR's) principal steady-state plan, the campaign support plan (CSP) serves the same purpose for the COMAFFOR. Responsibility for planning, execution, and assessment of the COMAFFOR CSP is typically aligned with the AFFOR staff, with the air operations center (AOC) in support. Although elements of the AFFOR staff may take lead for various parts of the planning and execution cycle, the steady state requires coordinated effort by the entire component staff. As in contingency planning, the transition from steady-state planning, typically led by the AFFOR/A5, to steady-state execution, typically led by the AFFOR/A3, should be clearly defined, documented as a key staff process, and closely followed.

The COMAFFOR CSP and country plans are unlike other deliberate plans in that they normally transition into execution. As such, the key constraint on the execution of the plan is the availability of resources—forces, funding, authorities, time, effort, etc. This is arguably the most significant difference between steady-state and contingency planning. Without addressing resource procurement, funding, and so on, steady-state tactical-level operations (and thus the achievement of desired effects) in support of the COMAFFOR CSP are not possible.

Another steady-state planning consideration, closely related to resources, is the need to take a multi-year approach to planning, execution, and assessment. By aligning the steady-state planning cycle with Department of Defense (DOD) and Air Force institutional processes for resource allocation (e.g., program objective memorandum development), Air Force planners acknowledge the vital linkage between steady-state operations and resources.

All plans developed in the steady state are considered deliberate plans, only one of which addresses crisis or contingency operations. The Air Force's Service component plan serves this need, summarizing the Air Force component's support to the CCDR's overarching operation plan (OPLAN) or concept plan (CONPLAN). Whereas the AFFOR/A5 often takes the lead in the development of deliberate Service component plans, expertise from the AOC may be necessary for its development.

Steady-state plans are normally developed using an effects-based approach, ensuring planned steady-state operations support COMAFFOR-established strategy, objectives, effects, and tasks. Geographic or functional objectives at the operational level are the centerpiece of these plans, enabling all subordinate planning and assessment. Objectives should be specific, measurable, achievable, relevant, results-oriented, and time-bound.

The Air Force CSP supports the Service's overall Title 10 responsibility to organize, train, and equip Air Force forces for employment by combatant commands. As such, this plan's primary value is to articulate a steady-state demand signal to sequentially inform institutional force planning, capabilities development, and resource allocation.

Steady-state planning products:

- COMAFFOR campaign support plan CSP
- COMAFFOR Country Plans
- Air Force Service component plan (deliberate)
- Air Force campaign support plan CSP

III. Deliberate and Crisis Action Planning

Under the larger APEX "umbrella," joint operation planning for contingencies is divided into deliberate and crisis action planning (CAP). Deliberate planning in the context of APEX is a process that is used to develop global and theater campaign plans, which operationalize a CCDR's ongoing theater or functional strategies in peacetime, as well as joint operation plans for contingencies identified in joint strategic planning documents. "Traditional" contingency plans (the type that have been developed by the JPEC for decades) are now often considered branches of ongoing CCDR theater or functional strategies. During deliberate planning, the Secretary of Defense, the Chairman of the Joint Chiefs of Staff (CJCS), and CCDRs determine the level of detail required and participate in in-progress reviews of each respective plan. This process prepares for possible contingencies based on the best available information and using forces and resources apportioned in strategic planning documents. It relies heavily on design assumptions about political and military circumstances that may prevail when the plan is implemented. Plan production generally takes six or more months and involves the entire JPEC. The Air Force Service component (the COMAFFOR's staff) usually develops supporting plans following the same process used by the JFC. During the steady state, this plan is a campaign support plan. During contingencies, this plan is the COMAFFOR's component plan.

CAP procedures are used in time-limited situations to adjust previously prepared operation plans (OPLANs) or otherwise conduct design and planning for military action Here, the crisis may occur with little or no warning, the situation will be dynamic, and time for planning may be very limited. Operational design and planning should revalidate or correct the majority of the assumptions made during deliberate planning, if accomplished. In some cases, however, commanders and their strategists must start the process with a "blank slate," accomplishing design and planning based on assumptions made in the absence of facts or the products of previous deliberate planning. An adequate and appropriate military response in a crisis demands flexible procedures keyed to the time available, rapid and effective communications, and use of previous planning and detailed databases and region analyses whenever possible. CAP often entails the positioning of forces, or at least the start of that process. CAP generally produces joint operation orders and other orders associated with the time-sensitive execution of operations.

JOPES technology and processes are still a vital, necessary part of Air Force planning, even though the joint operation planning process (JOPP) and joint operation planning process for air (JOPPA) are often accomplished separately from APEX system processes. JOPES helps planners focus on the identification and flow of resources and sequencing required to support a given course of action (COA) determined by APEX processes. Once a COA is selected, JOPES helps create detailed time-phased force and deployment data (TPFDD) to support the JFC's plan of operations. This entails reconciliation of the TPFDD with the requirements of the operation's major tasks and phasing. The areas in which the joint operation planning and JOPES processes overlap are shown in the shaded area in the figure, "The Cyclical Nature of Strategy, Design, and Planning."

There are no separate joint or Air Force procedures for deliberate and crisis action planning beyond some internal coordination and staffing procedures at the various component headquarters. When developing supporting plans, some of the steps may not be as in-depth, as they may reiterate work already done by the JFC and staff.

The contribution of JOPES processes extend beyond the TPFDD and other deployment considerations. These processes also provide a whole series of staff estimates and coordination steps, conducted by national-level agencies down through Air Force major command staffs carrying out force-provider responsibilities. Further, only JFC and Service component (e.g., the COMAFFOR's) staffs possess the information technology infrastructure to interface with many JOPES processes, thus

Service Component Planning during Contingencies

Ref: Annex 3-0, Operations & Planning (4 Nov 16), pp. 114 to 115.

There are three types of Service component plans that concern Air Force commanders and their staffs at the operational- and tactical-levels:

- Deliberate plans supporting ongoing, steady-state campaigns
- Deliberate plans supporting plans for a particular contingency
- Crisis action plans and orders supporting an imminent contingency

The first type is deliberate planning performed in support of the combatant commander's (CCDR) steady-state campaign plan. Air Force component planners, in turn, develop campaign support plans (CSP) and country plans that operationalize the commander, Air Force forces' (COMAFFOR's) theater or functional strategy in addition to supporting the CCDR's campaign plan. The second type is deliberate planning performed in support of a CCDR's operation plan (OPLAN) or concept plan (CONPLAN) for a contingency. The third type is crisis action planning performed in a contingency in support of a joint force commander (JFC) and, when applicable, a joint force air component commander (JFACC). A component-developed Air Force Service component plan is used to support both the second and third plan types. When a JFACC and associated joint air operations center (JAOC) are designated and active, the component-developed Air Force Service component plan supports the JFACC's joint air operations plan. When time available for planning is constrained, crisis action planning may produce an operation order (OPORD) rather than Air Force Service component plan.

How these plans are developed is significantly influenced by two distinct responsibilities of the COMAFFOR: operational and administrative. The operational side reflects the COMAFFOR's role as a Service component commander to a CCDR or other JFC with assigned responsibility to achieve operational objectives, effects, and tasks associated with the JFC's operations plan. This operational responsibility applies to the first plan type (deliberate steady-state), second plan type (deliberate contingency), and may apply for the third plan type (crisis action contingency). When a JFACC and JAOC are designated for crisis action contingency (as is normally the case, but not required), the operational hat is worn by the COMAFFOR in his role as the JFACC.

The Joint Operation Planning Process for Air (JOPPA)

The Air Force creates plans for contingencies and other operations using the process known as the joint operation planning process for air (JOPPA). The JOPPA is a seven-step process that essentially recapitulates the joint operation planning process (JOPP) at the component level. It culminates in the production of the joint air operations plan (JAOP) or a Service component plan, as well as supporting plans and orders. The JOPPA is the process by which commanders, Air Force forces (COMAFFORs) and their staffs create the detailed plans they require to effectively employ airpower. Since the COMAFFOR is normally also the JFACC, the JOPPA is also the joint force air component's equivalent of the joint force commander's (JFC's) JOPP and can be performed in parallel with it.

If the COMAFFOR anticipates the need for such planning, he or she may direct preparations before formal tasking is received. The JOPPA produces the JAOP and the COMAFFOR's component plan, and, as part of an ongoing battle rhythm, the guidance that helps create the air operations directive, which guides the tasking cycle.

See chap. six (pp. 6-5 to 6-18), "Planning for Joint Air Operations", for detailed discussion of the joint operation planning process for air (JOPPA).

Operations & Planning

Contingency & Crisis Action Planning

Ref: AFI 10-401, AF Operations Planning & Execution (w/Chg 4, 13 Mar '12), pp. 49 to 50

DCAPES, Force Modules, and UTCs

Air Force planners, regardless of organization, will use Deliberate and Crisis Action Planning and Execution Segments (DCAPES), force modules, and unit type codes (UTCs) during the planning process. DCAPES is the Air Force feeder to JOPES. DCAPES use is directed because it provides a variety of capabilities to Air Force planners and agencies not found in JOPES that are necessary for management and oversight of Air Force planning and execution. Force modules and UTCs are the building blocks of AEWs, AEGs, and AESs -the way the Air Force presents and sources capabilities to the JPEC.

Air Force Instruction 10-401 (w/Chg 4, 13 March 2012).

Planning, whether legacy or Adaptive has contingency and crisis action components.

1. Contingency Planning (formerly "Deliberate Planning")

Combatant commanders, their components, and supporting commands accomplish contingency planning during peacetime conditions. Planners use scenarios and threats identified in national guidance, such as the JSCP, along with the combatant commander's evaluation of their AOR, to develop a series of plans that span a wide range of operations. This formal process develops responses to potential crises, determines forces required to achieve objectives, prepares deployment plans, and continually evaluates selected courses of action (COAs). This process results in a series of formal plans within each theater that contain lists of apportioned forces and their time-phased deployment schedules. The process for contingency planning is cyclic and continual and is almost identical whether the resulting operation plan is a fully developed OPLAN, CONPLAN, or FUNCPLAN. Operations plans remain in effect until canceled or super ceded by another approved plan. While in effect, they are continuously maintained and updated.

2. Crisis Action Planning (CAP)

Crisis action planning is driven by current events in real time and normally occurs in emergencies and in the context of time-sensitive situations. Planners base their efforts on the actual circumstances that exist when crisis action planning occurs. Detailed guidance and instructions are located in JOPES Volumes I-III. Ideally, an existing contingency plan addresses the crisis situation. If there is not a contingency plan that can be used or modified to respond to the crisis, planners must start from scratch. Each MAJCOM must establish complementary procedures and must ensure adequate procedures exist for subordinate command and agency use. These procedures must be periodically exercised during joint and unilateral command post exercises and field training exercises to ensure the required capability is available. The JPEC's Global Force Management (GFM) process developed policy and procedures in support of Commander, U.S. Joint Forces Command (CDRUSJFCOM) as the DOD primary joint force provider (JFP). Commander, Air Combat Command (COMACC), as the Air Force component commander to USJFCOM, is the Air Force's primary Service force provider. MAJCOM and AEFC roles in sourcing crisis requirements will mature under GFM. The GFMB and CDRUSJFCOM will establish complementary procedures to determine sourcing recommendations and issues related to risk to sourcing other requirements, sustainability assessment and issues identified by other combatant commanders and JFCOM Service components. The AEFC and each MAJCOM must establish complementary procedures. MAJCOMs must ensure adequate procedures exist for subordinate command and agency use.

Crisis Action Planning (CAP) Orders

Ref: AF1 10-401, Air Force Operations Planning & Execution (Apr '07), pp. 53 to 54.

The Warning Order (WARNORD)

The CJCS Warning Order initiates COA development and applies to the supported command and supporting commands. It is normally published by the CJCS during Phase II planning. The WARNORD establishes command relationships (designating supported and supporting commanders) and provides the mission, objectives, and known constraints. It establishes a tentative C-day and L-hour. It may apportion capabilities for planning purposes or task the combatant commander to develop a list of forces required to confront the crisis. A warning order does not authorize movement of forces unless specifically stated. If the crisis is progressing rapidly, a planning order or alert order may be issued instead. When a WARNORD is issued, the Air Force component headquarters commander prepares a TPFDD in DCAPES for the Air Force portion of the supported commander's TPFDD in JOPES in accordance with CJCSM 3122.01A and CJCSM 3122.02B. The AEFC sources for Air Force requirements.

The Planning Order (PLANORD)

The CJCS can send a PLANORD to the supported commander and JPEC to direct execution planning before a COA is formally approved by the SecDef and President of the United States (POTUS). If the PLANORD is used in lieu of a WARNORD, the PLANORD will include a COA, provide combat forces and strategic lift for planning purposes, and establish a tentative C-day and L-hour. The PLANORD will not be used to deploy forces or increase readiness unless approved by the SecDef. When a PLANORD is issued, the Air Force component headquarters commander prepares a TPFDD in DCAPES for the Air Force portion of the supported commander's JOPES TPFDD in accordance with CJCSM 3122.01A and CJCSM 3122.02B. The AEFC sources Air Force requirements.

The Alert Order (ALERTORD)

The SecDef approves and transmits an ALERTORD to the supported commander and JPEC announcing the selected COA. This order will describe the COA sufficiently to allow the supported commander and JPEC to begin or continue the detailed planning necessary to deploy forces. If the ALERTORD is used in lieu of a WARNORD, the PLANORD will include a COA, provide combat forces and strategic lift for planning purposes, and establish a tentative C-day and L-hour. In a time-sensitive crisis, an Execute Order may be issued in lieu of an ALERTORD.

The Execute Order (EXORD)

This order is issued by the authority and direction of the Sec-Def and directs the deployment and/or employment of forces. If the EXORD was preceded by a detailed Alert Order or PLANORD, then the EXORD simply directs the deployment and employment of forces. If nature of the crisis results in an EXORD being the only order dispatched, then the EXORD must include all the information normally contained in the warning, alert, and planning orders.

The Prepare to Deploy Order (PTDO), Deployment Order (DEPORD) and Redeployment Order

Issued by the SecDef, these orders are used to prepare forces to deploy or deploy forces without approving the execution of a plan or OPORD. Prior to issuance, JFCOM develops a draft DEPORD with recommended sourcing solutions. The Joint Staff coordinates the draft DEPORD with agencies and OSD then forwards the proposed DEPORD to SecDef for approval. When a PTDO or DEPORD is issued, the AEFC, through ACC, sources Air Force requirements. Upon receipt of the CJCS Orders, the HAF Crisis Action Team (AFCAT) (or Air Force Operations Group (AFOG) if the CAT is not stood up) will transmit an order to all U.S. Air Force components and commands.

Operations & Planning

the air operations center's (AOC's) contribution to JOPES is dependent upon the COMAFFOR's staff. Specifics concerning the products of the deliberate and crisis action planning processes can be found in the JOPES/APEX manuals.

Absorbing lessons learned and adapting to them appropriately is critical to operational success. Observations should be captured after every operation in the form of lessons learned. Events should be documented in detail to provide information that improves planning and execution of future actions.

IV. The Relationship Between Operational Design and Planning

In many respects, operational design constitutes a necessary "front end" of planning, since the commander should frame the problem he or she seeks to solve and determine its scope and parameters. It logically forms the first steps of campaign, deliberate, crisis action, and other operational planning. It makes sense to determine an operation's overall end state before detailed steady-state or employment planning begins (or, for that matter, before many aspects of force deployment and sustainment planning begin). In other respects, design and planning are complementary and even overlap: Design may begin before initiation of the JOPP or JOPPA, but some portions of the mission analysis stage of the JOPP and JOPPA may provide insights needed to properly frame an operational problem. Design often begins with step 1 of the JOPP ("Initiation"), but certain formal products of contingency planning (such as warning and planning orders) may be issued after design efforts have begun but before more detailed planning has started. Design often also continues after completion of initial JOPP and JOPPA planning. There is no clear demarcation between when design ends and planning begins (or vice versa), especially during the "first round" of design and planning. Strategists often also identify possible branches and sequels at various points based on planning assumptions. In doing so, they must often make assumptions in the absence of facts in order to allow planning to continue. The need for many assumptions is typical of designing and planning for ill-structured problems.

Later, during plan execution and assessment, operational design may be conducted in concert with planning to adapt to emerging situations or behaviors. In this part of the process, commanders and strategists determine whether to implement preplanned branches or sequels, or even initiate complete re-design of an operation.

Lines of Effort

It is very helpful during design and planning to have a tool that depicts the relationship of effects to decisive points (DPs), centers of gravity (COGs), objectives, and other events and concepts, using the logic of purpose–cause and effect. Such a tool is usually arranged in proper time sequence to help commanders and strategists visualize how operations evolve and interact over time. Lines of Effort (LOE) provide just such a tool. Commanders and strategists may use LOEs to link multiple actions and effects on nodes and DPs with COGs and objectives to enhance effects-based planning efforts.

V. An Effects-Based Approach to Planning

The effects-based approach to operations (EBAO) informs every aspect of how the Air Force designs, plans, executes, assesses, and adapts operations. The effects-based approach applies as well to steady-state planning (such as campaign support plans and country plans) as it does for planning the employment of forces (as in the joint air operations plan).

See following pages (pp. 5-23 to 5-25) for further discussion of EBAO to include an overview of effects (direct, indirect, intended and unintended), objectives and actions.

Effects, Objectives & Actions (EBAO)

Ref: Annex 3-0, Operations & Planning (4 Nov 16), pp. 66 to 68.

Effects

"Effect" refers to "the physical or behavioral state of a system that results from an action, a set of actions, or another effect." Effects are elements of a causal chain that consists of tasks, actions, effects, objectives, and the end state(s), along with the causal linkages that conceptually join them to each other. "Tasks" refer to an action or actions that have been assigned to someone to be performed. Actions are the results of assigned tasks. Actions produce specific direct effects, those effects produce other, indirect effects that influence the adversary and other actors within the operational environment, and this chain of cause and effect creates a mechanism through which objectives and ultimately the end state are achieved. The end state is a set of conditions that needs to be achieved to resolve a situation or conflict on satisfactory terms, as defined by appropriate authority.

Objectives

Objectives at one level may be seen as effects at other, higher levels. Effects, however, comprise all of the results of actions, whether desired or undesired, intended or unintended, immediate or ultimate. From a military planning perspective, operations should be planned "from the top down," starting with the desired military end state, determining subordinate objectives needed to bring about that end state, then deriving the effects and causal linkages needed to accomplish the objectives, and finally determining the actions and resources necessary to create those effects. The end state should explain the operation's ultimate purpose—the outcome that is sought. The objectives and effects should explain what results are required to attain that outcome. The task and their resultant actions should explain the steps needed to achieve the required results.

Perspective is important here. What may seem like an action to the operational-level warfighter may seem like an objective to warfighters at tactical units. Conversely, what may be an objective for a component commander may seem like an action to the President of the United States.

Planners should maintain awareness of the "big picture"—how the component's effects and objectives support the joint force commander's (JFC's) effects and objectives. This is especially important during execution, where it is easy to get caught up in the details of daily processes and lose sight of the end state. For example, "gain and maintain air superiority to X degree in and over area Y for Z period" may be an objective for the joint force air component commander (JFACC), but will likely be one of the effects the JFC directs the JFACC to deliver (often stated as an execution task) in support of the notional objective "defeat enemy A's offensive into region B." In turn, the JFACC's objective may seem like an action to the President, who has given the JFC the desired effect of "defeating A's offensive" in order to accomplish his national strategic objective of restoring stability and maintaining political order in the applicable global region.

Actions

An action is performance of an activity to create desired effects. In general, there are two broad categories of actions that are relevant at the tactical and operational levels: Kinetic and nonkinetic. Examples of kinetic actions include the use of explosive munitions and directed energy weapons. Examples of nonkinetic actions include use of cyberspace weapons, an information operations radio broadcast to encourage enemy surrender, and employment of electronic warfare capabilities.

Effects

Ref: Annex 3-0, Operations & Planning (4 Nov 16), pp. 68 to 75.

There are four broad categories of effects, which often overlap. These categories are: direct,indirect, intended, and unintended. Within these categories, especially within the realm of indirect effects, there are many subcategories. Understanding some of these special types of effects is vital to an effects based approach to war.

1. Direct Effects

Direct effects are the results of action with no intervening effect or mechanism between act and outcome. They are also known as "first-order effects." In most cases they are physical, often immediate, and easy to recognize. They can usually be assessed empirically and can often be meaningfully quantified.

2. Indirect Effects

Direct effects trigger additional outcomes—intermediate effects or mechanisms that produce additional outcomes or results. These are indirect effects, sometimes also known as "second-," "third-," or "higher-order effects." Indirect effects can be categorized many ways, including physical, psychological, and behavioral. They may also occur in a cumulative or cascading manner, can occur sequentially or in parallel (since they are caused by direct effects that may be applied sequentially or in parallel), and may be intended or unintended and lethal or nonlethal. They are usually displaced from direct effects in time and space, and often can be hard to quantify or measure empirically. They are often assessed or evaluated in qualitative terms. Generally, the less direct the effect—the further removed it is in the causal chain or in time from the initial action—the harder it is to predict before the fact and measure after. Historically, it has proven extremely difficult to predict beyond third-order effects with any degree of certainty.

- **Cumulative and Cascading Effects.** Indirect effects can be achieved in a cumulative or cascading manner. Those effects that result from the aggregation of many direct and indirect effects are said to be cumulative. These effects typically flow from lower to higher levels of employment and occur at the higher levels, but they can occur at the same level as a contributing lower-order effect. Some indirect effects ripple through the adversary system, usually affecting other systems. These are called cascading effects. Typically, they flow from higher to lower levels of employment and are the result of affecting nodes that are critical to many related systems or sub-systems.
- **Physical effects** are the results of actions or effects that physically alter an object or system. Most physical effects are direct, but some may be indirect. Often, unintended or undesirable physical effects, like "collateral damage" can be major concerns in a campaign.
- **Psychological effects** are the results of actions or effects that influence the emotions, motives, and reasoning of individuals, groups, organizations, and governments. These result in changes in the outward behavior of the individual, group, organization, or government known as behavioral effects. While it is seldom possible to measure psychological effects directly, their behavioral results can be measured. Nonetheless, the intermediate psychological states leading to behaviors can be important to understanding causal mechanisms during planning. In most cases, friendly targeting actions are intended to produce—and accomplishment of objectives requires—some change in enemy behavior. Unless the enemy is destroyed outright, all such changes entail a change in the enemy's emotions, motivations, and/or reasoning.
- **Sequential and Parallel Effects.** Sequential, or serial, effects are the results of actions or effects that are imposed one after another. Parallel effects are the results

Operations & Planning

of actions or effects that are imposed at the same time or near-simultaneously. In general, it is often better to impose effects in parallel rather than sequentially. Parallel effects have greater potential for causing system-wide failures by placing stress on the enemy system in a manner that overwhelms its capacity to adapt. This is common sense—everyone is better at handling problems coming one after another from a single source than from many different sources or directions simultaneously. Some of the advantages parallel attack confers are purely physical, but many are psychological. Simultaneous stress from many sources is a major cause of psychological strain or breakdown and an effects-based approach seeks changes in enemy behavior more than it seeks simple destruction of enemy capability. Thus, as a rule, effects-based targeting should attempt to place the enemy under maximum psychological stress at all times through parallel efforts. Even if one is seeking predominantly physical effects, the psychological strain will act in synergy with the physical to have more impact than the physical effects could on their own. Another advantage of parallel operations is that they take less time to achieve desired effects and objectives.

- **Functional and Systemic Effects.** Functional effects are the direct or indirect effects of an action on the ability of a target or target system to function properly. Analysis and assessment of functional effects answer the questions, "in what ways and to what extent has the system the target is part of been affected by action taken against it?" Targets are usually elements of larger systems, even if they are systems themselves. Effects that relate to how well the targeted system functions as part of larger systems of which it is a part are systemic effects. Analysis and assessment here answer the question, "in what ways and to what extent has the system the target is part of been affected by action taken against it?"

3. Intended Effects

Intended effects are the desired, planned, and predicted outcomes of an action or set of actions. They can be direct or indirect. They should always represent a net gain in terms of accomplishing objectives or the end state.

4. Unintended Effects

Unintended effects are outcomes of an action that are not part of the original intent. These effects may be undesired or desired, presenting opportunities for exploitation. Almost all actions produce some unintended effects. These can be direct, but are usually indirect. If unplanned, they can also be desirable or undesirable from the friendly point of view, leading to outcomes that help or hinder achievement of friendly objectives. The case of the enemy commander being replaced by a more capable officer is an illustration of an undesired unintended effect. Unwanted civilian injuries or collateral damage to civilian property are examples of unintended effects that are planned, or for which risk is accepted, but which are undesired. Collateral civilian damage, of course, is a major concern for commanders today.

There is another aspect of unintended effects that is easy to overlook in planning. Even successful operations carry a cost in terms of lost opportunities. For example, destroying certain C2 or communications nodes in order to degrade enemy cohesion can remove valuable sources of friendly intelligence, or prevent transmission of surrender guidance by the adversary government. Likewise, destroying transportation nodes like bridges in order to impede enemy movement may interfere with future friendly schemes of maneuver or recovery efforts accompanying conflict resolution.

Effective planning should account for these "opportunity costs." Effective air, space, and cyberspace planning should also account for other components' schemes of maneuver, so that effects created by the airpower component are not undesired effects for the other components.

VI. Types of Plans

Ref: AFI 10-401, AF Operations Planning & Execution (w/Chg 4, 13 Mar '12), pp. 48 to 49.

Planning is a continuous, iterative, and highly structured process that allows for an orderly transition from peace to war and post-hostilities operations. The JPEC accomplishes planning. The JPEC includes the CJCS, Joint Staff and Services, the supported and supporting commands and their components, defense agencies, non-DOD departments and agencies, and Allied Commands and agencies. The resulting plans are implemented through a series of universally understood orders. These orders provide the mechanism for bringing together the resources, equipment, and personnel needed in a military response. Both contingency and crisis planning are driven by joint processes and conducted within JOPES.

Planning results in different types of plans depending on the level of detail required:

1. Operations Plan (OPLAN)

An OPLAN is a written description of the combatant commander's concept of operations to counter a perceived threat. An OPLAN includes all required annexes, appendices, and a supporting TPFDD. It may be used as the basis of a campaign plan (if required) and then developed into an operations order (OPORD).

2. Contingency Plan (CONPLAN)

A CONPLAN is an operational plan in an abbreviated format that may require considerable expansion or alteration to convert it into an OPLAN. The objective of concept planning is to develop sound operational and support concepts that can be rapidly expanded into an OPORD if necessary. Unless specified in the JSCP, detailed support requirements are not calculated and TPFDD files are not prepared. A CONPLAN may or may not have an associated TPFDD.

3. Functional Plans (FUNCPLAN)

Functional plans involve military operations in a peacetime or permissive environment. These types of plans are tasked by the JSCP for humanitarian assistance, disaster relief, peacekeeping, or counter-drug operations. Functional plans are written using the JOPES procedures and formats specified for a CONPLAN without a TPFDD.

4. Supporting Plans

Supporting plans are prepared as tasked by the supported combatant commanders in support of their contingency plans. They are prepared by supporting combatant commanders, subordinate joint force commanders, component commanders, or other agencies. These commanders may, in turn, assign their subordinates the task of preparing additional supporting plans.

5. Operation Order (OPORD)

OPORDs are prepared under joint procedures in prescribed formats during crisis action planning. They appear in the form of a directive issued by a commander to subordinates to effect the coordinated execution of an operation (e.g. air operations, training exercises, etc.). Normally, a combatant commander issues OPORDs to the Service component headquarters to effect the coordinated execution of an operation.

III. Execution

Ref: Annex 3-0, Operations & Planning (4 Nov 16), p. 78 and Volume 3, Command (22 Nov 16), pp. 33 to 35.

Plans describe the ways and means through which given ends (objectives and end states) can be achieved. Plans are carried out through a process called "execution," which involves putting into effect any courses of action, orders, or subordinate plans needed to achieve the ends specified by the governing plan. Execution takes place within the timeframe specified in the governing plan and usually encompasses some mechanism through which forces are tasked or ordered to carry out specific missions. Assessment of ongoing operations usually takes place during execution.

The process of tasking forces and generating orders for specific missions itself entails cycles of planning, execution, and assessment. Generally speaking, planning refers to activities intended to govern future operations and execution refers to actions taking place inside the timeframe spanning from whenever an order is given to carry out the governing plan to the point when the commander has decided that the operation can be terminated. Execution encompasses the commander's "battle rhythm"—the deliberate cycle of command, staff, and unit activities intended to integrate and synchronize current and future operations. The tasking cycles that govern execution vary greatly between steady-state conditions (one or two fiscal years) and contingency or crisis operations (days or hours).

I. Executing Operations

Execution of operations is an integral part of the overarching effects-based approach construct. Many Air Force operations are executed by means of a tasking cycle. The cycle is used with some modifications for tasking operations in the air, space, and cyberspace and is the heart of the Air Force battle rhythm.

Once execution begins, the commander continues to guide and influence operations through the air operations directive (AOD) (and, in some cases, equivalent space and cyberspace operations directives).

The Tasking Cycle

Many Air Force operations are executed by means of a tasking cycle. The tasking cycle creates a daily articulation of the overall airpower strategy and planning efforts. The tasking cycle is the means Airmen use to accomplish deliberate and dynamic targeting, among other requirements.

The tasking cycle develops the products needed to build and execute an air tasking order (ATO) and related products, and accomplish assessment.

Although it is presented below as six separate, sequential stages, in reality the tasking process is bi-directional, iterative, multidimensional, and sometimes executed in parallel. It is built on a foundation based on thorough joint intelligence preparation of the operational environment (JIPOE). The cycle typically consists of the following stages performed at various levels of command (illustrated in the figure on the following page, Typical Tasking Cycle).

See following pages (pp. 5-28 to 5-29) for an overview of the tasking cycle. See pp. 6-23 to 6-28 for detailed discussion of the joint air tasking cycle.

II. Contingency and Crisis Execution: The Tasking Cycle

Ref: Annex 3-0, Operations & Planning (4 Nov 16), pp. 116 to 120.

Many Air Force operations are executed by means of a tasking cycle. The cycle is used with some modifications for tasking operations in the air, space, and cyberspace and is the heart of the Air Force commander's battle rhythm. Once execution begins, the commander continues to guide and influence operations through the air operations directive (and, in some cases, equivalent space and cyberspace operations directives).

The tasking cycle creates a daily articulation of the overall airpower strategy and planning efforts. The tasking cycle is the means Airmen use to accomplish deliberate and dynamic targeting, among other requirements. The following discussion touches on targeting only as it relates to the tasking cycle and other aspects of an ongoing rhythm of operations. Conceptually, the tasking cycle–its people, processes, and products–forms the connecting link that transitions most airpower planning from the operational to the tactical level.

Typical Tasking Cycle

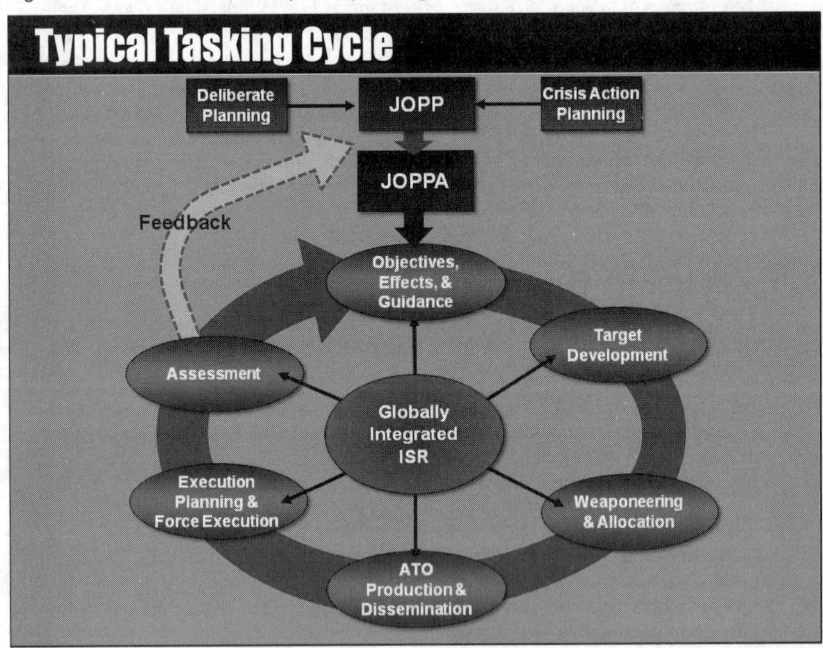

Ref: Volume 3, Command (22 Nov 16), p. 34. Typical Tasking Cycle.

The cycle is built around finite time periods that are required to plan, integrate and coordinate, prepare for, conduct, and assess operations in air, space, and cyberspace. These time periods may vary from theater to theater and much targeting effort may not be bound specifically to the cycle's timeframe, but the tasking cycle and its constituent processes drive the air operations center's (AOC's) battle rhythm and thus help determine deadlines and milestones for related processes, including targeting.

The tasking cycle develops the products needed to build and execute an air tasking order (ATO) and related products, and accomplish assessment. Although it is presented below as six separate, sequential stages, in reality the tasking process is iterative, multidimensional, and sometimes executed in parallel. It is built on a foundation based on thorough joint intelligence preparation of the operational environment. The cycle typically consists of:

- Assigning objectives, effects, and guidance
- Target development
- Weaponeering and allocation
- ATO production and dissemination
- Execution planning and force execution
- Assessment

Targeting and ATO production are essential to the tasking cycle. The tasking cycle encompasses the entire process of taking commanders' intent and guidance, determining when and where to apply force or other actions to fulfill that intent. It matches available capabilities and forces with targets (integrating this effort with the ongoing targeting cycle); puts this information into an integrated, synchronized, and coordinated order; distributes that order to all users; monitors execution of the order to adapt to changes in the operational environment; and assesses the results of that execution. The cycle is built around finite time periods that are required to plan, integrate and coordinate, prepare for, conduct, and assess operations in air, space, and cyberspace. These time periods may vary from theater to theater and much targeting effort may not be bound specifically to the cycle's timeframe, but the tasking cycle and its constituent processes drive the air operations center's (AOC's) battle rhythm and thus help determine deadlines and milestones for related processes, including targeting.

A principal purpose of the tasking cycle is to produce orders and supporting documentation that places an effective array of capabilities in a position to create desired effects in support of joint force objectives. This cycle is driven by the constraints of time and distance. For example, it takes time for ground crews to prepare aircraft for flight, for aircrews to plan missions, and for those crews to fly to the immediate area of operations from distant airfields. Likewise, commanders should have enough visibility on future operations to ensure sufficient assets and crews are available to prepare for and perform tasked missions. These requirements drive the execution of a periodic, repeatable tasking process that allows commanders to plan for upcoming operations.

The ATO (usually 24 hours in duration) and the process that develops it (usually 44-96 hours in duration) are a direct consequence of these physical constraints.

The ATO articulates tasking for joint air, space, and cyberspace operations (unless there are separate space and cyberspace tasking orders) for a specific period, normally 24 hours. Detailed planning generally begins 72 hours prior to the start of execution to properly assess the progress of operations, anticipate enemy actions, make needed adjustments to strategy, and enable integration of all components' requirements. The actual length of the tasking cycle may vary from theater to theater. Length should be based upon joint force commander (JFC) guidance, the commander, Air Force forces' (COMAFFOR's) direction, and theater needs. The length should be specified in theater standard operating procedures or other directives. If the length is modified for a particular contingency, this should be specified in the JFC's operation plan or operation order, in the joint air operations plan, or the COMAFFOR's component plan. The key to both the flexibility and versatility of the tasking process (and both deliberate and dynamic targeting and collection) is a shared understanding among the functional components of anticipated operations in all domains during the period the relevant orders and directives cover. Misperceptions may arise because other components may not have visibility on the wide variety of missions tasked to the COMAFFOR in support of the JFC's objectives and because airpower assets are often tasked to simultaneously conduct missions supporting overlapping operational phases. This shared understanding is largely accomplished by ensuring component liaisons are properly positioned during planning and execution.

See pp. 6-23 to 6-28 for detailed discussion of the joint air tasking cycle. See chap. seven for discussion of targeting and the targeting cycle.

III. Steady-State Execution

Ref: Annex 3-0, Operations & Planning (4 Nov 16), pp. 102 to 103.

Just as with combatant commanders' campaigns and country plans, the commander, Air Force forces' (COMAFFOR's) campaign support plans (CSPs) and country plans normally transition into execution annually. Because of the multi-year nature of these plans, at least three COMAFFOR CSPs may be in various levels of a life cycle at any given time: one in development, one preparing to execute, and one being executed during the current fiscal year (FY). There may also be one or more past FYs being assessed in order to appropriately adapt COMAFFOR's strategy.

COMAFFOR CSP "Battle Rhythm"

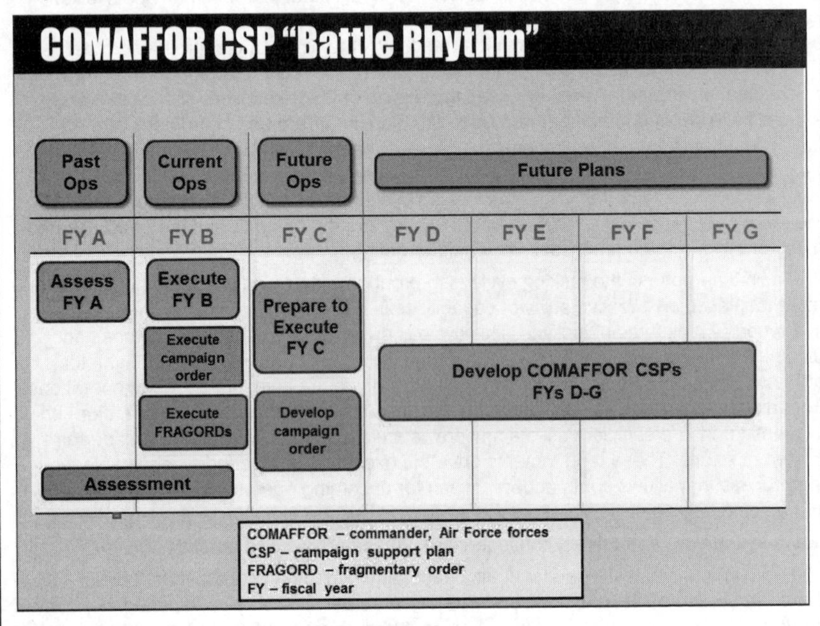

Ref: Annex 3-0, Operations & Planning (4 Nov 16), p. 102.

This is called a "battle rhythm" in this context because it is analogous to the day-to-day battle rhythm managed by the joint force air component commander during combat or contingency operations, even though its cycles are measured in years rather than days or hours.

Execution is normally preceded by preparation activities, facilitating the formal transition from planning led by the Air Force forces (AFFOR)/A5 to execution led by the AFFOR/A3. Since execution of steady-state operations largely consists of tactical-level actions, the prepare stage also defines the interface between operational-level entities (e.g., AFFOR staff) and tactical-level units (e.g., mobility support advisor squadron, air advisors). During preparation, tactical planning occurs for the missions and activities that support the achievement of operational-level effects as defined in the COMAFFOR CSP and COMAFFOR country plans. The final actions in the prepare stage occur at the operational-level. Resources requested in the planning stage must be received and distributed before execution can occur. Finally, the COMAFFOR authorizes mission execution with an execute order (EXORD) or a campaign order (also often known as a theater campaign order). The fragmentary order may be used to modify or amend an EXORD or campaign order. During execution, the AFFOR/A3 staff or air operations center monitor mission execution in real time.

IV. Assessment

Ref: Annex 3-0, Operations & Planning (4 Nov 16), pp. 79 to 94.

Assessment is a continuous process that measures the overall effectiveness of employing joint force capabilities during military operations. It is also the determination of the progress toward accomplishing a task, creating an effect, or achieving an objective (Joint Publication [JP] 3-0, Joint Operations). The purpose of assessment is to support the commander's decision-making process by providing insight into the effectiveness of the strategy and accompanying plans. Many types of assessment exist, and may be used in support of operations, but assessment in this document refers to activities that support the commander's decision-making process. In an effects-based approach, assessment should provide the commander with the answers to these basic questions:

- Are we doing things right?
- Are we doing the right things?
- Are we measuring the right things?

The first question addresses the performance of planned airpower operations by assessing the completion of tasks. The second question addresses the level at which the commander's desired effects are being observed in the operational area and prompts examination of the links between performance and effects. The third question addresses the process of assessment itself and the importance of understanding how one chooses to measure the links between performance, cause, and effect. When determined properly, the answers to these questions should provide the commander with valid information upon which to base decisions about strategy.

While often depicted as a separate "stage" of the tasking cycle for conceptual clarity, assessment is actually interwoven throughout operational design, planning, and execution. The assessment process should begin as the broad strategy is laid out (including development of assessment criteria), continue through detailed planning (with the development of metrics and data sources), and extend to evaluation of measures during and after execution. This process is iterative as assessment results influence future strategy and planning.

Assessment consolidates data from many sources and summarizes that data clearly, concisely, and in context. It should follow a rigorous, defensible analytical process that provides commanders and planners the ability to view details of methods used and results produced. It communicates relevant uncertainty in the data and the associated risks. In short, assessment provides analytically supportable judgments on a commander's strategy.

Editor's Note: See pp. 7-53 to 7-58 for related discussion of tactical-level assessment (combat assessment) as related to targeting from Annex 3-60.

I. Levels of Assessment

Ref: Annex 3-0, Operations & Planning (4 Nov 16), pp. 80 to 82.

Assessors perform many types of assessment across the strategic, operational, and tactical levels to inform a wide array of decisions. The figure, "Common Levels and Types of Assessment" displays some common types of assessment and, broadly, the levels where each would most likely be applied (the depiction is not all-inclusive). The figure also shows the level of commander who commonly directs a given type of assessment (e.g., the joint force commander [JFC] and joint force air component commander [JFACC]). At all levels–but especially at the operational level–the commander, Air Force forces (COMAFFOR), JFACC, and respective staffs should observe how the JFC takes information "on board" and craft assessment products that convey the Airman's perspective without seeming "air-centric" or presenting a biased view.

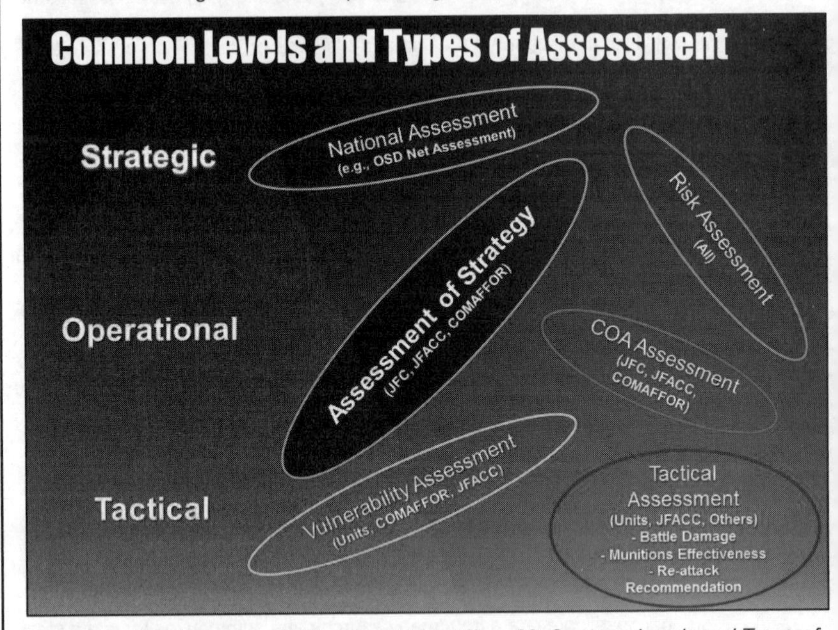

Ref: Annex 3-0, Operations & Planning (4 Nov 16), p. 80. Common Levels and Types of Assessment.

A. Tactical-Level Assessment

Tactical-level assessment is generally performed at the unit or joint force component level and typically measures physical, empirical achievement of direct effects. Combat assessment (CA) is an umbrella term covering battle damage assessment (BDA), munitions effectiveness assessment (MEA), and recommendations for re-attack (RR).

BDA is the estimate composed of physical and functional damage assessment, as well as target system assessment, resulting from the application of lethal or nonlethal military force. BDA consists of three phases. Phase I BDA consists of reporting physical damage (kinetic) or other changes (nonkinetic) to the target and, if possible, evaluating the physical damage or change to the target quantitatively or qualitatively. Phase II BDA measures what effect the weapon had on that individual target and to what extent it can perform its intended function. Phase III BDA then measures the effect of striking a particular target on the overall target system (e.g., what effect does taking out a command and control [C2] node have on the overall combat capability of an integrated air defense

Operations & Planning

system? This might relate to the overall effect of gaining and maintaining air superiority). MEA evaluates whether the selected weapon or munition functioned as intended. MEA is fed back into the planning process to validate or adjust weaponeering and platform selections. RR and future targeting recommendations merge the picture of what was done (BDA) with how it was done (MEA), comparing the result with predetermined measures of effectiveness, to determine the degree of success in achieving objectives and to formulate required follow-on actions, or indicate readiness to move on to new tasks.

Another assessment consideration at the tactical level is estimated damage assessment (EDA). EDA is a type of physical damage assessment; it anticipates damage using the probability of weapon effectiveness to support estimated assessments and allows commanders to accept risk in the absence of other information.

Tactical-level assessment should also be accomplished following tactical employment of nonkinetic actions and non-offensive capabilities. Examples include military information support operations (MISO; e.g., Commando Solo missions), public affairs (PA; e.g., media engagements), cyberspace operations (e.g., temporary utility outages), operations security (OPSEC; e.g., signature management), etc.

See pp. 7-53 to 7-58 for related discussion of tactical-level assessment (combat assessment) as related to targeting from Annex 3-60.

B. Operational-Level Assessment

Operational-level assessment is the component's evaluation of whether its objectives—at the tactical and operational levels—are being achieved. Operational assessment addresses effects, operational execution, environmental influences, and attainment of success indicators for the objectives to help the COMAFFOR/JFACC decide how to adapt the component's portion of the joint force strategy. Assessment at this level begins to evaluate complex indirect effects, track progress toward operational and strategic objectives, and make recommendations for strategy adjustments and future action extending beyond tactical re-attack. Assessment at this level often entails evaluation of course of action (COA) success, assessment of the progress of overall strategy, and joint force vulnerability assessment. Operational assessment should also include evaluation of changes to key parameters of adversary force performance, changes in adversary capabilities, and what the adversary is doing to limit the effects of friendly actions and to overcome friendly strategy. These are commonly performed by joint force component commanders and the JFC and their staffs.

Operational-level assessment evaluates a wide range of data: Quantitative and qualitative, objective and subjective, observed and inferred. Some measures can be expressed empirically (with quantitative measures); others, like psychological effects, may have to be expressed in qualitative or subjective terms. Both rely on extensive data and analysis from federated intelligence partners, including other US government agencies and multinational partners.

C. Strategic-Level Assessment

Strategic-level assessment addresses issues at the joint force ("theater strategic," as in bringing a particular conflict to a favorable conclusion) and national levels (enduring security concerns and interests). It involves a wide array of methodologies, participants, and inputs. The President and Secretary of Defense rely on progress reports produced by the combatant commander or other relevant JFC, so assessment at their levels often shapes the nation's, or even the world's, perception of progress in an operation. This places a unique burden on assessors, planners, strategists, and commanders to be accurate, meaningful, and to complete their analysis and communicate results clearly and logically.

The time frames considered by the various assessment types may vary widely, from rather short intervals at the tactical level to longer time horizons at the strategic level, even reaching well beyond the end of an operation, as lessons learned are determined and absorbed. The relationship among the various assessment types is not linear, with outputs from one type often feeding multiple other types and levels.

II. Assessing Strategy

The purpose of assessing strategy is to give commanders dependable insights into whether their strategy is effective and to measure progress toward the end state the commander is tasked to deliver. This type of assessment does not just entail assessment at the strategic level, but can be conducted for any commander from the tactical through the strategic level and should address the four main components of a strategy:

- **Ends**—The commander's end state and the objectives required to obtain it. These are generally derived from the commander's intent statement.
- **Ways**—The tasks or actions undertaken to help create the effects that achieve the ends, as generated during the detailed planning process.
- **Means**—The resources put toward accomplishing the ways. The doctrine, organization, training, materiel, leadership and education, personnel, and facilities (DOTMLPF) construct is often a useful source for examining and developing the means.
- **Risk**—The cost and amount of uncertainty and vulnerability the commander is willing to accept in executing the strategy.

It is critical to integrate the assessment process with strategy design and plan development. The assessment process begins with a review and analysis of lessons learned from previous operations, continues through operational design (where broad assessment criteria are often decided upon), detailed planning (where specific measures and indicators are usually selected to accompany objectives, effects, and tasks), and extends to evaluation of measures and indicators once tasks have been accomplished. Some forms of assessment continue long after the particular conflict or operation has concluded, supporting, for example, munitions effectiveness assessment and the lessons learned process.

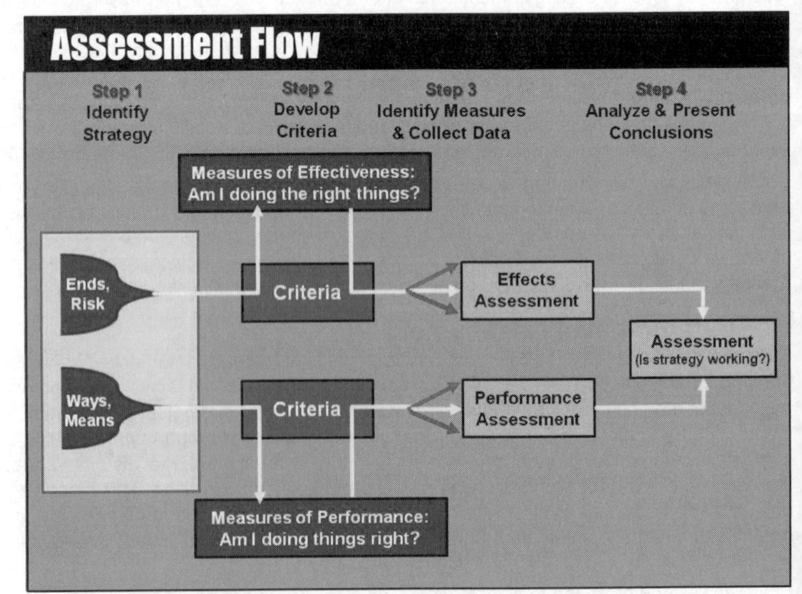

Ref: Annex 3-0, Operations & Planning (4 Nov 16), p. 84. Assessment Flow.

Assessment considers all these components, with the goal of developing insights into whether a strategy is working and what areas may need to be re-evaluated if that strategy is not working. The figure, "Assessment Flow" depicts this strategy-centric approach to assessment, which applies to operations during steady-state conditions, as well as during contingencies and major operations.

Throughout the assessment process, the assessor's focus should remain on informing the relevant commander's decisions. Even though planners may document various forms of guidance, including commander's intent, the assessment team should work to derive assessable effects from these statements. Often the commander's intent is written in terms of what operations the commander plans to undertake and not in terms of the conditions that they hope will result from these operations. Thus, planning for assessment should begin in dialogue with the commander during the design process. Assessment is also iterative, working to converge on a reasonably assessable commander's end state. In addition, understanding the objectives and tasks of the commander's boss is crucial in forming a comprehensive assessment.

Given the fluid nature of complex military operations involving higher-order effects, judgment should be an intrinsic part of any assessment. Instead of developing criteria or measures that take all judgment out of the process, the goal is to build a framework for the development of logically defensible judgments, which often involve qualitative (unquantifiable) and even subjective elements.

III. Assessment Criteria

Criteria define the attributes and thresholds for judging progress toward the end state and accomplishment of required tasks. Development of assessment criteria is the critical component of the assessment process and should be accomplished before specific measures or data requirements are defined. Developing measures without a clear understanding of how they fit into a judgment of the effectiveness of an overall strategy often leads to laborious data collection and analysis processes that provide little to no value to decision-makers. Spending additional time to thoroughly consider and develop meaningful and relevant assessment criteria help avoid this pitfall.

Criteria help focus data collection by ensuring that assessment measures relate clearly to the elements of the strategy being assessed. As data are collected, the criteria translate those data into meaningful insights on the commander's strategy, which may be presented in a variety of ways to visually display progress (or lack thereof) to the commander. Criteria should objectively indicate trends of significance and should be things that can be measured by known means. Determining them prior to commencement of operations allows for the establishment of baseline values for friendly and adversary forces and actions, which will facilitate objective reporting of changes, as well as rates of change.

All criteria have strengths and limitations. Which is used will depend in some part on the personality and preferences of the Commander. However, a variety of means should be used to comprehensively display progress toward (or away from) objectives and avoid losing relevant data by artificial form limitations. Criteria should be developed for the ends, ways, and means at each level of assessment. Well-written criteria should adhere to some basic attributes:

- Relevant to the effect or action being assessed
- Mutually exclusive across the assessment categories (e.g., good, marginal, poor) for a given effect or action assessed
- Collectively exhaustive across the range of outcomes for a given effect or action.
- Well-defined. Specific and relevant definitions should be developed for any confusing or ill-defined terms used in the criteria. Planners should attempt to define success thresholds and the boundaries between assessment categories objectively whenever possible (e.g., what are the criteria for transition between the 'good' and 'marginal' categories?). Nonetheless, judgment is always necessary when assessing the overall strategy.

IV. Assessment Measures

Ref: Annex 3-0, Operations & Planning (4 Nov 16), pp. 88 to 90.

Assessment measures are simply the data elements that, via the criteria, provide insight into the effectiveness of the commander's strategy. Assessment measures are commonly divided into two types:

- **Measure of Performance (MOP)** — A criterion used to assess friendly actions that are tied to measuring task accomplishment.
- **Measure of Effectiveness (MOE)** — A criterion used to assess changes in system behavior, capability, or operational environment that is tied to measuring the attainment of an end state, achievement of an objective, or creation of an effect.

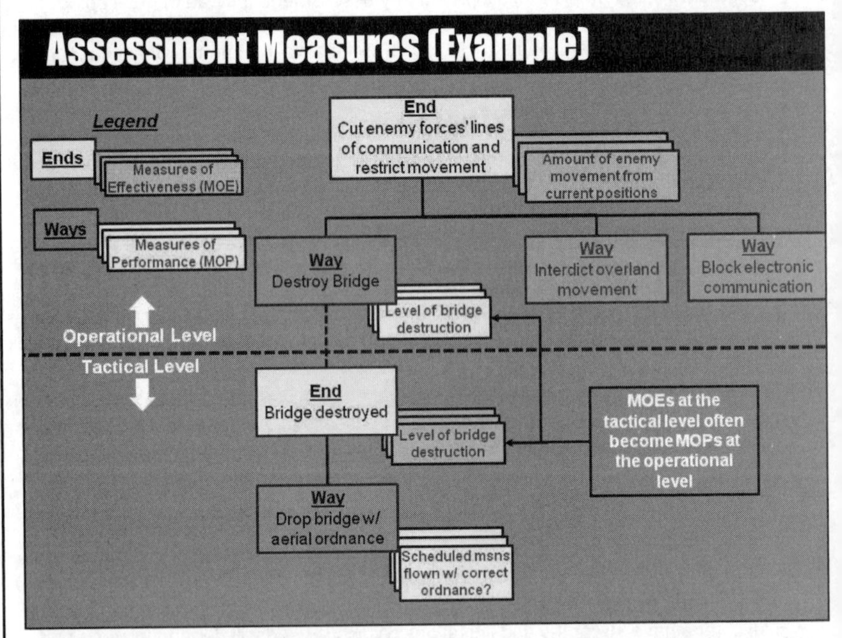

Ref: Annex 3-0, Operations & Planning (4 Nov 16), p. 88. Assessment Measures – An Example.

MOPs address the ways and means employed during execution to help achieve desired effects; they indicate progress toward accomplishing planned tasks or actions. MOEs assess progress toward creating desired effects and thus achieving the objectives and end state (Simply put, MOPs help tell us if we are doing things right; MOEs help tell us if we are doing the right things).

The distinction between MOEs and MOPs can depend on their context within the commander's strategy. The exact same measure can be an MOP for one commander and an MOE for another, lower echelon commander. The figure, "Assessment Measures—An Example" illustrates a practical application of this delineation.

Developing good measures is an art, though there are some general guidelines that can aid in developing high-quality measures:

Measures should be relevant and necessary. Measures should relate to the effect or task they are intended to describe and should feed directly into the already-es-

Operations & Planning

tablished criteria. Collection of irrelevant measures that do not shed light on the effectiveness of the commander's strategy is a misuse of valuable time and resources. Focusing primarily on collecting the data necessary to apply to the developed criteria should help avoid the creation of superfluous measures.

- **Bad Measure**: no friendly fighter losses.
- **Better Measure**: number of friendly fighters destroyed or damaged by enemy air defenses.

Measures should represent a scale, not a goal or objective. Metrics developers may be tempted to write a goal or criterion as a measure. Instead, the goal should be included in the criteria in accordance with the commander's risk tolerance and thresholds. Operators and planners should establish these goals (objectives) in coordination with the assessors.

The data satisfying a measure should be observable, or at least inferable. The measurements can be quantitative (numerical) or qualitative (non-numerical). In general, the more objectively measurable the better, but commanders and planners should avoid "the numbers trap:" blindly using rates, numbers, and other quantitative metrics, especially in assessing effects, since their seemingly "empirical" and quantified elements may be based on wholly subjective assumptions and the number may be meaningless—thus they may often lack direct linkages to the objectives or ends outlined in the strategy, while sometimes also imparting an illusion of "scientific validity" merely because they are quantified. Examples:

- **Bad Measure**: civilian populace attitude toward stability forces.
- **Better Measure (Quantitative)**: percentage of surveyed civilian population giving "favorable" rating to stability forces; number of riots and civil disturbances in response to friendly force activities; amount of enemy propaganda, graffiti, and the like discovered; and so on.
- **Bad Measure**: progress towards opening new air base.
- **Better Measure (Qualitative)**: current phase of air base stand-up (secured land, runway operational, 30-day sustainment capability in place, long-term sustainment capability in place).

Measures should be clear and concise. They should be written in plain language so that someone with no prior knowledge of the measures can still understand the data requirements. Examples:

- **Bad Measure**: status of enemy fighters.
- **Better Measure**: number or percentage of enemy fighters confirmed destroyed or rendered combat-ineffective.

Measures should be drafted during planning so that associated intelligence collection needs may inform surveillance and reconnaissance requirements. Measures may need to be refined or amended during the tasking cycle as the operational situation changes. Selection of assessment measures is an iterative, ongoing effort. Measure the entire plan, but do not overdo it. All elements of the strategy should be measured, and there may be multiple measures required to fully address the relevant criteria. However, attempting to assess too many measures can paralyze the assessment effort. Consider the value to the end result before adding more measures. Also consider what measures are readily available through immediate analysis of mission reports and planned collection tasking, rather than addressing new collection requirements. After assessors have built the entire set of measures, they should conduct a final review to identify those measures that have less relative importance/contribution or take inordinate effort relative to the insight provided, and remove them from the set. In general, assessment teams should prioritize their efforts to best support the commander's decision-making needs.

V. Steady-State Assessment

Ref: Annex 3-0, Operations & Planning (4 Nov 16), pp. 104 to 105.

The assessment of operations during steady-state conditions informs the commander concerning progress toward closing the gap between the commander, Air Force forces' (COMAFFOR's) security cooperation and other steady-state objectives and the associated baselines established in planning. This assessment influences COMAFFOR decision making with respect to resource allocation, prioritization, future planning guidance, future strategy revisions, interaction with partner nations, risk management, force protection, and other potential issues involving the commander and senior leadership. Operational-level assessment is informed by tactical-level assessments.

Assessment begins in planning, with the key being the establishment of objectives that adhere to "SMART" criteria (specific, measurable, achievable, relevant and results-oriented, and time-bound) in the steady-state plan. Effects describe the conditions necessary to achieve COMAFFOR objectives. Tasks describe friendly actions to create effects. Measures of effectiveness are used to assess end states (if specified), effects, and objectives. Measures of performance are used to assess the accomplishment of tasks.

Experience has shown that developing an assessment approach after the plan is complete, or while in execution, is ineffective. The development of an assessment annex to the steady-state plan is the preferred technique to support the commander's operation assessment process.

The AFFOR/A5 normally provides the COMAFFOR a comprehensive campaign support plan (CSP) assessment on a periodic basis. As the lead developer of the steady-state plans, the AFFOR/A5 is best qualified to summarize progress toward the achievement of COMAFFOR objectives, effects, and tasks associated with the CSP or country plans. The entire AFFOR staff should support the A5 in this effort.

Steady-state assessment products:

- **After-action report.** After-action reports summarize an entire military operation or a steady-state activity following its completion.
- **Periodic operation assessment updates to the COMAFFOR**. As steady-state operations are executed, the commander receives periodic updates on mission execution. These updates influence the commander's ongoing informal assessment of steady-state operations.
- **COMAFFOR CSP operation assessment**. On a recurring, predictable schedule, the commander requires a formal assessment of plan execution. The AFFOR staff presents its analysis of progress toward the accomplishment of plan objectives; however, only the commander can make the final assessment. The formal assessment results in COMAFFOR decisions related to future steady-state operations.
- **COMAFFOR country plan operation assessments**. Country plans are typically assessed at a level below the COMAFFOR, but these assessments may inform the COMAFFOR CSP assessment.
- **Tactical-level (mission, event, etc.) operation assessments**. Operation assessment, as a bottom-up process, initiates with tactical-level assessments. Mission, events, and activities at the tactical level are assessed conceptually in the same way as operational-level plans. These assessments help determine progress toward the achievement of tactical-level objectives, effects, and tasks. Tactical-level assessments inform operational-level assessments, which, in turn, inform strategic-level assessments.

VI. Operation Assessment during Contingencies and Crises

Ref: Annex 3-0, Operations & Planning (4 Nov 16), pp. 129 to 130.

Assessment is a vital part of any operation. Commanders, assisted by their staffs, and subordinate commanders, interagency and multinational partners, and other stakeholders, should continuously monitor the operational environment and assess the progress of ongoing operations toward the desired end state.

Operation assessment is a continuous process that supports decision making by measuring the progress toward accomplishing tasks, creating desired effects, and achieving objectives. It supports judicious allocation of resources in order to make operations more effective. It also analyzes risks, opportunities, gaps, and trends in ongoing operations. In general, any operations assessment framework should organize intelligence and operational data, analyze that data, and communicate recommendations to a decision maker.

The operation assessment process helps to frame the clear definition of tasks, desired effects, objectives, and end states, and gives the commander's staff a method for selecting the information and intelligence requirements (including commander's critical information requirements) that best support decision making.

The process consists of the following steps:

- **Identify information and intelligence requirements**. During planning, acquiring a baseline understanding assists in setting objectives and determining thresholds for success and failure.
- **Develop or modify the assessment plan**, which should link information and intelligence requirements to appropriate measures and indicators, and contain a collection plan to gather appropriate data.
- **Collect Information and intelligence**. During execution, forces use the collection plan and defined reporting procedures to gather information about the environment and ongoing operations.
- **Conduct periodic or event-based assessment**. Commanders and their staffs normally conduct assessment based on events or at specified intervals in the course of an operation.
- **Conduct change reporting**. Commanders are especially interested in learning how either friendly or adversary behavior has changed from expectations or norms established earlier in this or other operations. While many initial reports may prove false or are regarded as outliers, once a pattern of change is reliably perceived or discerned by assessment, it should be highlighted for commanders' attention.
- **Provide feedback and recommendations**. Assessment reports inform the commander and other stakeholders about current conditions and communicate progress toward desired objectives and end states.

For more in-depth information on this emerging area of joint doctrine, refer to Joint Doctrine Note 1-15, Operation Assessment (15 January 2015).

Refer to JFODS5: The Joint Forces Operations & Doctrine SMARTbook (Guide to Joint, Multinational & Interorganizational Operations) for further discussion. Topics and chapters include joint doctrine fundamentals (JP 1), joint operations (JP 3-0), joint planning (JP 5-0), joint logistics (JP 4-0), joint task forces (JP 3-33), information operations (JP 3-13), multinational operations (JP 3-16), interorganizational cooperation (JP 3-08), plus more!

VII. Assessment Interpretation

The purpose of assessment is not merely to report on the measures, but rather to provide analytically supported insights into the effectiveness of the commander's strategy and information with which to make decisions. There are numerous analytic techniques available to summarize data analysis in performing effective assessment. The technique chosen should be tailored to the operational environment, taking into account such factors as the pace of operations, available expertise, and reachback support capabilities. Assessors should also take into account the level of warfare and the commander's primary concerns. The figure, "Relation Between Performance and Effects Assessment," provides a framework with which to compare the effect and performance assessments when determining the level of objective achievement.

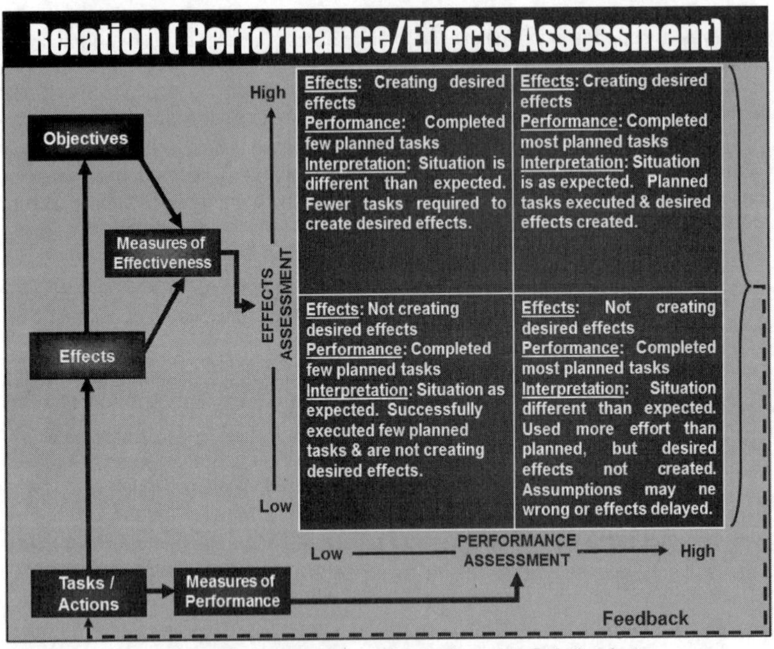

Ref: Annex 3-0, Operations & Planning (4 Nov 16), p. 92. Relation between Performance and Effects Assessment.

Overall, assessment interpretation can be broken into two major types: effects and performance assessment. Effects assessment, based on measures of effectiveness, should provide the commander with the overall picture of progress toward objective or end state achievement. Performance assessment, based on measures of performance, should provide commanders with an overall picture of how well their forces are executing the strategy's ways and means.

The relationship between effects assessment and performance assessment can be characterized in several basic ways. The scores may be similar, the performance assessment may be higher than the effect assessment, or the effects assessment may be higher than the performance assessment.

A significant consideration when interpreting effectiveness and performance results is that complex systems often begin internal change without showing outward signs that are measurable to observers. It is thus often necessary for commanders, planners, and strategists to counsel patience in following a particular course of action to allow time for desired changes to work their way through targeted systems and manifest themselves as desired behaviors in the operational environment.

Chap 6

Planning for Joint Air Operations

Ref: JP 3-30, Command and Control of Joint Air Operations (Feb '14), exec. summary.

The JFC's estimate of the operational environment and articulation of the objectives needed to accomplish the mission form the basis for determining components' objectives. The JFACC uses the JFC's mission, commander's estimate and objectives, commander's intent, CONOPS, and the components' objectives to develop a course of action (COA). When the JFC approves the JFACC's COA, it becomes the basis for more detailed joint air operations planning—expressing what, where, and how joint air operations will affect the adversary or current situation.

The Joint Air Estimate

The joint air estimate is described as a process of reasoning by which the air component commander considers all the circumstances affecting the military situation and decides on a COA to be taken to accomplish the mission. The joint air estimate reflects the JFACC's analysis of the various COAs that may be used to accomplish the assigned mission(s) and contains the recommendation for the best COA.

See pp. 6-3 to 6-4.

The Joint Operation Planning Process for Air (JOPPA)

The JFACC is responsible for planning joint air operations and uses the joint operation planning process for air (JOPPA) to develop a JAOP that guides employment of the air capabilities and forces made available to accomplish missions assigned by the JFC. JOPPA follows the joint operation planning process found in *Joint Publication 5-0, Joint Operation Planning,* with specific details for joint air operations. JOPPA drives the production of the JAOP and supporting plans and orders.

See pp. 6-5 to 6-18.

Joint Targeting

The JFC will normally delegate the authority to conduct execution planning, coordination, and deconfliction associated with joint air targeting to the JFACC and will ensure that this process is a joint effort. Targets scheduled for attack by component air capabilities and forces should be included on an ATO for deconfliction and coordination.

See pp. 6-19 to 6-22.

The Joint Air Tasking Cycle

The joint air tasking cycle process provides an iterative, cyclic process for the planning, apportionment, allocation, coordination, and tasking of joint air missions and sorties within the guidance of the JFC. The joint air tasking cycle is synchronized with the JFC's battle rhythm. The full joint air tasking cycle, from JFC guidance to the start of ATO execution, is dependent on the JFC's and JFACC's procedures. The precise timeframes should be specified in the JFC's operation plan or the JFACC's JAOP.

See pp. 6-23 to 6-28.

Planning Considerations (Joint Air Operations)

Ref: JP 3-30, Command and Control of Joint Air Operations (Feb '14), pp. xiii to xiv.

Intelligence, Surveillance, and Reconnaissance (ISR) Considerations

The JFC will normally delegate collection operations management for joint airborne ISR to the JFACC to authoritatively direct, schedule, and control collection operations for use by the J-2 [intelligence directorate of a joint staff] in associated processing, exploitation, and reporting. The JAOC should request ISR support from the JFC or another component if available assets cannot fulfill specific airborne ISR requirements.

Air Mobility Considerations

Integrating air mobility planning into the JAOP and monitoring mission execution is normally the responsibility of the AMD chief supported by a team of mobility specialists in the JAOC. Intratheater airlift and theater refueling assets may be attached to a joint task force, with operational control normally delegated down to the appropriate Service component commander (usually the commander, Air Force forces). The director of mobility forces (DIRMOBFOR) is normally a senior officer who is familiar with the AOR and possesses an extensive background in air mobility operations. Operationally, the DIRMOBFOR exercises coordinating authority for air mobility with commands and agencies within and external to the joint force. Specifically, the DIRMOBFOR coordinates with the JFACC's JAOC, Air Mobility Command's 618 Operations Center Tanker Airlift Control Center, and joint movement center/joint deployment and distribution operations center to expedite the resolution of air mobility issues.

Unmanned Aircraft Systems (UASs) Considerations

Unmanned aircraft systems (UASs) should be treated similarly to manned systems with regard to the established doctrinal warfighting principles. Several characteristics of UASs can make C2 particularly challenging:

- UAS communication links are generally more critical than those required for manned systems.
- UASs may be capable of transferring control of the aircraft and/or payloads to multiple operators while airborne.
- Most larger UASs have considerably longer endurance times than comparable manned systems.
- Compliance with the airspace control order is critical. Unlike manned aircraft, UASs cannot typically "see and avoid" other aircraft.

Personnel Recovery (PR) Considerations

Since personnel recovery (PR) often relies on air assets to accomplish some of the PR execution tasks, coordination between the joint personnel recovery center (JPRC) and JAOC is essential. The JPRC is responsible for providing the information that goes into the PR portion of the ATO special instructions.

Command and Control of Space Forces

Space forces typically operate in general or direct support to other JFCs. Geographic combatant commanders may designate a SCA and delegate appropriate authorities for planning and integrating space requirements and support for the theater.

I. The Joint Air Estimate

Ref: JP 3-30, Command and Control of Joint Air Operations (Feb '14), chap. 3.

The joint air estimate is described as a process of reasoning by which the air component commander considers all the circumstances affecting the military situation and decides on a COA to be taken to accomplish the mission. The joint air estimate is often produced as the culmination of the COA development and selection stages of the joint operation planning process described below. The joint air estimate reflects the JFACC's analysis of the various COAs that may be used to accomplish the assigned mission(s) and contains the recommendation for the best COA.

Joint Air Estimate

Operational Description

- **Purpose of the operation**
- **References**
- **Description of military operations**

Narrative -- Five Paragraphs

- **Mission**
- **Situation and courses of action**
- **Analysis of opposing courses of action (adversary capabilities and intentions)**
- **Comparison of friendly courses of action**
- **Recommendation or decision**

Remarks

- **Remarks -- Site plan identification number of the file where detailed requirements have been loaded into the Joint Operation Planning and Execution Systemn (JOPES)**

Ref: JP 3-30, Command and Control of Joint Air Operations, fig. III-2, p. III-3.

The JFACC's estimate of the situation is often produced as the culmination of the air COA development and selection stages of the JOPPA. It can be submitted in response to or in support of creation of a JFC's estimate of the situation. It should also be used to assist in creation of the JAOP and daily AODs (as required). It reflects the JFACC's analysis of the various air COAs that may be used to accomplish the assigned mission(s) and contains the recommendation for the best air COA. The estimate may contain as much supporting detail as needed to assist further plan development, but if the air estimate is submitted to the JFC or CCDR for a COA decision, it will generally be submitted in greatly abbreviated format, providing only the information essential to the JFC for arriving at a decision.

See following page (p. 6-4) for a joint air estimate of the situation template. It is a notional example of a joint air estimate in paragraph format. Use of the format is desirable, but not mandatory and may be abbreviated or elaborated where appropriate. It is often published in message format.

Joint Air Estimate of the Situation Template

Ref: JP 3-30, Command and Control of Joint Air Operations (Feb '14), app. B.

A. Mission. State the assigned or deduced mission and its purpose.

- JFC's mission statement (from the JFC's estimate), or other overarching guidance if the latter is unavailable
- JFACC's mission statement. Include additional language indicating how overarching guidance will be supported, as required

B. Situation and Courses of Action.

1. Commanders' Intent

- JFC's intent statement, if available (or other overarching guidance stipulating the end state, as required)
- JFACC's intent statement

2. Objectives. Explicitly state air component objectives and the effects required to support their achievement. Include as much detail as required to ensure that each objective is clear, decisive, attainable, and measurable.

3. Summary of the Results of JIPOE. Include a brief summary of the major factors pertaining to the characteristics of the operating environment and the relative capabilities of all actors within it that may have a significant impact on alternative air COAs.

4. Adversary Capability. Highlight, if applicable, the adversary capabilities and psychological characteristics that can seriously affect the accomplishment of the mission, giving information that would be useful in evaluating the various air COAs. This section should describe, at a minimum, the enemy's most likely and most dangerous potential COAs.

5. Force Protection Requirements. Describe potential threats to friendly forces, including such things as the threat of terrorist action prior to, during, and after the mission that can significantly affect accomplishment of the mission.

6. Own Courses of Action. List air COAs that offer suitable, feasible, and acceptable means of accomplishing the mission. If specific air COAs were prescribed in the WARNING ORDER, they must be included. For each air COA:

- Combat forces required. List capabilities needed, and, if applicable, specific units or platforms. For each, list the following, if known:
    1. Force provider
    2. Destination
    3. Required delivery date(s)
    4. Coordinated deployment estimate
    5. Employment estimate
    6. Strategic lift requirements, if appropriate
- ISR forces required. List capabilities needed, and, if applicable, specific units or capabilities
- Support forces required. List capabilities needed, and, if applicable, specific units or capabilities

C. Analysis of Opposing Courses of Action. Highlight adversary capabilities and intent (where known) that may have significant impact on friendly COAs.

D. Comparison of Own Courses of Action. For submission to the JFC, include only the final statement of conclusions and provide a brief rationale for the favored air COA. Discuss the relative advantages and disadvantages of the alternative air COAs if this will assist the JFC in arriving at a decision.

E. Recommended Course of Action. State the JFACC's recommended COA.

II. Joint Air Operation Planning (JOPPA)

Ref: JP 3-30, Command and Control of Joint Air Operations (Feb '14), chap. 3.

This discussion reflects that the JFC has designated a JFACC. Planning for joint air operations begins with understanding the JFC's mission and intent. The JFC's estimate of the operational environment and articulation of the objectives needed to accomplish the mission form the basis for determining components' objectives. The JFACC uses the JFC's mission, commander's estimate and objectives, commander's intent, CONOPS, and the components' objectives to develop a course of action (COA). When the JFC approves the JFACC's COA, it becomes the basis for more detailed joint air operations planning—expressing what, where, and how joint air operations will affect the adversary or current situation.

Joint Operation Planning Process for Air

1. Initiation
2. Mission Analysis
3. COA Development
4. COA Analysis and Wargaming
5. COA Comparison
6. COA Approval
7. Plan or Order Development

Ref: JP 3-30, Command and Control for Joint Air Operations, fig. III-4, p. III-5.

The Joint Operation Planning Process for Air (JOPPA)

The JFACC is responsible for planning joint air operations and uses the joint operation planning process for air (JOPPA) to develop a JAOP that guides employment of the air capabilities and forces made available to accomplish missions assigned by the JFC.

JOPPA follows the joint operation planning process found in JP 5-0, Joint Operation Planning, with specific details for joint air operations. JOPPA drives the production of the JAOP and supporting plans and orders. JOPPA may be utilized during deliberate planning, producing JAOPs that support OPLANs or concept plans. JOPPA may also be utilized as part of crisis action planning. It must always be tied closely to the overall joint planning being done by the JFC's staff and other Service or functional component staffs. While the steps are presented in sequential order, work on them can be concurrent or sequential. Nevertheless, the steps are integrated and the products of each step should be checked and verified for coherence and consistency.

Joint Air Operations Planning Overview

Ref: JP 3-30, Command and Control of Joint Air Operations (Feb '14), pp. III-2 to III-3.

The JFACC's role is to plan joint air operations. In doing so, the JFACC provides focus and guidance to the JAOC staff. The amount of direct involvement depends on the time available, preferences, and the experience and accessibility of the staff. The JFACC uses the entire staff during planning to explore the full range of adversary and friendly COAs and to analyze and compare friendly air capabilities with the adversary threat. The JFACC must ensure that planning occurs in a collaborative manner with other components. Joint air planners should meet on a regular basis with the JFC's planners and with planners from other joint force components to integrate operations across the joint force. Planning is a continuous process and only ends with mission accomplishment.

Joint Air Operations Planning

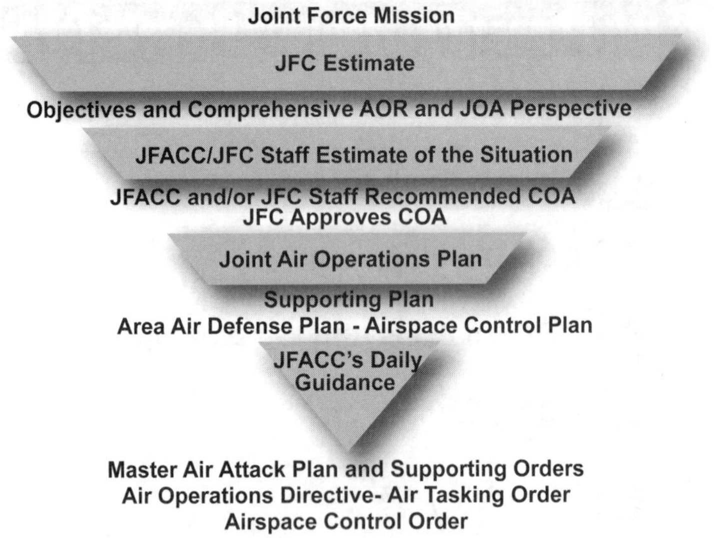

Joint Force Mission

JFC Estimate

Objectives and Comprehensive AOR and JOA Perspective

JFACC/JFC Staff Estimate of the Situation

JFACC and/or JFC Staff Recommended COA
JFC Approves COA

Joint Air Operations Plan

Supporting Plan
Area Air Defense Plan - Airspace Control Plan

JFACC's Daily Guidance

Master Air Attack Plan and Supporting Orders
Air Operations Directive- Air Tasking Order
Airspace Control Order

Ref: JP 3-30, Command and Control of Joint Air Operations, fig. III-1, p. III-2.

The Joint Air Operations Plan (JAOP)

The JAOP is the JFACC's plan for integrating and coordinating joint air operations and encompasses air capabilities and forces supported by, and in support of, other joint force components. The JFACC's planners must anticipate the need to make changes to plans (e.g., sequels or branches) in a dynamic and time-constrained environment. Planners should include representatives from all components providing air capabilities or forces to enable their effective integration.

See pp. 6-16 to 6-18 for a sample joint air operations plan (JAOP) format.

Joint Air Operations Planning Process

Ref: Adapted from JP 3-30, Command and Control of Joint Air Operations (Jan '10), fig. III-13, p. III-25 (not provided in Feb '14 edition).

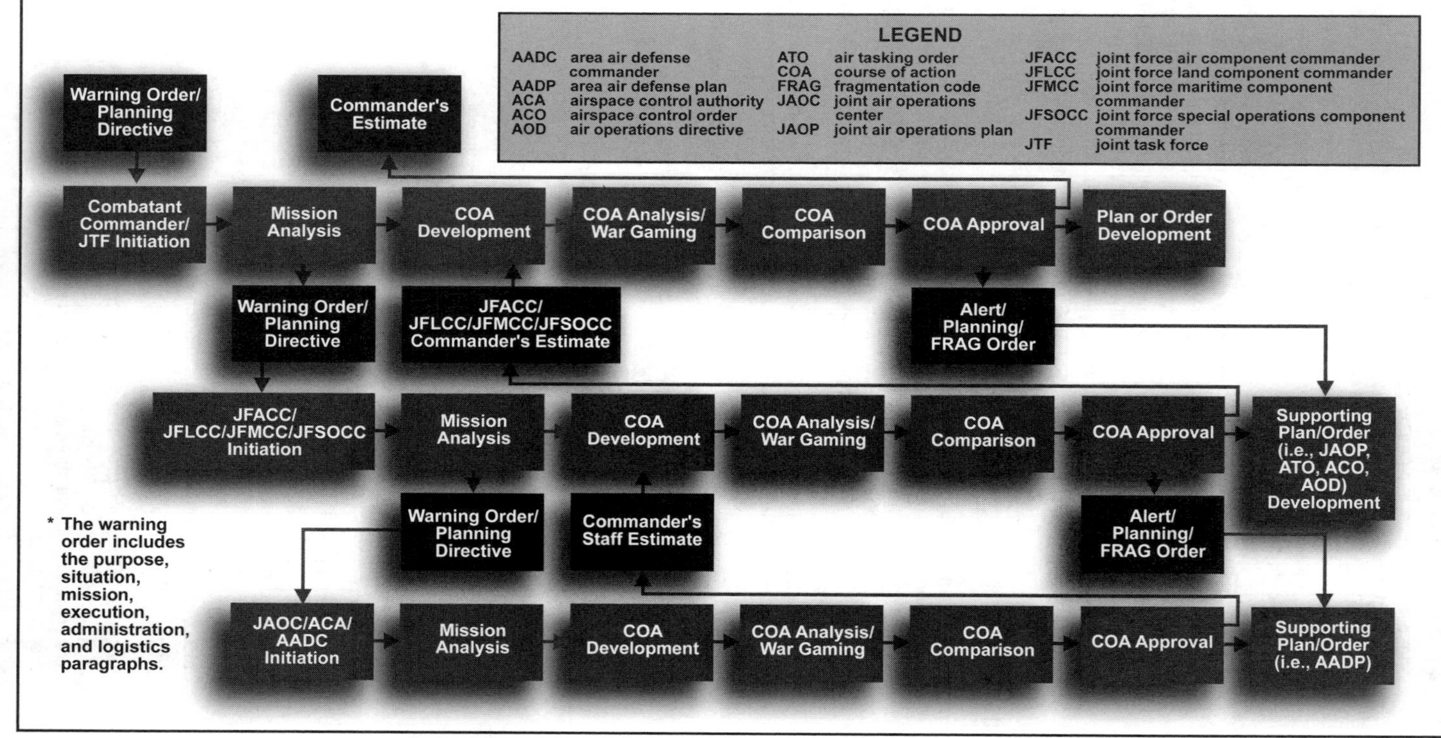

Joint Air Operations

Step 1. Initiation

Planning is usually initiated by direction of a JFC, but the JFACC may initiate planning in anticipation of a planning requirement not directed by higher authority, but within the JFACC's authority. Joint air operations should be coordinated with space and cyberspace operations. Military air options are normally developed in combination with the other military and nonmilitary options so the JFC can appropriately respond to a given situation.

The JFACC and staff perform an assessment of the initiating directive to determine how much time is available until mission execution, the current status of intelligence products and staff estimates, and other relevant factors that influence the planning situation. The JFC and JFACC typically provide initial guidance that may specify time constraints, outline initial coordination requirements, authorize movement of key capabilities within the commanders' authority, and direct other actions as necessary. The JFACC may produce an initial commander's intent during this step.

See facing page to see a sample JFACC mission statement and commander's intent.

Step 2. Mission Analysis

Mission analysis is critical to ensure thorough understanding of the task and subsequent planning. It results in the JFACC's final mission statement that describes the joint air component's essential tasks. It should include the "who, what, when, where, and why" for the joint air operation, but seldom specifies "how." At the end of mission analysis, the JFACC should issue his intent for the overall joint air operation, that is, the JFACC's contribution to the JFC's military end state. The JFACC's intent should express the end state to be produced by joint air operations and the purpose for producing them. It should also include the JFACC's assessment of where and how much risk is acceptable during the operation. While the commander's intent for the overall operation is needed at the end of mission analysis, the JAOP will eventually contain the commander's intent for each phase of the operation, and the AODs will contain the JFACC's intent for a specific ATO or period of time. Hence the commander's intent articulates a desired set of conditions for a given point in time and the purpose those conditions will support.

See facing page to see a sample JFACC mission statement and commander's intent.

Anticipation, prior preparation, and a trained staff are critical to timely mission analysis. Staff estimates generated during mission analysis are continually revisited and updated during the course of planning, execution, and assessment.

Mission analysis includes developing a list of critical facts and assumptions.

Facts

Facts are statements of known data concerning the situation.

Assumptions

Assumptions are suppositions on the current situation or a presupposition on the future course of events, either or both assumed to be true in the absence of positive proof, necessary to enable the commander in the process of planning to complete an estimate of the situation and make a decision on the COA. Assumptions may also become commander's critical information requirements or drive the development of branch plans to mitigate the risks of a wrong assumption. Assumptions must be continually reviewed to ensure validity. Once an assumption is proven correct, it becomes a fact; or if proven incorrect, a new fact or assumption is determined. They are necessary to enable commanders to complete estimates of the situation, influence commander's critical information requirements, drive branch planning, and make decisions on COAs.

Sample Mission & Intent Statements

Ref: JP 3-30, Command and Control of Joint Air Operations (Feb '14), fig. III-5 and fig. III-6.

Sample Joint Force Air Component Commander's Mission Statement

When directed, the joint force air component commander (JFACC) will conduct joint air operations to deter aggression and protect deployment of the joint force.

Should deterrence fail, the JFACC, on order, will gain and maintain air superiority to enable joint operations within the operational area. Concurrently, the JFACC will support the joint force land component commander (JFLCC) in order to prevent enemy seizure of vital areas (to be specified).

On order, the JFACC, in conjunction with the JFLCC and joint force maritime component commander (JFMCC), will render enemy fielded military forces combat ineffective and prepare the operational environment for a counteroffensive. Concurrently, the JFACC will support the JFMCC in gaining and maintaining maritime superiority. The JFACC, on order, will support JFLCC and joint force special operations component commander (JFSOCC) ground offensive operations, degrade the ability of enemy national leadership to rule the country as directed, and destroy enemy weapons of mass destruction, in order to restore territorial integrity, and enemy military threat to the region, support legitimate friendly government, and restore regional stability.

Sample Joint Force Air Component Commander's Intent Statement

***Purpose**. The purpose of this joint air operation will be initially to deter enemy aggression. Should deterrence fail, I will gain air superiority, render enemy fielded forces ineffective with joint airpower, degrade enemy leadership and offensive military capability as directed, and support joint group and special operations forces in order to restore territorial integrity and ensure the survival or restoration of legitimate government in a stable region.*

***End State**. At the end state of this operation: Enemy military forces will be capable of limited defensive operations, will have ceased offensive action, and will have complied with war termination conditions. The succeeding state will retain no weapons of mass destruction capability; I will have passed air traffic control to local authorities, territorial integrity will be restored, and joint force air component commander operations will have transitioned to support of a legitimate and stable friendly government.*

Identify and Analyze Centers of Gravity (COGs)

IPB should identify and analyze adversary and friendly centers of gravity (COGs) at the operational and tactical levels and contribute to the joint intelligence preparation of the operational environment (JIPOE). JIPOE is the analytical process used by joint intelligence organizations to produce intelligence assessments, estimates, and other intelligence products in support of the JFC's decision-making process. The process is used to analyze the physical domains; the information environment; political, military, economic, social, information, and infrastructure systems; and all other relevant aspects of the operational environment, and to determine an adversary's capabilities to operate within that environment. The IPB effort must be fully coordinated, synchronized, and integrated with the JIPOE effort of a joint intelligence center. A COG is a source of power that provides moral or physical strength, freedom of action, or will to act. In coordination with the JFC, the joint air component may focus on strategic and operational COGs as well as tactical-level details of adversary forces because air power can often directly or indirectly affect COGs through application of lethal and nonlethal force and through peaceful means.

See JP 2-01.3, Joint Intelligence Preparation of the Operational Environment, for greater detail on JIPOE.

The JFACC and staff prioritize the analyzed **adversary and friendly critical vulnerabilities (CVs)** associated with COGs based on their impact on achieving the objectives most effectively, in the shortest time possible, and at the lowest cost. The analyses of adversary and friendly critical vulnerabilities are incorporated into the various COAs considered during COA development.

The JFACC, supported by the staff, determines the joint air objectives and the specified, implied, and essential tasks. The JFACC typically includes essential tasks in his mission statement. Essential tasks are specified or implied tasks that the JFACC must perform to accomplish the mission. The JFACC and staff examine readiness of all available air capabilities and forces to determine if there is enough capacity to perform all the specified and implied tasks. **The JFACC identifies additional resources needed for mission success to the JFC.** Factors to consider include available forces (including multinational contributions), command relationships (joint force, national, and multinational), force protection requirements, ROE, law of war, applicable treaties and agreements (including existing status of force agreements), base use (including land, sea, and air), over flight rights, logistic information (what is available in theater ports, bases, depots, war reserve material, host nation support), and what can be provided by other theaters and organizations.

End State

The end state is the set of required conditions that defines achievement of the commander's objectives and specific criteria for mission success. By articulating the joint air component's purpose, the JFACC provides an overarching vision of how the conditions at the end state support the joint operation and follow-on operations.

Step 3. COA Development

COA development is based on mission analysis and a creative determination of how the mission will be accomplished. The staff develops COAs. **A COA represents a potential plan the JFACC could implement to accomplish the assigned mission.** All COAs must meet the JFACC's intent and accomplish the mission.

A COA consists of the following information: what type of military action will occur; why the action is required (purpose); who will take the action; when the action will begin and how long it will last (best estimate); where the action will occur; and how the action will occur (method of employment of forces). COAs may be broad or detailed depending on available planning time and JFACC's guidance. The staff should assess each COA to estimate its success against all possible adversary COAs. The staff converts the approved COA into a CONOPS. COA determination consists of four primary activities: COA development, analysis and wargaming, comparison, and approval. Air COAs will often require input from other component commanders to synchronize them with land and maritime operations.

When time is limited, the JFACC should determine how many COAs the staff will develop and which adversary COAs to address. A complete COA should consider, at a minimum:

- The JFACC's mission and intent (purpose and vision of military end state)
- Desired end state
- Commander's critical information requirements
- C2 structure
- Essential tasks
- Available logistic support
- Available forces
- Available support from agencies and organizations
- Transition strategies between each phase
- Decision points

COAs should include the following specifics:

- Operational and tactical objectives and effects and their related tactical tasks, in order of accomplishment
- Forces required and the force providers
- Force projection concept
- Employment concept
- Sustainment concept

The speed, range, persistence, and flexibility of air assets are their greatest advantages, and their employment location and purpose may change in minutes. Air strategists and planners deal with objective sequencing and prioritization, operational phasing, employment mechanisms, and weight of effort. In some cases, there may be flexibility in how to attain the JFACC's objectives. For example, an objective may be to "destroy WMD capability," but an alternate objective may be to "destroy WMD delivery means." In addition, COAs may vary by the phase in which an objective is achieved or the degree to which an objective is achieved in each phase.

Air COAs may be presented in several ways. They may be presented as text and may discuss the priority and sequencing of objectives. Air COAs may also be depicted graphically—displaying weights of effort, phases, decision points, and risk. One helpful way to depict an air COA graphically is to depict it as one or more logical lines of operations (LOOs), as described in Joint Operation Planning. Any quantitative estimates and assessment criteria presented should clearly indicate common units of measurement in order to make valid comparisons between COAs. For example, a sortie is not a constant value for analysis—one F/A-18 sortie does not equate to one B-2 sortie. Air COAs should avoid numerical presentation. Ultimately, the JFACC will direct the appropriate style and content of the COA.

The first step in COA development is to determine the measures that will accomplish the JFACC's mission and support achievement of the JFC's objectives. The framework of objectives, effects, and tasks provides a clear linkage of overall strategy to task. While the JFC normally provides operational objectives to the JFACC, they may also emerge through mission analysis or COA development, developed by the JFACC and the JAOC SD staff in consultation with the JFC. An objective should be clearly defined, decisive, and state an attainable goal. JFACC support to other components should also be expressed in terms of objectives. Resulting objectives can then be prioritized with other JFACC objectives in accordance with the JFC's CONOPS. **Supporting objectives should describe what aspect of the adversary's capability the JFC or other component wants to affect.**

Linking Objectives, Effects, and Tasks

Commanders plan joint operations by developing objectives supported by measurable effects and assessment indicators. Analysis of effects (desired and undesired) and determination of measures of effectiveness during planning for joint air operations is usually conducted by the JAOC strategy plans and operational assessment teams, assisted by all other planning elements of the JAOC.

To clarify, objectives prescribe friendly goals. Effects describe system behavior in the operational environment. Desired effects are the conditions related to achieving objectives. Tasks, in turn, direct friendly action. Objectives and effects are assessed through measures of effectiveness (MOEs). Empirically verifiable MOEs may help ensure the JFACC knows when the desired ends have been achieved. Accomplishment of friendly tasks is assessed through measures of performance (MOPs). MOEs help answer questions like, "are we doing the right things, or are alternative actions required?" MOEs also help focus component operational assessment efforts, inform processing, exploitation, and dissemination (PED) priorities, and identify ISR requirements. MOPs help answer questions like, "are we doing things right: were the tasks completed to standard?"

Linking Objectives, Effects, and Tasks Examples

Objective:	Gain and maintain air superiority over Sector X.
Effect 1:	Multinational air forces capable of conducting air operations over Sector X without prohibitive interference from Red counter-air forces.
MOE:	Proportion of multinational air missions ineffective due to prohibitive interference from Red counter-air forces.
Task:	Destroy all Red SAM radars covering Sector X.
MOP:	Proportion of known Red force SAM radars covering Sector X destroyed.
Task:	Degrade Red air C2 capacity by 90%.
MOP:	% of Red air C2 links severed.
Effect 2:	Coalition ground forces capable of conducting operations in Sector X without prohibitive interference from Red forces air or missile attack.
MOE:	% of coalition ground force phase objectives achieved without delay or losses from Red air or missile attack.
Task:	Destroy known fixed Red SRBM launch facilities within range of Sector X.
MOP:	Proportion of known Red SRBM launch facilities within range of Sector X destroyed.
Task:	Deny Red attack aircraft sortie generation.
MOP:	Proportion of Red attack airfields denied sortie generation capability.

Ref: JP 3-30 (2014), fig. III-7. Linking Objectives, Effects, and Tasks Examples.

Once strategists and planners define the joint air objectives and supporting effects and tasks, they further refine potential air COAs based on the objective priority, sequence, phasing, weight of effort, and matched resources. This is one method of differentiating COAs. Other methods include varying time available, anticipated adversary activities, friendly forces available, and higher-level guidance. For air planning, a single COA may be developed with several branches and sequels that react to possible adversary activities.

Planners should determine the validity of each air COA based on suitability, feasibility, acceptability, distinguishability, and completeness.

The relationship between resources and COA development is critical. COA development must take into account the resource constraints of the joint force at large. Competing requirements for limited airlift will often result in deployment orders less than ideal for all components but optimal for the joint force as a whole.

See facing page for an overview and further discussion of combat support requirements. Additional risk management (operational factors) can be found on the following page.

Combat Support Considerations

Ref: JP 3-30, Command and Control of Joint Air Operations (Feb '14), pp. III-10 to III-13.

The relationship between resources and COA development is critical. COA development must take into account the resource constraints of the joint force at large:

- Basing
- Force protection
- Petroleum, oils and lubricants availability
- Armaments/precision-guided munitions
- After-action reports
- Reachback
- Out-of-theater staging
- Long-range assets
- Sustainment (airlift/sealift)
- Communications systems
- Aerial port and seaport of debarkation location

During air COA development, the JFACC and staff help the commander identify risk areas that require attention. These will vary based on the specific mission and situation and may be divided into two broad areas: combat support and operational considerations. Combat support includes TPFDD planning that will critically affect the joint force strategy and execution. Also considered with the TPFDD are basing, access, logistic support available, and force protection requirements. However, since TPFDD execution, basing, and logistic support are the responsibility of the JFC and Service components, the JFACC's planning effort needs to focus on the limitations and constraints imposed by them.

Decisions related to operational assumptions may drive changes in how the JFACC operates. These changes range from JOPPA process changes to targeting and weaponeering methods. One of the first considerations for the JFACC is air superiority. The JFACC is responsible for considering the risk related to air defense planning when designated as the AADC. The commander's operational assumptions will determine the resources committed, force posturing, and structure of the air and missile defense plan.

The JFC's assumptions will also affect the operational assumptions made by the joint force air strategists and planners. The joint force structure and campaign or OPLAN directly influence the JFACC's risk estimate and guidance.

Risk Management (Combat Support Factors)

Minimizing the risk of fratricide and collateral damage are operational factors in risk management. The commander must balance the potential for fratricide and collateral damage with mission success. When the risk becomes unacceptable, the commander should consider changes in operational employment. Risk management considerations include:

- Time-phased force and deployment data
- Basing
- Regional access
- Logistics support
- Airbase defense
- Reachback operations
- Just-in-time logistic considerations
- Communications/bandwidth
- Host-nation support

Risk Management: Operational Factors

Minimizing the risk of friendly fire and collateral damage are operational factors in risk management. The commander must balance the potential for friendly fire and collateral damage with mission success. When the risk becomes unacceptable, the commander should consider changes in operational employment.

- Joint force commander assumptions concerning the operation
- Friendly fire
- Collateral damage
- Force protection
- Information assurance
- Multinational considerations
- Command and control architecture

Step 4. COA Analysis and Wargaming

COA analysis involves wargaming each COA against the adversary's most likely and most dangerous COAs. Wargaming is a recorded "what if" session of actions and reactions designed to visualize the flow of the conflict or operation and evaluate each friendly COA in the light of adversary adaptation. War gaming is a valuable step in the planning process because it stimulates ideas and provides insights that might not otherwise be discovered. It also provides initial detailed planning while also determining the strengths and weaknesses of each COA. This may alter or create a new COA based on unforeseen critical events, tasks, or problems identified. Wargaming is often a sequential process, but planning groups should adjust their wargame style based on JFACC guidance, time available, situation, and staff dynamics. Wargaming begins by assembling all the tools and information planners require and establishing the general rules to follow. Recording the activity is vital and directly contributes to identifying the advantages and disadvantages of a COA and providing sufficient detail for future JAOP development. Planners may use a synchronization matrix to detail the results of war gaming. Time permitting, the staff should:

- Consider all facts and assumptions in the estimate and their possible effects on the action
- Consider active and passive measures to decrease the impact of adversary counteractions
- Consider conflict termination issues and the end state
- Think through one's own actions, adversary reactions, and friendly counteractions

COA analysis and wargaming concludes when planners have refined each plan in detail and identified the advantages and disadvantages of each air COA. Automation in the planning process and joint analysis centers may provide additional modeling support to wargaming, increasing the accuracy and speed of COA analysis.

Step 5. COA Comparison

Comparing air COAs against predetermined criteria provides an analytical method to identify the best employment options for air forces and capabilities. The same method used in *JP 5-0, Joint Operation Planning*, is used in air COA comparison.

Another technique for air COA comparison involves developing an objective-risk timeline. Logical LOOs may help to elucidate the relationships between objectives, effects, time, and risk. In logical LOOs, objectives, decisive points, or other significant events are plotted against a timeline that identifies when certain objectives or actions will occur. Risk for each air COA based on the logical LOO is identified. The resulting graphical representation may form the basis for the staff's recommendation and presentation to the JFACC.

Step 6. COA Approval

The staff determines the best air COA to recommend to the commander. The staff presents their recommended air COA usually in the form of a briefing. This briefing includes a summary of the operational design and planning process that led to the recommended air COA. Ideally, the JFACC should be involved in the process, especially in the early operational design stages. Depending on the level of JFACC involvement and the degree of parallel planning the commander accomplishes, air COA selection will vary from choosing among various alternatives to directly approving the staff-recommended air COA. The air COA is identified, adjusted (if required), and selected by the JFACC for presentation to the JFC. Once the JFC approves an air COA, the JOPPA contributes directly to JAOP preparation.

Step 7. Plan or Order Development

For the joint air component, this step concentrates on the preparation of the JAOP. The JAOP details how the joint air effort supports the JFC's overall operation or campaign plan. JAOP development is a collaborative effort of the JFACC staff, the JFC staff, other joint force and Service component staffs, and outside agencies. Once the total force structure is determined, force availability, deployment, timing, basing availability, and sustainment requirements are matched with logistic and planning requirements. With this information, the JFACC's ability to accomplish the assigned mission is reevaluated and adjusted as necessary. The JAOP should accomplish the following:

- Integrate the efforts of joint air capabilities and forces and, where applicable and appropriate, space and cyberspace capabilities and/or support mechanisms/enablers
- Identify objectives, effects, and tasks
- Identify MOEs and indicators used to determine whether air operations are creating desired effects and achieving objectives.
- Account for current and potential adversary COAs
- Integrate and synchronize the phasing of operations with the JFC's plan
- Indicate what capabilities and forces are required to achieve joint air objectives. In addition to air capabilities and forces, planners should include land, maritime, space, cyberspace, and information-related capabilities required to meet joint air objectives.
- Develop specific procedures for allocating, tasking, exercising, and transitioning C2 of joint air capabilities and forces.

In addition to building the plan for the employment of air forces, the JAOP should also include considerations for phase transitions, decision points, conflict termination, redeployment (if applicable), and procedures to capture lessons learned. Incomplete planning for conflict termination and the end state can result in the waste of valuable resources, aggravate a tenuous peace, cause a return to hostilities, or lead to numerous other unintended consequences. The list of considerations for conflict termination is specific to each situation and is never formulated in a vacuum or without extensive consultations with national leadership. This part of the plan should also address the prospect of the "surge" of air forces to accomplish phases of the operation, based on projected operating tempo.

See following pages (pp. 6-16 to 6-18) for a sample joint air operations plan (JAOP) format.

Sample Joint Air Operations Plan (JAOP)

Ref: JP 3-30, Command and Control of Joint Air Operations (Feb '14), app. C.

The JAOP format uses the same format as the JFC's OPLAN but from an air power point of view. Each air operations plan will differ with the JOA, situation, and capabilities of the joint force. A sample template follows:

Copy No.
Issuing Headquarters
Place of Issue
Date/Time Group of Signature

JOINT AIR OPERATIONS PLAN:
(Number or Code Name)

REFERENCES: Relevant documents, maps, and charts. This should generally include CJCSM 3122.03C, JOPES Volume II, Planning Formats.

1. SITUATION

Briefly describe the situation that the plan addresses (see the JFC's estimate and the template below as a guide). Related OPLAN(s) should be identified, as appropriate.

a. **General Guidance**. Summarize the operational environment and overall JFC mission, guidance, intent, prioritized effects, operational limitations, and specified tasks for the JFACC and established support relationships among components that are relevant to that guidance.

b. **Area of Concern**. Applicable boundaries, as of the operational area(s), area(s) of interest, and etc. Include maps as appropriate.

c. **Deterrent Options**. Describe air power's role in these JFC options, if applicable.

d. **Adversary Forces**. Overview of the hostile threat, to include:

(1) Composition, location, disposition, movement, of major adversary forces and capabilities that can influence action in the operational environment

(2) Adversary strategic concept (if known): should include adversary's perception of friendly vulnerabilities and adversary's intention regarding those vulnerabilities

(3) Major adversary objectives (strategic and operational)

(4) Adversary commanders' motivations, thought patterns, idiosyncrasies, and doctrinal patterns (to the extent known)

(5) Operational and sustained capabilities (all relevant adversary forces, not just air and counterair)

(6) Adversary COGs and decisive points

(a) Analysis of critical capabilities (CCs), critical requirements (CRs), and critical vulnerabilities (CVs) for each

(b) Description using logical LOOs and LOEs, if appropriate

e. **Friendly Forces.** Overview of friendly (US and coalition partner), to include:

(1) Forces available according to TPFDD considerations

(2) Forces required, based on employment CONOPS. Highlight shortfalls

(3) Intent of higher, adjacent, and supporting US and coalition forces and commands

(4) Friendly COGs

(a) Analysis of CCs, CRs, and CVs

(b) Steps to be taken to protect friendly CVs

f. **Assumptions**. List, as required.

g. **Legal Considerations**. List those of critical importance to operations, such as legal restrictions and guidance on targeting. Refer to Annexes, as required.

2. MISSION

JFACC's Mission Statement

3. EXECUTION

a. **CONOPS for Joint Air Operations**. A statement of the JFACC's intent, objectives, desired effects, and broad employment concepts, to include logical LOOs for the desired end state. Phase plans for each phase of the operation.

(1) Operational concept for the phase, including objectives (ongoing and specific to the phase), intent, desired effects, risk, logical LOOs, plan of operations, timing, and duration

(2) General guidance for subordinate units and component's supported and supporting requirements. Ensure that all subordinates' missions are complementary.

(3) Forces or capabilities required by objective

(4) "Be prepared to" missions; phase branches

(5) Reserve capabilities and/or forces, if applicable – reserve in this sense meaning capabilities held in operational reserve, not Reserve Component elements of the joint force

(6) Mobility considerations for the phase – transportation, lines of communications, over flight, basing considerations, and the like that are unique to this phase

(7) Information-related capabilities that contribute to IO that are unique to the phase of the operation.

b. **Tasks**. State the component's supporting and supported requirements for the operation in general. Include implied tasks and guidance to subordinates that are not specific to a given phase.

c. **Coordinating Instructions.** Explain operational terms required for complete understanding of the operation, but which are not defined in current JPs.

d. **Exchange of LNOs**. Explain and direct any liaison requirements here, including the role of the JACCE.

4. ADMINISTRATION AND LOGISTICS

a. **Concept of Sustainment**. A broad statement of the functional areas of logistics, transportation, personnel policies, and administrations, if required.

b. **Logistics**. Broad sustainment concept for air operations. Phase considerations (synchronized with execution phases – may not be required if already explained in phase plans).

(1) Basing and Overflight. Explain any unique clearance and buildup requirements in this section, if not already explained in phase plans.

(2) Lines of Communications. Explain any requirements relevant to the operation.

(3) Base Opening and Development. Explain any general base opening requirements for the operation. Information may also be included in phase plans.

(4) Maintenance and Modification. Use as required.

(5) Host Nation Considerations. Explain any unique requirements for the operation.

(6) Reconstitution of Forces. Use as required.

(7) Inter-Service, Interagency, and Inter-Component Requirements. Use as required.

(8) Foreign Military Assistance. Use as required.

c. **Personnel**. Use as required.

d. **Public Affairs**. Identify key public affairs requirements necessitated by major event (may also be identified in phase plans).

Continued on next page

Continued on next page

Sample Joint Air Operations Plan (JAOP)

Ref: JP 3-30, Command and Control of Joint Air Operations (Feb '14), app. C.

Continued from previous page

e. **Civil Affairs**. Use as required.

f. **Meteorological and Oceanographic**. Explain factors like climate and terrain, and how they will likely affect air operations.

g. **Geospatial Information**. Explain common geospatial reference system requirements and plans here.

h. **Medical Services**. Use as required.

5. COMMAND AND CONTROL

a. **Command**

(1) Command Relationships. Specify command relationships for all organizations relevant to the JFACC operations. Be as specific as possible

(2) Memoranda of Understanding. As applicable

(3) Command Headquarters. Designation and location of all air-capable command headquarters

(4) Continuity of Operations. Any general considerations unique to the operation

(5) Command Posts. List the designations and locations of each major headquarters

(6) Succession to Command. Designate, in order of succession, the commanders responsible for assuming command of the operation in specific applicable circumstances

b. **C2 and Communications Systems**. General overview of C2 and communication systems required to support air operations.

6. ANNEXES

JAOP annexes should be written for a functional domain-specific audience and contain technical details necessary for C2 of all air organizations and capabilities across the joint force. They should contain any details not considered appropriate for the relevant section of the main plan.

A. Task Organization
B. Intelligence
C. Operations
D. Logistics
E. Personnel
F. Public Affairs
G. Civil Affairs
H. Meteorological and Oceanographic Operations
I. Force Protection
J. Command Relationships
K. Joint Communications System
L. Environmental Considerations
M. Geospatial Information and Services
N. Space Operations
P. Host-Nation Support
Q. Medical Services
S. Special Technical Operations
V. Interagency Coordination

Continued from previous page

(Signed) (Commander)

DISTRIBUTION:

SECURITY CLASSIFICATION.

III. Joint Targeting

Ref: JP 3-30, Command and Control of Joint Air Operations (Feb '14), chap. III.

Targeting is the process of selecting and prioritizing targets and matching the appropriate response to them, considering operational requirements and capabilities. Targeting is both a joint- and component-level function to create specific desired effects that achieve the JFC's objectives. Targeting selects targets that, when attacked, can create those effects, and selects and tasks the means to engage those targets. Targeting is complicated by the requirement to deconflict unnecessary duplication of target nominations by different forces or different echelons within the same force and to integrate the attack of those targets with other components of the joint force. An effective and efficient target development process coupled with the joint air tasking cycle is essential for the JFACC to plan and execute joint air operations. The joint targeting process should integrate the intelligence databases, analytical capabilities, and data collection efforts of national agencies, combatant commands, subordinate joint forces, and component commands.

I. Joint Targeting Cycle

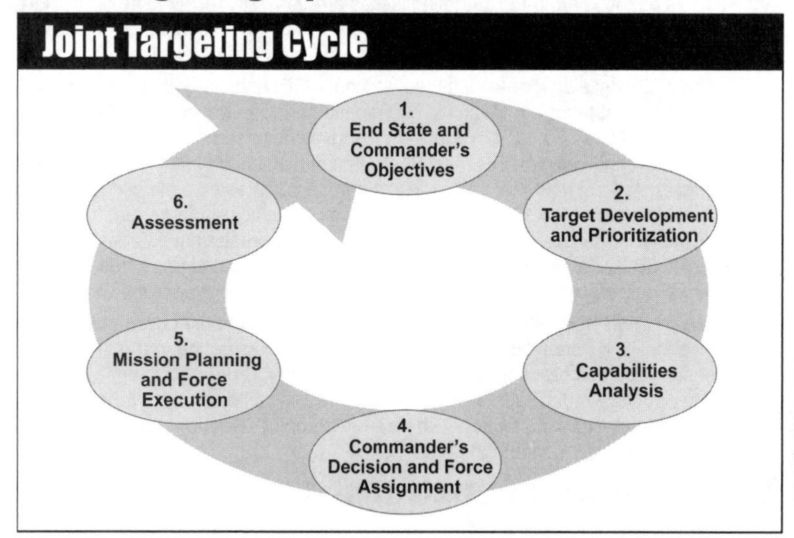

Ref: JP 3-60, Joint Targeting (Jan '13), fig. II-2. Joint Targeting Cycle.

The joint targeting cycle is an iterative process that is not time-constrained, and steps may occur concurrently, but it provides a helpful framework to describe the steps that must be satisfied to successfully conduct joint targeting. The deliberate and dynamic nature of the joint targeting process is adaptable through all phases of the air tasking cycle. As the situation changes and opportunities arise, steps of the joint targeting process can be accomplished quickly to create the commander's desired effects. There are six phases to the joint targeting cycle: end state and commander's objectives, target development and prioritization, capabilities analysis, commander's decision and force assignment, mission planning and force execution, and assessment.

See chap. 7, Targeting, for further discussion from Annex 3-60. See pp. 7-6 to 7-7 for an overview of the targeting cycle.

Targeting Mechanisms

Targeting mechanisms should exist at multiple levels. The President, Secretary of Defense, or headquarters senior to JFCs may provide guidance, priorities, and targeting support. Joint force components identify requirements, nominate targets that are outside their operational area or exceed the capabilities of organic and supporting assets, and conduct execution planning. After the JFC makes final targeting decisions, components plan and execute assigned missions.

II. Joint Targeting Coordination Board (JTCB)

Typically, the JFC organizes a joint targeting coordination board (JTCB). The JTCB's focus is to develop broad targeting priorities and other guidance in accordance with the JFC's objectives as they relate operationally. The JFC normally appoints the deputy JFC or a component commander to chair the JTCB. If the JFC so designates, a JTCB may be an integrating center to accomplish broad targeting oversight functions or a JFC-level review mechanism to refine or approve the joint integrated prioritized target list (JIPTL). The JTCB needs to be a joint activity comprised of representatives from the staff, all components, and, as required, other agencies, multinational partners and/or subordinate units.

The JFC defines the role of the JTCB. The JTCB provides a forum in which all components can articulate strategies and priorities for future operations to ensure they are integrated and synchronized. The JTCB normally facilitates and coordinates joint force targeting activities with the components' schemes of maneuver to ensure that the JFC's priorities are met. Targeting issues are generally resolved below the level of the JTCB, by direct coordination between elements of the joint force, but the JTCB and/or JFC may address specific targeting issues not previously resolved.

The JFC may also form a joint fires element (JFE) and/or a joint targeting working group (JTWG), both of which aid coordination and integration of the joint targeting process. The JFE is an optional staff element comprised of representatives from the JFC's J-3, the components, and other elements of the JFC staff. The JFE is an integrating staff element that synchronizes and coordinates fires planning and coordination on behalf of the JFC. The JTWG is an action-officer-level venue, typically chaired by the JFE chief, J-2 (chief of targets), or similar representative. The JTWG supports the JTCB by conducting initial collection, consolidation, and prioritization of targets and synchronization of target planning and coordination on behalf of the JFC.

The JFC will normally delegate the authority to conduct execution planning, coordination, and deconfliction associated with joint air targeting to the JFACC and will ensure that this process is a joint effort. The JFACC must possess a sufficient C2 infrastructure, adequate facilities, and ready availability of joint planning expertise. A targeting mechanism tasked with detailed planning, weaponeering, and execution is also required to facilitate the process.

III. Targeting Functions

The JFACC develops a JAOP that accomplishes the objectives directed by the JFC. Integration, synchronization, deconfliction, allocation of air capabilities and forces, and matching appropriate weapons against target vulnerabilities are essential targeting functions for the JFACC. National agencies, higher headquarters, JTFs, and task forces subordinate to the JFC, supporting unified commands, and functional/ Service components may nominate targets to the JFC for processing and inclusion on the JIPTL through a target nomination list (TNL). Targets scheduled for attack by component air capabilities and forces should be included on an ATO for deconfliction and coordination. All component commanders within the joint force should have a basic understanding of each component's mission and general scheme of maneuver. All components should provide the JFACC a description of their air plan to minimize the risk of friendly fire, assure deconfliction, avoid duplication of effort, and provide visibility to all friendly forces. This basic understanding allows for integration of targeting efforts between components and within the JFC staff and agencies.

IV. The Targeting Effects Team (TET)

Ref: JP 3-30, Command and Control of Joint Air Operations (Feb '14), pp. III-17 to III-18.

The JFACC may establish a targeting effects team (TET) as part of the JAOC. The TET's responsibilities are varied but key to the targeting process. The TET validates targets to be engaged by joint air forces per the JFC's targeting guidance, links targets to appropriate tactical tasks in the AOD, weaponeers targets to create desired effects, and verifies MOEs/MOPs. It also deconflicts and coordinates target nominations based on estimates of what targets can be attacked and provides other targeting support requiring component input at the JFACC level. If the JFC delegates joint targeting oversight authority to the JFACC, that commander should possess or have access to sufficient C2 infrastructure, adequate facilities, and joint planning expertise to effectively manage and lead the JFC's joint targeting operations. When the JFACC has JFC approval, the TET receives all target nominations (that cannot be addressed at lower echelon levels) and prioritizes them in accordance with the objectives and tactical tasks set forth in the AOD to form the draft JIPTL. Common organizational guidelines of the TET include the following:

- Chaired by the deputy JFACC or designated representative.
- Senior component liaison officers (LNOs) and key JFACC staff members comprise the TET.
- The JAOC CPD provides the staff support to the TET during the joint air tasking cycle.

Draft JIPTL Construction

The draft JIPTL is formed from a prioritized listing of targets based on JFC and component target priorities. To be a true integrated target list, the TET must present the JFACC with a draft JIPTL that includes targets and associated effects for engagement by both lethal and nonlethal means. In the case of a theater JFACC supporting multiple JFCs (e.g., two or more JTF commanders), the draft JIPTL should be constructed to meet the requirements of each supported JFC. Members consider the estimated available capabilities and their ability to engage the targets on the list. A draft JIPTL "cut line" is normally established. The draft JIPTL "cut line" should reflect which nominated targets will most likely be serviced (barring technical problems with aircraft, weather, retasking for higher priority targets, or other operational circumstances) with the projected apportionment of assets assigned or made available to the JFACC. Component LNOs and JAOC staff members should be ready to justify and/or prioritize target nominations among all the priorities of the joint operation. The JFACC may also recommend to the JFC that additional assets from other components be used against targets on the draft JIPTL. However, only the JFC can approve the use of other components' assets and forces. Close coordination must continue with the development of the JIPTL and with the development of the joint integrated prioritized collection list (JIPCL) to ensure effective and efficient use of assets that may be used to address targets on both the JIPTL and/or the JIPCL.

V. Target Development and Prioritization

Ref: JP 3-60, Joint Targeting (Jan '13), pp. II-5 to II-6.

Target development is the analysis, assessment, and documentation processes to identify and characterize potential targets that, when successfully engaged, support the achievement of the commander's objectives. A fully developed target must comply with national and command guidance, law of war, and the applicable ROE to be engaged. Phase 2 is comprised of three steps:

(a) Target system analysis;

(b) Entity-level target development; and

(c) Target list management (TLM).

Target developers systematically examine the enemy to the entities to the elements utilizing the targeting taxonomy, which hierarchically orders the adversary, its capabilities, and the targets which enable the capabilities into a clarifying framework.

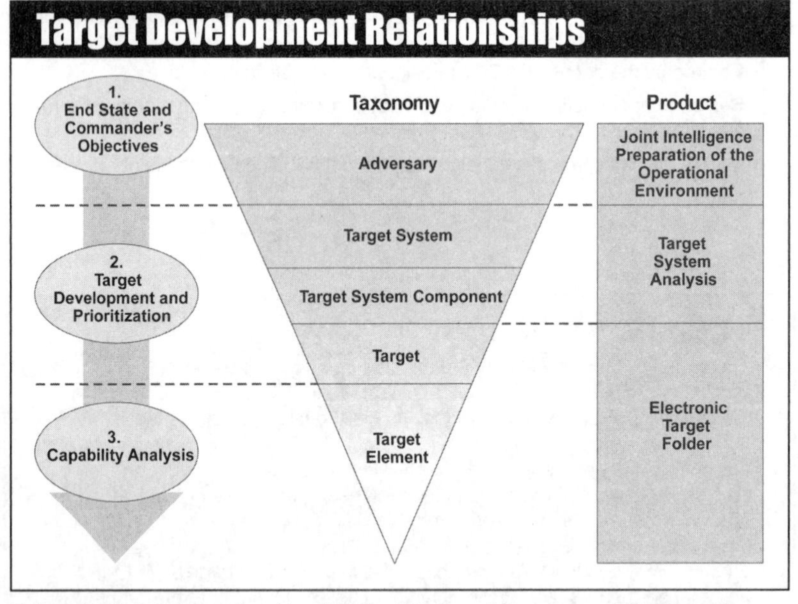

Ref: JP 3-60, Joint Targeting (Jan '13), fig. II-3. Target Development Relationships.

Target systems are typically a broad set of interrelated functionally associated components that generally produce a common output or have a shared mission. Target development always approaches adversary capabilities from a target systems perspective. This includes physical, logical, and complex social systems, and the interaction among them. While a single target may be significant because of its own characteristics, the target's real importance lies in its relationship to other targets within an operational system. A target system is most often considered as a collection of assets directed to perform a specific function or series of functions. While target systems are intra-dependent to perform a specific function, they are also interdependent in support of adversary capabilities. System-level target development links these multiple target systems and their components to reflect both their intra- and interdependency that, in aggregate, contribute to the adversary capabilities. JIPOE helps target developers prioritize an adversary's target systems based on how much each contributes to the adversary's ability to wage war.

IV. The Joint Air Tasking Cycle

Ref: JP 3-30, Command and Control of Joint Air Operations (Feb '14), chap. III.

The joint air tasking cycle provides for the effective and efficient employment of joint air capabilities and forces made available. This process provides an iterative, cyclic process for the planning, apportionment, allocation, coordination, and tasking of joint air missions and sorties within the guidance of the JFC. The cycle accommodates changing tactical situations or JFC guidance as well as requests for support from other component commanders. The joint air tasking cycle is an analytical, systematic cycle that focuses joint air efforts on accomplishing operational requirements. Much of the day-to¬day tasking cycle is conducted through an interrelated series of information exchanges and active involvement in plan development, target development, air execution, and assessment (through designated component LNOs and/or messages), which provide a means of requesting and scheduling joint air missions. A timely ATO is critical—other joint force components conduct their planning and operations based on a prompt, executable ATO and are dependent on its information.

The Joint Air Tasking Cycle

1. **Objectives, Effects, and Guidance**
2. **Target Development**
3. **Weaponeering and Allocation**
4. **ATO Production and Dissemination**
5. **Execution Planning and Force Execution**
6. **Assessment**

Ref: JP 3-30, Command and Control for Joint Air Operations, chap. III.

The joint air tasking cycle begins with the JFC's objectives, incorporates guidance received during JFC and component coordination, and culminates with assessment of previous actions. The ATO articulates the tasking for joint air operations for a specific execution timeframe, normally 24 hours. The joint air tasking cycle is synchronized with the JFC's battle rhythm. The JAOC normally establishes a 72- to 96-hour ATO planning cycle. The battle rhythm or daily operations cycle (schedule of events) articulates briefings, meetings, and report requirements. It provides suspense for targeting, AIRSUPREQs, friendly order of battle updates, etc., to produce the air battle plan (ABP) that includes the ATO message and other products. The battle rhythm is essential to ensure information is available when and where required to provide products necessary for the synchronization of joint air operations with the JFC's CONOPS and supporting other components' operations. Nonetheless, airpower must be responsive to a dynamic operational environment and the joint air tasking cycle must be flexible and capable of modification during ATO execution. The net result of the tasking process is a series of ATOs and related products in various stages of progress at any time (see Figure III-12 on previous page). The primary factor that drives the daily schedule for the development of the ATO is the battle rhythm. The battle rhythm is a very detailed timeline that lists series of briefings, meetings, etc., to produce specific products by a specified time.

The full joint air tasking cycle, from JFC guidance to the start of ATO execution, is dependent on the JFC's and JFACC's procedures. A 72-hour cycle, starting with objectives, effects, and guidance is fairly standard. The precise timeframes should be specified in the JFC's OPLAN or the JFACC's JAOP. Long-range combat air assets positioned outside the theater but operating in the JOA may be airborne before ATO publication or execution. These assets require the most current ATO information and updates. The JAOC, however, possesses the capability to retask such missions even during execution. Intertheater air mobility missions may not necessarily operate within the established tasking cycle. The AMD is an AOC division that assists the CPD with intertheater and intratheater air mobility missions that should be integrated into the ATO.

The ATO matches and tasks air forces and capabilities made available to the JFACC for tasking to prosecute targets and resource AIRSUPREQs and other requirements. Other component air missions should be on the ATO to improve joint force visibility and to assist with overall coordination and deconfliction. The other component air missions that appear on the ATO may not be under the control of the JFACC, and the JFACC will coordinate changes with all affected components.

JFACC Tasking Process Responsibilities

- Plan, integrate, coordinate, allocate, task, and direct the joint air effort in accordance with the joint force commander's (JFC's) guidance and joint force objectives
- Develop a joint air operations plan derived from the JFC's broader operation or campaign objective and guidance regarding the objectives, effects, tasks, and responsibilities of joint air capabilities and forces.
- After consulting with other component commanders, recommend apportionment of the joint air effort by priority that should be devoted to various air operations for a given period of time
- Translate air apportionment into allocations and develop targeting guidance into the air operations directive and air tasking order
- Direct and ensure deconfliction of joint air operations
- Synchronize joint air operations with space and cyberspace operations
- Coordinate with the appropriate components' agencies or liaison elements for integration and deconfliction with land and maritime operations
- Coordinate with the appropriate components' agencies or liaison elements or tasking of the air forces and capabilities made available
- Coordinate with the joint force special operations component commander's special operations liaison element for integration, synchronization, and deconfliction with special operations
- Monitor execution and redirect joint air operations as required
- Compile component target requirements and prioritize targets based on JFC guidance
- Accomplish tactical and operational assessment

Joint Air Operations

Joint Air Tasking Cycle

Ref: JP 3-30, Command and Control of Joint Air Operations (Feb '14), fig. III-13, p. III-21.

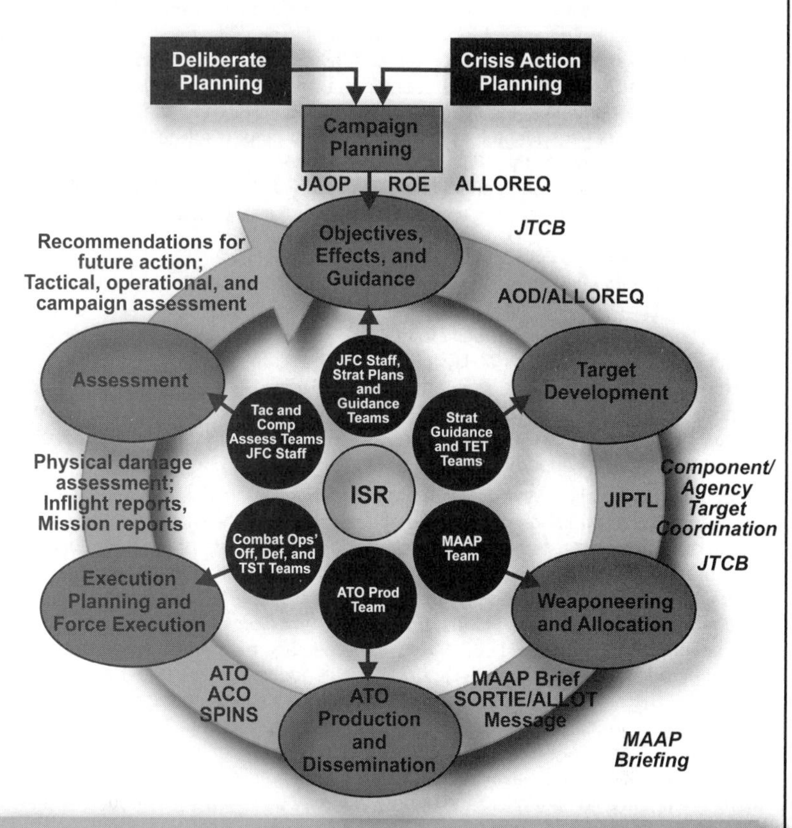

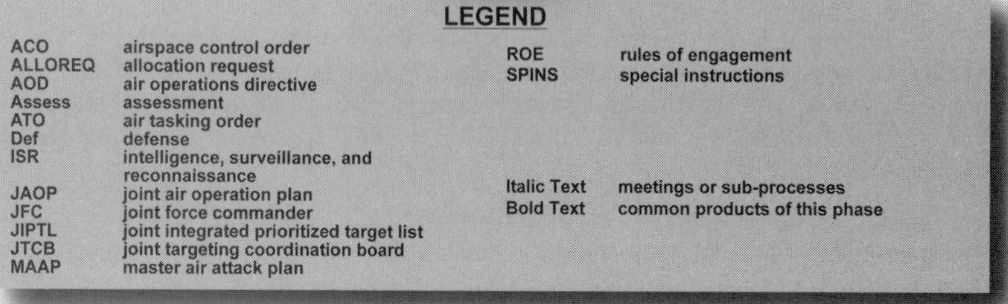

LEGEND

ACO	airspace control order
ALLOREQ	allocation request
AOD	air operations directive
Assess	assessment
ATO	air tasking order
Def	defense
ISR	intelligence, surveillance, and reconnaissance
JAOP	joint air operation plan
JFC	joint force commander
JIPTL	joint integrated prioritized target list
JTCB	joint targeting coordination board
MAAP	master air attack plan
ROE	rules of engagement
SPINS	special instructions
Italic Text	meetings or sub-processes
Bold Text	common products of this phase

II. Joint Air Tasking Cycle Stages

Ref: JP 3-30, Command and Control of Joint Air Operations (Feb '14), pp. III-19 to III-28.

The joint air tasking cycle consists of six stages. The joint air tasking cycle receives products from information developed during the joint targeting cycle and other joint force processes. Both the joint targeting cycle and joint air tasking cycle are systematic processes to match available capabilities and forces with specific targets to achieve the JFC's objectives. Unlike the joint targeting cycle, the joint air tasking cycle is time-dependent, built around finite time periods to plan, prepare for, and conduct joint air operations. There is a set suspense for product inputs and outputs for each stage of the joint air tasking cycle. Prior to the JFC and component commanders' meeting, the JFACC should meet with senior component liaisons and the JFC's staff to develop recommendations on joint air planning and apportionment for future operations.

Stage 1: Objectives, Effects, and Guidance *(see also pp. 7-32 to 7-33)*

The JFC consults often with component commanders to assess the results of the joint force's efforts and to discuss the strategic direction and future plans. This provides component commanders an opportunity to make recommendations, make support requirements known, and state their ability to support other components. The JFC provides updates to the guidance, priorities, and objectives based on enemy operations and the current/expected friendly order of battle. The JFC also refines the intended CONOPS. The JFC's guidance on objectives and effects will identify targeting priorities and will include the JFC's air apportionment decision.

Air apportionment allows the JFC to ensure the priority of the joint air effort is consistent with campaign or operation phases and objectives. Given the many functions that joint air forces can perform, its operational area-wide application, and its ability to rapidly shift from one function to another, JFCs pay particular attention to air apportionment. After consulting with other component commanders, the JFACC makes the air apportionment recommendation to the JFC. The methodology the JFACC uses to make the recommendation may include priority or percentage of effort devoted to assigned mission-type orders, JFC objectives, or other categories significant to the campaign or operation. The air apportionment recommendation is a vital part of the joint air planning and tasking process. The JAOC SD formulates the air apportionment recommendation that the JFACC submits to the JFC for upcoming iterations of the joint tasking cycle. With air capabilities made available to the JFACC, the strategy plans team can recommend the relative level of effort and priority that may be applied to various JFC and/or JFACC objectives. The end result is an air apportionment recommendation. This product is normally forwarded to the JTCB for coordination and approval by the JFC. In the case of a theater JFACC supporting multiple JFCs (e.g., two or more JTF commanders), the air apportionment recommendation (e.g., CAS, interdiction) referenced here is made to each supported JFC. The JFC is the final approval authority for the air apportionment decision.

Stage 2: Target Development *(see also pp. 7-35 to 7-44)*

This is the point in the joint targeting cycle and intelligence process, after analysts from other organizations have incorporated all source intelligence reports into a targeting database, where efforts of the joint air targeting cycle relate target development to air tasking and target aim points are selected, and these and other data are submitted to the TET. The TET correlates target nominations. It screens nominated targets and ensures that once attacked, they create the desired effects that meet JFC guidance as delineated in the AOD, and verifies that chosen MOEs will accurately evaluate progress and can be collected against. It prioritizes nominated targets based on the best potential for creation of the JFC's desired effects and components' priorities and timing requirements. The product of this effort, when approved by the JFC or his designated representative (e.g., JTCB), is the JIPTL.

Stage 3: Weaponeering and Allocation *(see pp. 7-45 to 7-49)*

During this stage, JAOC personnel quantify the expected results of the employment of lethal and non lethal means against prioritized targets to create desired effects. The JIPTL provides the basis for weaponeering assessment activities. All approved targets are weaponeered, to include recommended aim points, weapon systems and munitions, fusing, target identification and description, desired direct effects of target attack, probability of creating the desired effect, and collateral damage concerns. The final prioritized targets are developed and are then provided to the master air attack plan (MAAP) team. The TET may provide the MAAP team a draft JIPTL to begin initial planning. Once the JIPTL is approved by the JFC, the MAAP team can finalize force allocation (sortie flow plan). The force application cell can complete coordination with the supporting force enhancement cell to satisfy mission requirements to ensure the prioritized targets are planned to generate effects to achieve objectives while maximizing the combat effectiveness of joint air assets. The resulting MAAP is the plan for employment that forms the foundation of the ATO. The MAAP is normally a graphic depiction of capabilities required for a given period. The development of the MAAP includes review of JFC and JFACC guidance, component plans and their support requests, updates to target requests, availability of capabilities and forces, target selection from the JIPTL, and weapon system allocation. Components may submit critical changes to target requests and asset availability during this final phase of ATO development. The completed MAAP matches available resources to the prioritized target list. It accounts for air refueling requirements, suppression of enemy air defenses requirements, air defense, ISR, and other factors affecting the plan.

- **Air Allocation**. Following the JFC's air apportionment decision, the JFACC translates that decision into total number of sorties by weapon system type available for each objective and task. Based on the apportionment decision, internal requirements, and air support request messages, each air-capable component prepares an allocation request (ALLOREQ) message for transmission to the JFACC (normally not less than 36 hours prior to the start of the ATO execution period, thus coinciding with the beginning of the MAAP process).
- **Allotment**. The sortie allotment (SORTIEALOT) message confirms (and where necessary modifies) the ALLOREQ and provides general guidance to plan joint air operations. The JAOC reviews each component's allocation decision/ALLOREQ message and may prepare a SORTIEALOT message back to the components as required, in accordance with established operations plans guideline.

Stage 4: ATO Production and Dissemination *(see also pp. 7-48 to 7-50)*

JFC and JFACC guidance, including the AOD; target worksheets; the MAAP; and component requirements are used to finalize the ATO, SPINS, and ACO. Planners must develop airspace control and air defense instructions in sufficient detail to allow components to plan and execute all air missions listed in the ATO. These directions must enable combat operations without undue restrictions, balancing combat effectiveness with the safe, orderly, and expeditious use of airspace. Instructions must provide for quick coordination of task assignment and reassignment (redirection, retargeting, or change of type of mission) and must direct aircraft identification and engagement procedures and ROE appropriate to the nature of the threat. These instructions should also consider the volume of friendly and possibly neutral air traffic, friendly air defense requirements, identification-friend-or-foe technology, weather, and adversary capabilities. Instructions are contained in SPINS and in the ACO, and are updated as frequently as required. The AOD, ATO, ACO, and SPINS provide operational and tactical direction at appropriate levels of detail. The level of detail should be very explicit when forces operate from different bases and mult-component or composite missions are tasked. In contrast, less detail is required when missions are tasked to a single component or base.

Continued on next page

Joint Air Tasking Cycle Stages (Continued)

Ref: JP 3-30, Command and Control of Joint Air Operations (Feb '14), III-19 to III-28.

Continued from previous page

Stage 5: Execution Planning & Force Execution *(see also p. 7-51)*

The JFACC directs the execution of air capabilities and forces made available for joint air operations. Inherent in this is the authority to redirect joint air assets. The JFACC will co-ordinate with affected component commanders upon redirection of joint sorties previously allocated for support of component operations. Aircraft or other capabilities and forces not apportioned for joint air operations, but included in the ATO for coordination purposes, may be redirected only with the approval of the respective component commander or JFC.

- The JAOC must be responsive to required changes during the execution of the ATO. In-flight reports, discovery of time-sensitive targets (TSTs), and initial assessment (such as battle damage assessment [BDA]) may cause a redirecting of joint air capabilities and forces before launch or a redirection once airborne.
- During execution, the JAOC is the central agency for receiving the tasking of joint air capabilities and forces. It is also charged with coordinating and de-conflicting those changes with the appropriate control agencies and components.
- Due to operational environment dynamics, the JFACC may be required to make changes to planned joint air operations during execution. Employment of joint air assets against emerging targets requires efficient, timely information sharing and decision-making among components. It is critical that procedures be established, coordinated, and promulgated by the JFC before operations begin. The dynamic targeting portion of the joint targeting cycle is established to facilitate this process. The JFACC will coordinate with affected component commanders to ensure deconfliction of targets and to ensure those forces are out of danger relative to the new target area(s).
- During execution, the JFACC is responsible for redirecting joint air assets to respond to moving targets or changing priorities. Ground or airborne C2 platform mission commanders may be delegated authority from the JFACC to redirect sorties or missions made available to higher priority targets. The JAOC must be notified of all redirected missions.

Stage 6: Assessment *(see also p. 7-52)*

The JFC should establish a dynamic system to conduct assessment throughout the joint force and to ensure that all components are contributing to the overall joint assessment effort. Normally, the joint force J-3 is responsible for coordinating assessment, assisted by the J-2. Assessment is a continuous process that measures the overall effectiveness of employing joint force capabilities during military operations. It determines progress toward accomplishment of tasks, creation of effects, and achievement of objectives. The JFACC should continuously plan and evaluate the results of joint air operations and provide assessments to the JFC for consolidation into the overall assessment of the current operation.

Continued from previous page

Within the joint force, assessment is conducted at both the tactical and operational levels. At the tactical level, assessment is essential to decision making during ATO execution. However, the tactical assessment process continues over days or weeks to evaluate the effectiveness of weapons and tactical engagements as additional information and analysis become available from sources within and outside the operational area. This should also include a determination of actual collateral damage. Air planners should determine MOPs to evaluate task accomplishment and MOEs to assess changes in system behavior, capability, or the operational environment. Planners should ensure that they establish logical links between air objectives and tasks and the measures used to evaluate them early in the planning sequence.

At the operational level, assessment is concerned with gathering information on the broader results achieved by air operations and planning for future operations. In general, the assessment process at the tactical level provides one of the major sources of information for performing assessment at the operational level.

Chap 7

I. Targeting Fundamentals

Ref: Annex 3-60, Targeting (14 Feb '17), pp. 1-30.

Targeting is the process of selecting and prioritizing targets and matching the appropriate response to them, considering operational requirements and capabilities. This process is systematic, comprehensive, and continuous. Combined with a clear understanding of operational requirements, capabilities, and limitations, the targeting process identifies, selects, and exploits critical vulnerabilities of target systems and their associated targets to achieve the commanders' objectives and desired end state. Targeting is a command function requiring commander oversight and involvement to ensure proper execution. It is not the exclusive province of one type of specialty or division, such as intelligence or operations, but blends the expertise of many disciplines.

Targeting helps translate strategy into discrete actions against targets by linking ends, ways, means, and risks. It is a central component of Air Force operational art and design in the application of airpower to create lethal and nonlethal effects. Strategy allows commanders to choose the best ways to attain desired outcomes. Strategy forms the plans and guidance that can be used to task specific airpower capabilities through the tasking process. The processes of planning, tasking, targeting, and assessing effects provide a logical progression that forms the basis of decision-making and ensures consistency with the commander's objectives and the end state.

Too often targeting is tied just to the delivery of kinetic capabilities and the tasking cycle. However, achieving JFC objectives can be accomplished by creating lethal and nonlethal effects, using a variety of kinetic and non-kinetic capabilities. To optimize military action, targeting should integrate the full spectrum of capabilities beginning at the onset of planning. In addition, targeting should occur in peacetime well before hostilities and continue through post-hostilities. Targeting occurs at all levels of conflict (strategic, operational, and tactical), for all phases of operations (Phase 0 through Phase 5), across all domains, and across the range of military operations. Airmen tie the targeting process to creating specific desired effects that achieve objectives. Additionally, Airmen recognize that targeting is a systematic process of analyzing adversaries and enemies to determine critical vulnerabilities against which national capabilities can be applied to create specific desired effects that achieve objectives, taking into account operational requirements and capabilities.

A target is an entity or object considered for possible engagement or other actions. Joint doctrine describes entities as facilities, individuals, equipment, virtual, and organizations. Targets are identified for possible action to support the commander's objectives, guidance, and intent. It is a fundamental tenet of targeting that no potential target derives its importance or criticality merely by virtue of the fact that it exists, or even that it is a crucial element within a target system and other interdependent target systems. Any potential target derives importance, and thus criticality, only by virtue of the extent to which it enables enemy capabilities and actions that must be affected in order to achieve the commander's objectives. Military actions employed may produce lethal or nonlethal effects. Multiple actions may be taken against a single target, and actions may often be taken against multiple targets to achieve a single effect.

I. Target Fundamentals & Characteristics

Ref: Annex 3-60, Targeting (14 Feb '17), pp. 6 to 9.

Targeting is focused on achieving objectives. Through targeting courses of action, objectives and effects are translated into detailed actions against adversary targets. Every target nominated should in some way contribute to attaining the commander's objectives and end state.

Targeting is fundamentally effects-based. Targeting is in part accomplished by targeteers who have specialized training in analyzing targets and developing targeting solutions to support the commander's objectives. In performing their job, they use the targeting cycle and an effects based approach to operations (EBAO).

Targeting is more than just the selection of targets for physical destruction; however this is a limiting perspective. Destruction may be the best means to the end, but it is only one effect within a spectrum of possible options, that may include influence operations, electronic warfare operations and cyberspace operations. The underlying premise of an effects-based approach is that it is possible to direct the instruments of power -- diplomatic, information, military, economic (DIME) -- against targets in ways that cause effects beyond the mere destruction of targets. These effects will influence the adversary's political, military, economic, social, infrastructure, and information systems (PMESII). Targeting should consider all possible means to achieve desired effects, drawing from all available forces, weapons, and platforms. Target selection based upon desired effects includes consideration of second- and third-order consequences that may either positively contribute to campaign objectives or negatively outweigh the near-term results of the applied lethal or nonlethal capability.

Targeting is interdisciplinary. It requires the expertise of personnel from many functional disciplines. For example, strategists and planners bring knowledge of the context and integrated plans; operators bring experience gained from combat execution; intelligence personnel provide analysis of enemy strengths and vulnerabilities and targeting expertise; and judge advocates provide expertise in the application of the Law of War (LOW) and interpretation of rules of engagement (ROE) vital for mission planning and weapons delivery. An effects-based approach to targeting is fundamentally a team effort, requiring these specialties and many more.

Targeting is inherently estimative and anticipatory. Matching actions and effects to targets requires estimating and anticipating future outcomes. In some cases the outcome is straightforward, such as anticipating that disabling a fire control radar may significantly impact a surface-to-air missile battery's capabilities. In most cases, however, estimation is more complicated and planners should consider the following processes to aid in making estimates. The joint intelligence preparation of the operational environment (JIPOE) should yield insight on the enemy and his intentions. The target system analysis (TSA) yields understanding of how components of the enemy system interact and how the system functions as a whole. Intelligence, surveillance, and reconnaissance (ISR) processing, exploitation, and dissemination gathers and processes needed data and helps improve the accuracy and extent of estimation. Such analyses enable planners to select targets and methods of affecting them that increase the probability of desired outcomes and allow the most efficient use of limited airpower resources. This does not imply perfect knowledge or anticipation; uncertainty and friction still apply.

Targeting is systematic. In supporting the commander's objectives, the targeting process is designed to achieve effects in a systematic manner. Targeting, like other planning processes that it complements, is rational, iterative, and methodically analyzes, prioritizes, and assigns forces and capabilities against adversary targets to achieve the effects needed to meet campaign objectives. While targeting is systematic, it is not mechanical and does not assume that the same actions always produce the same effects. If the desired effects are not achieved, targets may be replanned for subsequent engagement, or different targets may be selected.

Targeting is integrated with other processes. Targeting is essential to creating the operational strategy and the joint air operations plan (JAOP), the daily tasking cycle that produces tasking orders, and combat and targeting assessment that measures progress toward operation and campaign objectives. It cannot be separated from the overarching set of processes without resulting in an inputs-based exercise in target servicing—taking a target list, and matching available resources to those targets. Integrating targeting within these overarching processes enables an effects-based approach.

Target Characteristics

In general, there are five characteristics by which targets can be defined: physical, functional, cognitive, environmental, and temporal. The features of each category are briefly described below.

Physical Characteristics

These are features that describe what a target is. These are discernible to one or more of the five senses or through sensor-derived signatures. These may greatly affect the type and number of weapons, the weapon systems, and the methods or tactics employed against the target.

Functional Characteristics

These are features that describe what the target does and how it does it. They describe the target's function within the enemy system, how the target or system operates, its level of activity, the status of its functionality, and, in some cases, its importance to the enemy. Functional characteristics are often hard to discern because they most often cannot be directly observed. Reaching plausible conclusions can often entail speculation and much deductive and inductive reasoning.

Cognitive Characteristics

Features that describe how some targets think, exercise control functions, or otherwise process information. These can be critical to how something is targeted and can be especially important from an effects-based perspective, where information related capabilities (IRC) are considered. These characteristics can also be critical to targeting an enemy system, since nearly every system possesses some central controlling function, and neutralizing this may be crucial to obtaining the desired behavior. As with functional characteristics, these are often difficult to discern or deduce.

Environmental Characteristics

These are features that describe the effect of the environment on the target and its surroundings. These characteristics may also affect the types and numbers of weapons, weapon systems, and the methods or tactics employed against the target.

Temporal Characteristics

The factor of time, as a characteristic of a target, describes the targets vulnerability to detection, attack, or other engagement in terms of time available. All potential targets and all targets nominated for attack continually change in priority due to the dynamic nature of the evolving operational environment. Many targets may be fleeting and some may be critical to friendly operations. Those that are both fleeting and critical present one of the biggest targeting challenges faced by the joint force. This time factor can help determine when and how to find or engage a target. By comparing this factor to information latency and knowledge of friendly capabilities, the staff can make better recommendations to the commander regarding possible actions.

II. Types Of Targeting

There are two categories of targeting: deliberate and dynamic. It is a mistake to associate deliberate targeting with fixed targets and dynamic targeting with mobile targets.

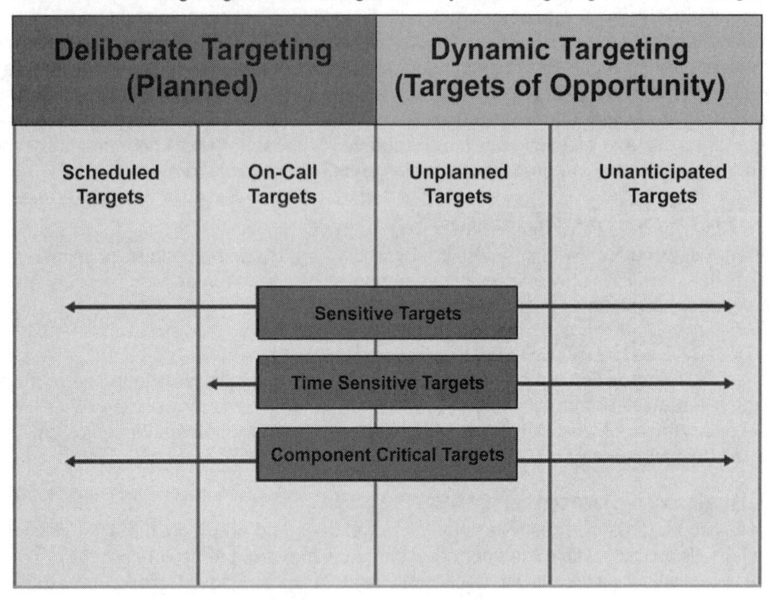

A. Deliberate Targeting

Deliberate targeting applies when there is sufficient time to add the target to an air tasking order (ATO) or other plan. Deliberate targeting includes targets planned for attack by on-call resources. The air tasking cycle is sufficiently flexible to allow for most mobile targets to be planned and attacked with deliberate targeting.

B. Dynamic Targeting

Dynamic targeting includes targets that are either identified too late, or not selected in time to be included in deliberate targeting, but when detected or located, meet criteria specific to achieving objectives. When plans change and planned targets must be adjusted, dynamic targeting can also manage those changes.

Sensitive and Time Sensitive Targets

Two subsets of targets that require special consideration are sensitive and time sensitive. **Sensitive targets** are targets where the commander has estimated the physical and collateral effects on civilian and/or noncombatant persons, property, and environments occurring incidental to military operations, exceed established national-level notification thresholds. Sensitive targets are not always associated with collateral damage. They may also include those targets that exceed national-level rules of engagement (ROE) thresholds, or where the combatant commander (CCDR) determines the effects from striking the target may have adverse political ramifications. **Time-sensitive targets (TSTs)** are joint force commander (JFC) validated targets or sets of targets requiring immediate response because they are highly lucrative, fleeting targets of opportunity or they pose (or will soon pose) a danger to friendly forces. These targets present one of the biggest targeting challenges.

III. Effects-Based Approach to Operations (EBAO)

In the most fundamental sense, effects-based approach to operations (EBAO) is defined as an approach in which operations are planned, executed, assessed, and adapted to influence or change systems or capabilities in order to achieve desired outcomes. Consequently, targeting personnel seek to understand and exploit the complex connections among individual actions, the effects—direct and indirect—that actions produce, how those effects influence the states and behaviors of complex systems in the operational environment, and how effects contribute to the accomplishment of ultimate desired outcomes.

See pp. 5-23 to 5-25 for related discussion of EBAO to include an overview of effects (direct, indirect, intended and unintended), objectives and actions.

Effects

Effects are the physical or behavioral state of a system that results from an action, a set of actions, or another effect. They are the full range of outcomes, events, or consequences of a particular cause. A cause can be an action, set of actions, or another effect. Effects join actions to objectives. The actions and effects in any causal chain can derive from any instrument of national power—diplomatic, informational, military, economic (DIME), and may occur at any point across the range of military operations from peace to global conflict. Properly understanding the relationship among effects at all levels is important to planning and conducting any campaign.

Effects can be intended or unintended, direct or indirect, lethal or nonlethal. Intended and unintended are straightforward in meaning. A direct effect is the first-order result of action with no intervening mechanism between act and outcome—usually immediate and empirically verifiable, like the results of weapons employment. Indirect effects are more complicated. An indirect effect is a second-, third-, or higher-order effect created through an intermediate effect or causal linkage following a tactical action—usually a delayed and/or displaced consequence associated with the action that caused the direct effect(s). Direct and indirect effects can be intended or unintended. For instance, a secondary or tertiary effect in the space domain can be the desired outcome of a weaponeered first order effect in the land domain, leading to achieving the objective. Objectives are achieved through an accumulation of direct and indirect effects, but the effects sought at the strategic and operational levels are almost invariably indirect. Broadly speaking, lethal effects are those effects intended to kill and/or destroy or cause great physical damage to the target, while nonlethal effects are not intended to cause death and destruction. Lethal and nonlethal effects are not necessarily tied to a capability. A capability normally used to create nonlethal effects may be used to create lethal effects and vice versa.

Effects are often categorized as physical or behavioral, and are assessed functionally or systemically. Physical effects materially alter a system or target and are most important at the tactical level. Behavioral effects are those that impact reasoning, emotion, and motivation and result in measurable changes in behavior. Functional effects relate how well a system performs its intended function(s) and systemic effects relate how well that system functions as a component of larger systems.

Effects can be imposed cumulatively or in a cascading manner, sequentially or in parallel. Effects can accumulate over time leading to gradual change, or can be cascading changes that occur catastrophically and ripple through related and subordinate systems. Often, there are both cumulative and cascading components to effects. Effects can also be imposed sequentially or in parallel. Effects imposed in series, one after another over time, are sequential. Those imposed near-simultaneously are parallel effects, which may place greater stress upon targeted systems and require faster adaptation. Full understanding of the types of effects and the principles of effects-based thinking can offer commanders more options, hasten success, and lead to success with lower cost in terms of lives, assets, and time.

IV. The Targeting Cycle

Ref: Annex 3-60, Targeting (14 Feb '17), pp. 11 to 12.

The targeting cycle supports the joint force commander's (JFC) joint operation planning and execution with a comprehensive, iterative and logical methodology for employing ways and means to create desired effects that support achievement of objectives. The targeting cycle described in joint doctrine is also used for Air Force targeting operations. Joint targeting selects and prioritizes targets and matches the appropriate means to engage them, considering operational requirements and capabilities. The **joint targeting cycle** is an iterative, non-linear process that provides a framework for successfully conducting joint targeting. The deliberate and dynamic nature of the joint targeting cycle supports all of the planning horizons of the joint operation planning process/joint operation planning process for air (JOPP/JOPPA) future plans, future operations, and current operations.

See pp. 6-19 to 6-22 for related discussion of joint targeting.

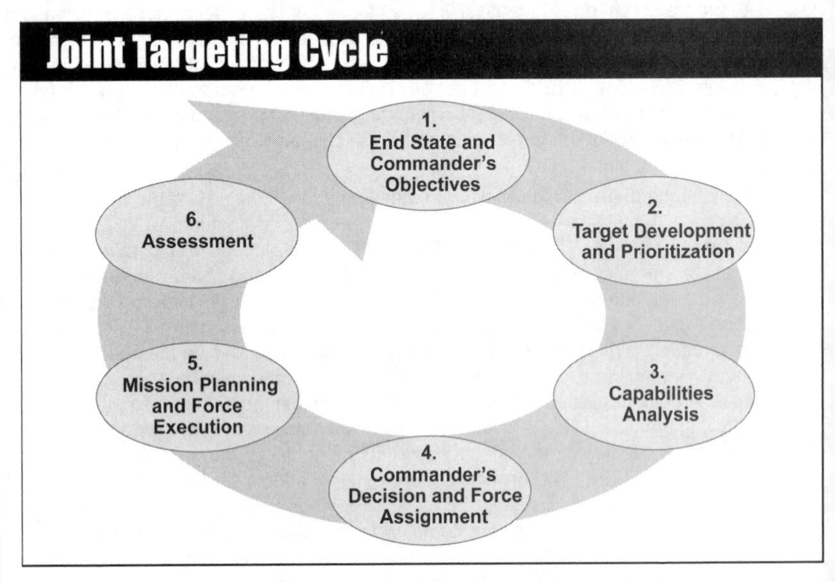

Ref: JP 3-60, Joint Targeting (Jan '13), fig. II-2. Joint Targeting Cycle.

1. The End State and Commander's Objectives

The military end state is the set of conditions that need to be achieved to resolve the situation or conflict on satisfactory terms, as defined by appropriate authority. The combatant commander (CCDR) typically may be concerned with the military end state and related strategic military objectives. The commander's objectives are developed during the mission analysis step of JOPP, or are derived from theater-strategic or national-level guidance. The commander, Air Force forces (COMAFFOR) staff, using the JOPPA, should establish the air component's objectives; the specified, implied, and essential tasks that support the CCDR's military objectives and contribute to achievement of the end state.

Objectives are the basis for developing the desired full spectrum effects and the scope of target development. Attainment of objectives is essential to the successful realization of the desired end state. Effective targeting is distinguished by the ability to generate the type and extent of effect necessary to achieve the commander's objectives. Integrating and employing the appropriate lethal and nonlethal capabilities creates the desired effects.

2. Target Development and Prioritization

Target development is the systematic examination of potential target systems to determine the type and duration of full spectrum action that should be exerted on each target to create desired effects that achieve the commander's objectives. Target development always approaches adversary capabilities from a systems perspective. Target vetting leverages the expertise of the national intelligence community to verify the fidelity of the intelligence and analysis used to develop the target(s). Target validation determines whether a target remains a viable element of a target system and whether prosecution of that target complies with the law of war and the rules of engagement. Validation is a continuous process that occurs until the target is serviced or removed from consideration for servicing. Once candidate targets are developed, vetted, and validated, they are added to the joint target list (JTL) or restricted target list (RTL). During execution, they are prioritized relative to all joint targets in a joint integrated prioritized target list (JIPTL), which is submitted to the commander for approval.

3. Capabilities Analysis

This portion of the joint targeting process involves evaluating the full spectrum of available capabilities (including forces, sensors, and weapons systems) against desired effects to determine the appropriate options available to the commander. Inputs to this stage include lethal and nonlethal considerations: target characteristics, desired damage criteria or probability of damage (Pd) calculations, delivery parameters, effects timing and effect duration. The outputs of this stage include the probability of effectiveness (Pe) which is the result of selected capabilities and target pairings required to create desired effects to inform the commander's estimate within the joint planning and execution system.

4. Commander's Decision and Force Assignment

The force assignment process integrates previous phases of joint targeting and fuses capabilities analysis with available forces, sensors, and weapons systems. It is primarily an operations function, but requires considerable intelligence support to ensure Intelligence, surveillance, and reconnaissance (ISR) assets are integrated into the plan. The process of resourcing joint integrated prioritized target list (JIPTL) targets with available forces or systems and ISR assets lies at the heart of force assignment. Once the JFC has approved the JIPTL, either entirely or in part, tasking orders are prepared and released to the executing components and forces. The joint targeting process facilitates the publication of tasking orders by providing amplifying information necessary for detailed force-level planning of operations. Coordination with other services and special programs at this point in the process is essential to ensure that targets are not serviced by multiple or conflicting resources.

5. Mission Planning and Force Execution

Upon receipt of tasking orders, detailed planning should be performed for the execution of operations. The joint targeting process supports this planning by providing tactical-level planners with direct access to detailed information on the targets, supported by the nominating component's analytical reasoning that linked the target with the desired effect (conducted in Phase 2 of the joint targeting cycle). This may provide the background information necessary for the warfighter to focus on the JFC's objectives as the operation unfolds.

6. Assessment

Assessment measures whether desired effects are being created, objectives are achieved, and next steps are evaluated. Effective planning and execution require continuing evaluation of the effectiveness of friendly and enemy action. Consequently, assessment is much more than "battle damage" or "combat assessment," as it has traditionally been presented—and more than just an intelligence function that takes place after execution has concluded. Planning for it begins prior to commencement of operations, takes place throughout planning and execution, and continues after the conflict is over. The assessment phase is common to both deliberate and dynamic targeting. The assessment of deliberate and dynamic target engagement results must be integrated to provide the overall targeting assessment.

V. Targeting-Related Responsibilities

Ref: Annex 3-60, Targeting (14 Feb '17), pp. 17 to 21.

Targeting occurs from the combatant command level to the tactical unit level. Across this organizational span, Air Force targeting focuses on a wide variety of targeting issues both within and outside of the targeting cycle such as: target planning, target materials production, targeting database maintenance, target systems analysis, targeting automation and support to weapons acquisition. The **air component** is responsible for enacting the targeting process for the **joint force commander (JFC)** and servicing approved targets, regardless of which service or functional component nominates them. Within this command structure the targeting cycle of planning, execution and assessment occurs under a very structured process, and normally under a compressed timeline.

The **COMAFFOR** establishes a close working relationship with the JFC. This relationship extends through the JFC and COMAFFOR staffs and other component staffs with a role in supporting the JFC with targeting capabilities. The COMAFFOR, as the Air Force's warfighting commander, directs execution of Air Force capabilities. If a **JFACC** is appointed, that commander directs execution of air component capabilities and forces made available for joint or combined operations. The COMAFFOR normally operates from an air operations center (AOC). The AOC and the COMAFFOR's staff are manned with subject matter experts who reflect the capabilities/forces available to the COMAFFOR for tasking and include appropriate component representation.

During day-to-day operations, the **air operations center (AOC)** plans, directs, and monitors theater air component operations including integrated targeting activities.

The **theater air ground system (TAGS)** is a system of systems that consists of component C2 elements for the purpose of working together in planning and executing operations. TAGS enables employment of the air targeting cycle from the operational to the tactical level. Comprised of airborne and ground elements, the Theater Air Control System (TACS) is the Air Force component of TAGS and the mechanism for C2 of airpower. The AOC is the senior C2 element of the TACS. *See pp. 3-32 to 3-33.*

The **air support operations center (ASOC)** is the tactical level organization that facilitates Air Force-Army integration and provides primary control of air power in support of the Army continuing down through the TACS Air Force component liaisons aligned with land combat forces. The ASOC's primary mission is to provide direction and control of air operations directly supporting Army ground forces. Within the targeting arena, this is a critical component in that it supports deliberate targeting requirements during planning and fulfills the dynamic targeting role where immediate targeting supports Army forces.

The COMAFFOR may establish one or more **joint air component coordination elements (JACCEs)** with other commanders' headquarters to better integrate joint air operations with their operations. The JACCE is a service/functional component level liaison that serves as the direct representative of the COMAFFOR/JFACC. The JACCE can be critical to targeting processes. For example, the JACCE located with the joint force land component commander (JFLCC) provides valuable assistance and liaison from the JFACC/COMAFFOR and assists the JFLCC in planning and synchronizing operational fires and the establishment and control of fire support coordination measures (FSCMs)

See pp. 2-21 and 3-13 for further discussion of the JACCE.

Commander, Air Force Forces (COMAFFOR)

These are the COMAFFOR's responsibilities unless the JFC has appointed a Joint Force Air Component Commander (JFACC).

The targeting responsibilities of the COMAFFOR are assigned by the joint force commander (JFC). As the air component facilitator for servicing of all targets nominated for airpower effects, the COMAFFOR is responsible for establishing a targeting process that meets the needs of the JFC and all represented components within the air operations center (AOC). Targeting processes are inherent to or affected by the COMAFFOR responsibilities listed below.

- Plan, coordinate, integrate, task, and direct the joint air effort in accordance with the JFC's guidance and joint force objectives.
- Develop a joint air operations plan (JAOP) derived from the JFC's broader objectives for the operation, and guidance regarding the roles, missions, tasks, and responsibilities of joint capabilities and forces.
- Recommend, after consulting with other component commanders, apportionment of the joint air effort that should be devoted to various air operations for a given period of time.
- Translate air apportionment into allocation and develop targeting guidance into the air tasking order, which may include specific aimpoints.
- Conduct target development and produce required target materials for operational and tactical level force employment planning and execution.
- Direct and ensure deconfliction of joint air operations.
- Integrate and synchronize joint air operations.
- Coordinate with the appropriate components, national agencies, and liaison elements for synchronization and deconfliction with land, maritime, space, cyberspace, and special operations.
- Coordinate with the appropriate components' agencies and liaison elements for tasking of the capabilities and forces made available.
- Monitor execution and redirect joint air component operations as required.
- Compile component target requirements and prioritize targets based on JFC guidance.
- Establish rules of engagement (ROE) and special instructions (SPINS) that state clear combat identification (CID) requirements; for example, which CID systems may be used, who can declare a track hostile, etc.
- Accomplish tactical and operational assessment and support accomplishment of campaign and national assessment.

Unit-Level Targeting Responsibilities

Individual units have targeting responsibilities that support and enhance air operating center (AOC) efforts and tactical-level execution. Commanders, mission planners, and intelligence specialists within these units should ensure the validity and accuracy of the targeting information provided to them for mission planning purposes. This responsibility may include verification of air tasking order (ATO) guidance coordinates and adjudication of problems with the AOC if errors or conflicts become evident. Specific data provided to mission planners should be checked for integrity, including verification of the joint desired point of impact (JDPI) coordinates and elevations, weapon azimuths and impact dive angles, fusing instructions, collateral damage considerations, target area graphics, etc. when direct electronic transfer of such data is not possible or fails.

Considerable benefits are realized when air and ground units work together directly to accomplish mission planning at the tactical level. Army ground liaison officers (GLO) working with tactical air units can provide insight into ground component plans and offer direct coordination for missions flown in support of the ground commander's intent. Air liaison officers (ALO) are aligned with tactical ground maneuver units and serve as advisors to ground commanders on targeting and other aspects of airpower. Such coordination is essential for joint operations.

VI. Targeting Coordination & Liaisons

Ref: Annex 3-60, Targeting (14 Feb '17), pp. 27 to 30.

Proper database management is necessary for effective targeting. Joint Targeting Toolbox (JTT) is the tool of record for the joint targeting community. However, there are still systems used in the field that are "stovepiped" and cannot talk to JTT or store data within MIDB. If support organizations lack appropriate interoperable systems and databases, it is the responsibility of the supported entity to work with the supporting entity and targeting and systems maintenance staffs to develop procedures (in peacetime) to overcome the difficulties associated with using systems that are not interoperable. There are many users of information in the AOC. Ideally, everyone should work from the same databases (i.e., data and imagery) to facilitate effective use of manpower and coordination.

Targeteers should coordinate with many different teams to ensure the flow and management of data and database information in the AOC is as seamless as possible. Those with whom targeteers should coordinate include (but are not limited to):

Analysis, Correlation, and Fusion Team (ACF Team). The ACF Team in the intelligence, surveillance, and reconnaissance division (ISRD) is responsible for updating enemy order of battle (EOB) databases. Targeteers should be able to pull from this database to ensure targeteers are using the most current EOB.

ISR Operations Team. The ISR Ops Team in the ISRD is responsible for planning and coordinating intelligence-gathering missions by air component assets. They also have insight into intelligence-gathering platforms that the air component does not own, including spacecraft. Ensuring targeting and collection management databases are the same may reduce the time required to task collection assets to support targeting efforts, especially in the case of dynamic targeting.

Targets and Tactical Assessment (TGT/TA). The TGT/TA team is comprised of two primary cells, the target development cell and the TA cell, which provide direct support and embedding of personnel to other AOC divisions to ensure continuity in the targeting effort. This team provides full-spectrum effects-based approach to operations (EBAO) based targeting development, solutions, and products/materials in support of the air tasking cycle. It is also responsible for assessing the immediate results and effects of capability employment during tactical operations. These assessments may lead to some type of follow-on action by friendly forces.

Senior Intelligence Duty Officer (SIDO) Team. The ISR Team in the combat operations division (COD), led by (and sometimes consisting only of) the SIDO, provides intelligence support to ATO execution in the areas of analysis, collection management, targeting, and assessment. Access to the Joint Targeting Database (JTDB) within the MIDB enables the seamless targeting support when the ATO requires modification. This access is magnified when supporting dynamic targeting operations, especially those involving time sensitive targets (TST).

Operational Assessment Team (OAT). The OAT in the SD is responsible for determining whether or not desired effects are being created and if those effects are leading to the attainment of COMAFFOR and JFC objectives. Since the JTDB within the MIDB is used by the OAT, specific targets can be tracked to specific effects and objectives.

Strategy Plans Team. The strategy plans team in the strategy division is responsible for building the overall air component strategy and is responsible for producing the Joint Air Operations Plan (JAOP). This phase of planning may involve a need to access the JTBD within MIDB in order to support JAOP creation.

The Strategy Guidance Team. The strategy guidance team is responsible for the AOC's transition from operational-level to tactical-level planning and culminates in the air operations directive (AOD). The guidance provided is typically short-range; 24-hours to

10 days from execution. This team develops operational guidance, prioritizes operational and tactical objectives, and determines tactical allocation.

Space Operations Specialty Team (SOST). The SOST consist of space operators assigned from multiple services to support theater operations. They are embedded into the Strategy, Plans, and Combat Operations Divisions to fulfill multiple roles to serve as theater advisors for space capabilities (national, military, civil, commercial, and foreign). *Refer to Annex 3-0, Operations and Planning, and AFTTP 3-1.AOC, 30 March 2016, for more information.*

Information Operations/Non-Kinetic Operations Team (IO/NKOT). The IO/NKOT provides the focal point for ensuring the synchronized planning, execution, and assessment of IO and nonlethal capabilities into the targeting cycle to create nonlethal effects. Its primary operations are electronic attack (EA), electronic warfare support (ES), space control, offensive cyberspace operations (OCO), and defensive cyberspace operations (DCO). Additionally, the IO/NKOT serves as the integration point employing information-related capabilities (IRC) via military support operations (MISO), operations security (OPSEC), and military deceptions (MILDEC). The IO/NKOT leads and develops IO and non-kinetic capability requirements as part of the effects-based approach to targeting for both preplanned situations via the ATO cycle and dynamic situations. *Refer to Annex 3-0, Operations and Planning, Annex 3-12, Cyberspace Operations, Annex 3-13, Information Operations, Annex 3-14, Space Operations, Annex 3-51, Electronic Warfare, and AFTTP 3-1.AOC, 30 March 2016, for more information.*

In addition, targeteers should also coordinate with the following **liaisons**. Each liaison represents their component or agency, and provides critical communication to the targeting process. This communication includes the submission of targets for consideration, coordination of targeting information, coordination of targeting capabilities, targeting support, and many other functions. Please reference AF Doctrine Annex 3-0 for more information. The following list is not all inclusive:

Battlefield Coordination Detachment (BCD). The BCD is the Army Forces (ARFOR) commander's liaison to the supporting air component commander's AOC. BCDs are assigned to Army service component commands of geographic combatant commands with duty at each numbered Air Force with a geographic AOC. The BCD expedites the exchange of information digitally and performs face-to-face coordination with elements in the AOC. *Refer to Annex 3-03, Counterland Ops, and AFTTP 3-1.AOC, 30 March 2016, for more information.*

Special Operations Liaison Element (SOLE). The SOLE is a joint element provided by the joint force special operations component commander (JFSOCC) or joint special operations task force (JSOTF) commander. SOLE personnel work with the various AOC functional areas to ensure that all SOF targets, SOF teams, and SOF air taskings/missions are deconflicted, properly integrated, and coordinated during all planning and execution phases. *Refer to Annex 3-0, Operations and Planning, Annex 3-05, Special Operations, and AFTTP 3-1.AOC, 30 March 2016, for more information.*

The main targeting database is the **modernized integrated database (MIDB)** with its associated data access layers, which can be accessed via the **joint targeting toolbox (JTT)** and command and control (C2) tools like the **Theater Battle Management Core System (TBMCS)**. Problems with compatibility between different versions of MIDB within the AOC weapons system versus the MIDB installed at combatant commands and DIA has forced targeteers in some theaters to utilize workarounds in order to transfer data between systems. Specialized databases also exist with functional tools like Joint Capabilities Analysis and Assessment System and the Space Integrated Planning Service (SIPS). Given the potential for incompatibility and diverging information, a thorough understanding of the interoperability and processes to maintain synchronicity between databases and C2 tools is necessary for successful execution of operations.

VII. Establishing Collaborative and Support Targeting Relationships

Targeting is a collaborative effort. Targeteers are consumers of multi-source intelligence data and operate across both the intelligence and operations functions. Manning and targeting resources at the joint task force (JTF), air operations center (AOC), and Joint Intelligence Operations Center (JIOC) are typically insufficient to support robust target planning and execution. The targeting process requires resources from many organizations to meet the commander's targeting demands. Targeting therefore requires reachback support via distributed and federated operations to be effective.

A. Reachback

Reachback is the process of obtaining products, services, and applications, or forces, or equipment for material, from organizations not forward deployed. For example, during crisis planning or contingency operations, the 363rd ISR Wing (363 ISRW) may stand up a crisis management element (CME) to provide direct targeting support to the commander, Air Force forces (COMAFFOR). Personnel assigned to the CME may operate in a supporting relationship to the COMAFFOR.

B. Distributed Operations

Distributed operations in support of targeting occur when independent or interdependent nodes or locations participate in the operational planning and/or operational decision-making process to accomplish goals/missions for engaged commanders. In some instances, the commander may establish a formal supported/supporting relationship between distributed nodes. In other instances, distributed nodes may have a horizontal relationship.

C. Split Operations

Split operations are a type of distributed operations. The term describes those distributed operations conducted by a single entity separated between two or more geographic locations. A single commander should have oversight of all aspects of a split operation. For example, sections of the air tasking order (ATO) may be developed from a rear area or backup operation center to reduce the deployed AOC footprint. In this case, the AOC is geographically separated and is a split operation. During split operations, the COMAFFOR has the same degree of authority over geographically separated elements as he or she does over the deployed AFFOR and AOC.

Although distributed operations are similar to reachback, there is one major difference. Reachback provides ongoing combat support to the operation from organizations that are not forward deployed, while a distributed operation indicates teaming with forward deployed independent or interdependent nodes. With distributed operations, some operational planning or decision-making may occur from outside the joint area of operations. The goal of effective distributed operations is to support the operational commander in the field; it is not a method of command from the rear. The concept of reachback allows functions to be supported by a staff at home station, to keep the manning and equipment footprint smaller at a forward location.

Federated operations are based on the needs of geographic combatant commanders, JFCs, or COMAFFOR. Joint targeting federation needs are coordinated with the larger joint community and national agencies through the JTF staff J-2's targeting directorate. Coordination should delineate specific duties to federated partners, establish timelines, and determine the methods of communication to be used.

While the COMAFFOR may have direct authority over some units, he/she may not have control over targeting organizations beyond the AOC and those units/personnel who augment the air component.

II. Target Planning

Ref: Annex 3-60, Targeting (14 Feb '17), pp. 31 to 37.

From guidance to assessment, targeting is a critical component in activities across the range of military operations. Air Force targeting principles may be applied to all instances in which military force is planned and executed. Air Force targeting personnel are involved in activities in all levels of command and operations. Targeteers and other planners should keep effects-based concepts in mind while building formal plans and conducting ongoing deliberate targeting once operations begin.

Planning encompasses all the means through which strategies and courses of action (COA) are developed, such as operational design, deliberate planning and crisis action planning. As a Service and as part of a joint or combined force, the Air Force uses the joint operations planning process for air (JOPPA). This process is the air component's equivalent of the joint force commanders' (JFC) joint operation planning process (JOPP) and is often performed in sequence or parallel with it. The JOPPA produces the joint air operations plan (JAOP) and the air operations directive (AOD), which guides the tasking cycle through its iterative execution as part of an ongoing battle rhythm. Since it sets the stage for all other actions, planning is where effects-based principles have the largest play and may have the greatest impact on operations. Plans should tie objectives, actions, and effects at all levels together into a logical, coherent whole strategy.

Targeting supports operational-level planning and validates that operational plans can be accomplished within the time and resources available. This support also helps create the detailed tactical-level products, usually appended to operational-level plans, for the opening phases of action. The objectives, guidance, and intent derived during planning guide all efforts, including targeting, throughout employment and assessment. This serves to inextricably tie planning, employment, and assessment together. Further, planning continues once operations commence and the battle rhythm is under way. Operational planning continues as adversary actions are evaluated or anticipated through revision of strategy and implementation of branches and sequels.

Types of Targeting

Deliberate Targeting

Dynamic Targeting

Targeting planning is divided into two categories, deliberate and dynamic. Deliberate targeting normally supports the future plans effort which is focused on all planning activities from 72/96 hours out to, but not including, the current air tasking order (ATO) execution day; whereas, dynamic targeting normally supports the current ATO execution with immediate targeting responsiveness to the active environment created by ongoing weapons employment and real-time, all-source identification of emerging and time sensitive targets (TSTs) (i.e., unplanned and unanticipated targets).

See pp. 7-17 to 7-20 for an overview and discussion of deliberate targeting and pp. 7-21 to 7-30 for an overview and discussion of dynamic targeting.

Targeting Considerations during JOPPA

Ref: Annex 3-60, Targeting (14 Feb '17), pp. 35 to 37.

Targeting During Formal Planning

Targeting supports every form of employment planning for joint operations. Joint operation planning employs an integrated process for orderly and coordinated problem solving and decision-making of JFC's desired objectives. In its peacetime application, the process is highly structured to support the thorough and fully coordinated development of contingency plans. In crisis, the process is shortened as needed to support the dynamic requirements of changing events. In wartime, the process adapts to accommodate greater decentralization of joint operation planning activities.

The JAOP is created through the seven step JOPPA and is normally developed in support of the JFC's plan or order. Almost all targeting support to pre-conflict planning is accomplished through the JOPPA.

See pp. 6-5 to 6-18 for an overview and complete discussion of the Joint Operations Planning Process for Air (JOPPA). *The discussion below only highlights certain targeting considerations during specific steps of JOPPA.*

Initiation

The commander, Air Force forces (COMAFFOR) and staff performs an assessment of the initiating directive to determine time available until mission execution, current status of intelligence products, and other factors relevant to the specific planning situation.

Mission Analysis

During this stage, joint intelligence preparation of the operational environment (JIPOE) begins. In order to fully support an effects-based campaign, the intelligence community should conduct robust JIPOE to inform planning. JIPOE provides a comprehensive framework for Intelligence, surveillance, and reconnaissance (ISR) support to planning and COA selection. Consequently, JIPOE should assist commanders in anticipating enemy intent and enable them in pre-empting enemy actions. The JIPOE process continues throughout planning by examining adversary and friendly capabilities, adversary intent, and the operational environment. Enemy and friendly centers of gravity (COG) are also identified during this initial stage of the JOPPA. As mission analysis is refined through later stages of the JOPPA, enemy COGs are analyzed, yielding critical vulnerabilities or other key system nodes. These are further examined through target system or nodal analysis to yield target sets, targets, critical elements, and aimpoints, as well as commander's critical information requirements (CCIRs) to support JIPOE and tactical assessment. Such analysis carries a considerable information-flow cost. In order to properly identify collection and exploitation requirements for targeting, target system analysis (TSA) and or targeting effects studies should begin well in advance of operations and should continue throughout them. It should begin during the initial stages of JIPOE and draw upon as much ongoing peacetime intelligence/targeting material as is available for the theater or area of operations. While space, cyberspace, and information operations should already be fully integrated into mission analysis, JIPOE, TSAs and target development should also ensure integration of specialized analysis in support of space, cyberspace, and information operations.

COA Development

JIPOE is refined during this stage and includes detailed analysis of COGs identified during mission analysis. COG analysis is important to targeting efforts because it identifies the enemy's sources of power and will to fight and tries to discover how and where those sources of power are vulnerable, where critical nodes within them are, and how they can be exploited by the full capabilities of the joint force (e.g., air, space, cyberspace, information operations, etc.). Critical vulnerabilities can be difficult to pick from critical requirements or to translate those vulnerabilities into explicit target sets. Techniques for translating vulnerabilities into targets can be used as the foundation for development of COAs or a selected COA may be directed by the JFC.

Plan or Order Development

This step and its ultimate product, the JAOP, describe how the air component may support the JFC's operational plan. The JAOP identifies objectives, desired effects, targets, and assessment measures in as much detail as available time and intelligence allow. Objectives and the end state are derived from commander's guidance, strategy development and planning. Targeting efforts should always aim toward achieving these objectives and the end state. During JAOP development, deliberate targeting is used to develop targets and target sets included in the JAOP and its attachments. Even if targeting information developed during planning is not included in the JAOP or its attachments, JAOP development may require considerable targeting effort in order to validate selected COAs, CONOPS, and other elements of the plan. Commanders and planners should know, at least approximately, how much effort and what resources are required to achieve the operation's desired effects. This knowledge can be gained by conducting some (at least notional) deliberate targeting systems analysis using existing TSA products, functional system products (i.e., power, roads, communications, chemical, etc.), targeting databases, and/or assessment of the total number of potential targets within the modern integrated database (MIDB) binned into functional categories (e.g., airfields, air defense, ballistic missile, WMD, C4I, etc.) before the conflict begins. Target selection should be based upon desired effects against enemy COGs, which in turn should be based upon the objectives for the conflict.

The JAOP should be effects-based, including lethal and/or nonlethal effects, as appropriate. It is the air component's main source of guidance. Targeting efforts play a major role in building an effects-based JAOP by relating effects to particular targets and target systems and helping validate whether planned resources can achieve those effects.

The JAOP should provide broad guidelines for prioritizing targets/target systems, as well as making clear which categories or sets are most important to the campaign. The JAOP should also provide guidance on the sequencing of targeting actions or effects, which is not the same thing as priority. Although parallel effects are generally best, sometimes some targets should be attacked first to enable effects against other targets. The JAOP, as well as subsequently published special instructions (SPINS), AOD, and ATOs, should clearly articulate the commander's rules of engagement (ROE) that ensure operations comply with the law of war (LOW).

Finally, the JAOP should establish guidelines for dynamic, especially time-sensitive, targeting. Dynamic targeting is one of the most labor-intensive and intellectually demanding challenges the air component faces. Anticipating as much of the challenge as possible and spelling out guidance and priorities in the JAOP may ease the burden on commanders and air operations center (AOC) combat operations division (COD) personnel once the daily battle rhythm begins. This may prevent mistakes from being made during employment or may at least mitigate their impact. Planners should address as broad a scope as possible in as much detail as time and planning resources allow. This should include robust ROE and related legal considerations.

Targeting support to formal operational planning, and the deliberate targeting conducted once operations begin, are both accomplished through the deliberate targeting process. Deliberate targeting is the procedure for prosecuting targets that are detected, identified, and developed in sufficient time to schedule actions against them in tasking cycle products such as the ATO. Deliberate targeting handles targets in one of two ways: 1) plans and schedules specific actions against specific targets and 2) creates on-call packages or missions that deal with targets through pre-determined concept of operations (CONOPS). Preplanned missions are typically used against fixed targets or targets that are transportable, but operate in fixed locations. However, deliberate targeting can be used against mobile targets. On-call missions can be used against fixed, transportable, and mobile targets. For instance, a fixed building may be watched, but does not become a target until some critical person, group, or equipment arrives, at which time the on-call mission is scheduled on the tasking order if intelligence arrives in sufficient time. Other potential targets that are detected or become significant during the current execution period (once all formal products of the planning and tasking processes are issued), including the JFC's TSTs, are dealt with using dynamic targeting.

Target Nomination

Target nomination processes remain unchanged when addressing offensive, non-lethal operations and should be leveraged appropriately by planners. That is, target development and selection are based on what the commander wants to achieve rather than on the available ways and means to achieve them. Therefore, nonlethal targets should be nominated, vetted, and validated within the established targeting processes. However, nonlethal operations may require parallel target development, selection, nomination, capability analysis and allocation, etc. that arise from unique authorities (i.e., cyber targets), which may extend the planning and execution approval timelines. Targeting personnel should work closely with the appropriate liaisons to these authorities to synchronize target planning within the ATO.

See p. 7-41 for discussion of the target nomination process.

Effects-Based Approach to Operations (EBAO)

The effects-based principles set forth in Annex 3-0, Operations and Planning, should guide all planning efforts, including targeting. An effects-based approach is even more critical for success in stability operations such as counterinsurgency and peace enforcement, because they may rely more on nonlethal means and less on types of effects for which cause and effect are well understood. Effects-based approach to operations (EBAO), and in particular targeting, ensures that every effect delivered can be linked to the JFC's end state, objectives, and plans. Within targeting, EBAO focuses on why we are taking an action rather than what action we are taking. To exploit the full range of possible effects in a given situation, planners should understand what effects are, how they relate to actions and objectives, how to measure different effects, and how various types of effects can be exploited to yield desired outcomes.

See pp. 5-23 to 5-25 for related discussion of EBAO to include an overview of effects (direct, indirect, intended and unintended), objectives and actions.

Chap 7

III. Deliberate Targeting

Ref: Annex 3-60, Targeting (14 Feb '17), pp. 38 to 41.

Deliberate targeting provides a systematic analytical approach that focuses targeting efforts on supporting operational requirements and the commander's objectives. It helps focus the appropriate capabilities against adversary targets at the right time and place to impose specific desired effects that achieve joint force objectives.

Deliberate targeting supports the air tasking cycle, which creates a daily conveyance of the overall air component strategy. Deliberate targeting within the tasking cycle is the means Airmen use to accomplish the COMAFFOR's non-dynamic targeting requirements. Therefore, this section discusses deliberate targeting within the context of the air tasking cycle. The air tasking cycle develops the products needed to build and execute an ATO and accomplish assessment. Although it is presented below as six separate, sequential phases, in reality the targeting process is bi-directional, iterative, multi-dimensional, sometimes executed in parallel, and part of a larger set of processes. It is built on a foundation laid by thorough JIPOE. Participants from the AOC's strategy, ISR, plans, and operations divisions accomplish various targeting responsibilities, integrating their products into all levels and stages of the air tasking cycle. The cycle consists of the following phases performed at various levels of command:

Deliberate Targeting

- **Objectives, effects, and guidance**
- **Target development**
- **Weaponeering and allocation**
- **ATO production and dissemination**
- **Execution planning and force execution**
- **Assessment**

The tasking cycle has usually been represented as a set of distinct processes that separately accomplish targeting, apportionment and allocation of joint air capabilities to produce the ATO. In fact, these processes are all closely interrelated, regarding them as distinct entities misses the central insight that they should work together as an integrated whole, if targeting and tasking are to be most effective. Targeting and ATO production are essential to the tasking cycle. Although the targeting and tasking cycles perform separate and distinct functions, they are highly intertwined and require close coordination, the two cycles run almost in parallel. Once a daily battle rhythm is established, the tasking cycle as a whole encompasses the entire process of taking commander's intent and guidance; determining where to apply force or other actions to fulfill that intent; matching available capabilities and forces with targets; putting this information into an integrated, synchronized, and coordinated order; distributing that order to all users; monitoring execution of the order to adapt to changes in the operational environment; and assessing the results of that execution. The cycle is built around finite time periods required to plan, integrate, coordinate, prepare, conduct, and assess air operations. These time periods may vary from theater to theater, but the tasking cycle and its constituent processes drive the AOC's battle rhythm and thus helps determine deadlines and milestones for related processes, including targeting.

A principal purpose of the air tasking cycle is to produce orders and supporting documentation to place a flexible array of capabilities in a position to create desired effects in support of the commander's intent. This cycle is driven by the tyranny of time and distance. It takes time for ground crew to prepare aircraft for flight, for aircrew to plan missions, and for aircrew to fly to the immediate theater of operations from distant airfields. Likewise, commanders should have enough visibility on future operations to ensure sufficient assets and crews are available to prepare for and perform tasked missions. These requirements drive the execution of a periodic, repeatable tasking process to allow commanders to plan for upcoming operations. The ATO execution period (usually 24 hours in duration) and the preceding process during which the ATO is developed (usually 72-96 hours in duration) are a direct consequence of these physical constraints.

In contrast to the misperception that targeting information should be provided to planners 72-96 hours in advance, it is evident targets can actually be struck within minutes from when information is first made available in the dynamic targeting process. The key to both the flexibility and versatility of deliberate and dynamic targeting is a shared understanding among the functional components. Misperceptions may arise because other components may not have visibility on the wide variety of missions tasked to the air component in support of the JFC's operation and because air component assets are often tasked to simultaneously conduct missions supporting overlapping operational phases. This important shared understanding is largely accomplished by ensuring component liaisons are properly positioned during planning and execution.

The ATO conveys tasking for joint air operations for a specific period of time, normally 24 hours. Detailed planning generally begins 72 hours prior to the start of execution to properly assess the progress of operations, anticipate enemy actions, make needed actual length of the tasking cycle may vary from theater to theater.

The tasking cycle length may be based upon JFC guidance, air component commander direction, and theater needs. The length should be specified in theater standard operating procedures or other directives. If it is modified for a particular contingency, this should be specified in JFC's operation plan (OPLAN) or the air component's JAOP. The net result of this part of the tasking cycle—and of deliberate targeting efforts—is that there are usually five ATOs in various stages of progress at any one time.

- One, or more, previously executed ATO undergoing assessment at various levels.
- Current ATO in execution.
- Next ATO in production.
- Next successive ATO in detailed planning (target development and weaponeering).
- Following successive ATO in strategy development (objectives and guidance).

Some assets may not operate within the established cycle. These include most space assets, which are tasked via the joint space tasking order (JSTO); cyberspace assets, which are tasked via the cyber tasking order (CTO); and airborne information operations (IO) assets, which are tasked via the ATO. However, some theater-specific space and cyberspace operations may be included in the daily ATO for the sake of situational awareness, integration, and synchronization. During major conventional operations, special operations function within a 96-hour planning cycle; however, during contingency operations they often operate within or drive the dynamic targeting process. Certain IO and other nonlethal capabilities operate within a

The Air Tasking Cycle

Ref: Annex 3-60, Targeting (14 Feb '17), pp. 38 to 41.

A principal purpose of the air tasking cycle is to produce orders and supporting documentation to place a flexible array of capabilities in a position to create desired effects in support of the commander's intent. This cycle is driven by the tyranny of time and distance. It takes time for ground crew to prepare aircraft for flight, for aircrew to plan missions, and for aircrew to fly to the immediate theater of operations from distant airfields. Likewise, commanders should have enough visibility on future operations to ensure sufficient assets and crews are available to prepare for and perform tasked missions. These requirements drive the execution of a periodic, repeatable tasking process to allow commanders to plan for upcoming operations. The ATO execution period (usually 24 hours in duration) and the preceding process during which the ATO is developed (usually 72-96 hours in duration) are a direct consequence of these physical constraints.

See pp. 6-23 to 6-28 for detailed discussion of the Joint Air Tasking Cycle from JP 3-30.

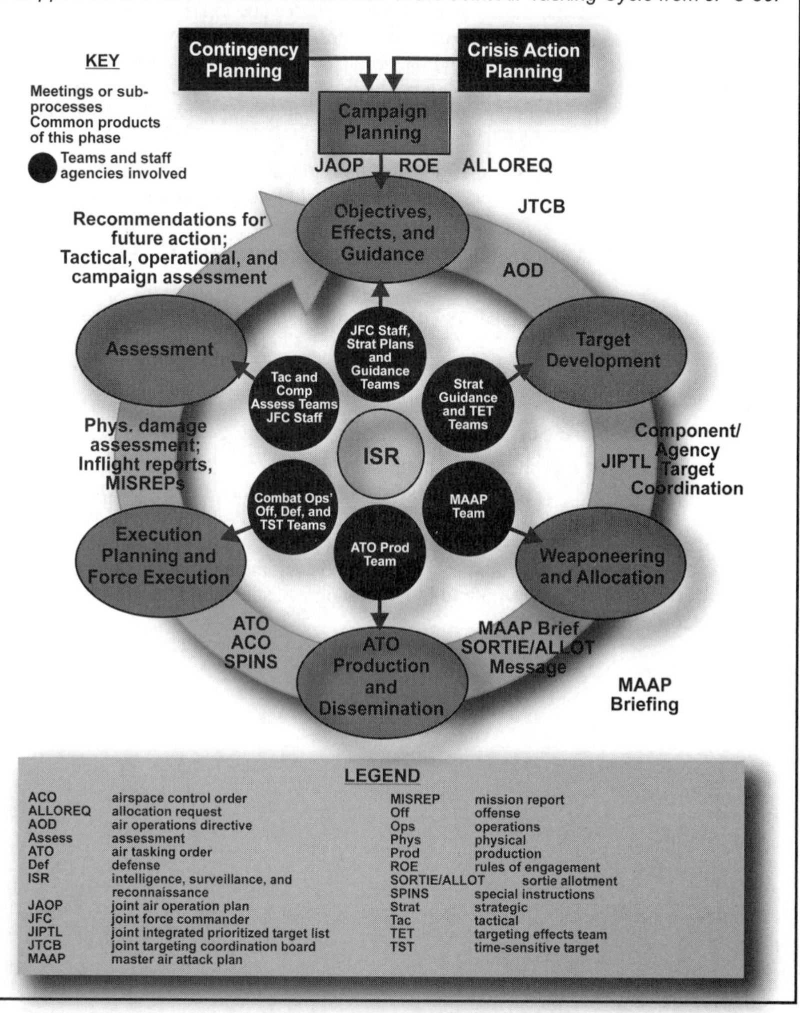

96-hour cycle as well, and it is critical for AOC planners to know if special operations forces (SOF) and IO personnel may assist with targeting. Intertheater air mobility assets also do not necessarily operate within the tasking cycle. In large operations, the existence of differing planning cycles among components can lead to increased complexity in the process. Most component planning cycles are approximately 72-96 hours. However, the requirement within the air tasking cycle to manage as many as five separate ATOs drives the requirement for discipline to manage defined inputs and outputs during particular slices of time.

Deliberate targeting supports every phase of the JOPPA and the joint air tasking cycle. It is interwoven throughout the phases up to and including ATO production and dissemination. Effective deliberate targeting comes at a high cost in terms of the volume and flow of information. Targeting and assessment, which are integrally related, impose most of the intelligence collection burden the joint force carries—to support deliberate targeting efforts before, dynamic targeting efforts during, and assessment during and after ATO execution. Successful targeting requires in-depth information on such things as enemy force posture; capabilities and movement; tactics, techniques, and procedures (TTPs); COGs and target vulnerabilities; enemy leadership's intentions, habits, and movement patterns; the flow and interconnections of enemy economic behavior; and the linkages and interconnections within major infrastructure systems, like electrical power and electronic communications webs. The process also takes into account such things as friendly objectives, concept of operations (CONOPS), ROE, target time constraints, and friendly force capabilities to create five general types of products:

- Target nominations and target lists intended to achieve desired effects which will accomplish commander's objectives while complying with the published guidance for the use of forces.
- Capability recommendations based upon effects chosen to achieve commander's objectives.
- Capability effectiveness estimates logically linked to effects specified during target development to support force application recommendations (may also include commensurate collateral damage estimates for targets of concern).
- Force/capabilities selection and planning.
- Target materials built to support current and future targeting efforts.

Once the ATO is published, adjustments are made in the COD and targeting decisions are handled through dynamic targeting. The final phase of the cycle is assessment, which is closely tied to ISR and may lag established battle rhythms and timelines due to its heavy dependence on planning and direction, collection, processing and exploitation, analysis and production, and dissemination. It is accomplished primarily by the ISR Division and the operational assessment team (OAT) within the Strategy Division (SD).

IV. Dynamic Targeting

Ref: Annex 3-60, Targeting (14 Feb '17), pp. 42 to 79.

Dynamic targeting complements the deliberate planning efforts, as part of an overall operation, but also poses some challenges in the execution of targets designated within the dynamic targeting process. Dynamic targets are identified too late, or not selected for action in time to be included in deliberate targeting. The doctrine of a deliberate targeting process controlled by strategy, law of war (LOW), and rules of engagement (ROE), etc. is equally applicable to the dynamic targeting process.

The Dynamic Targeting Process

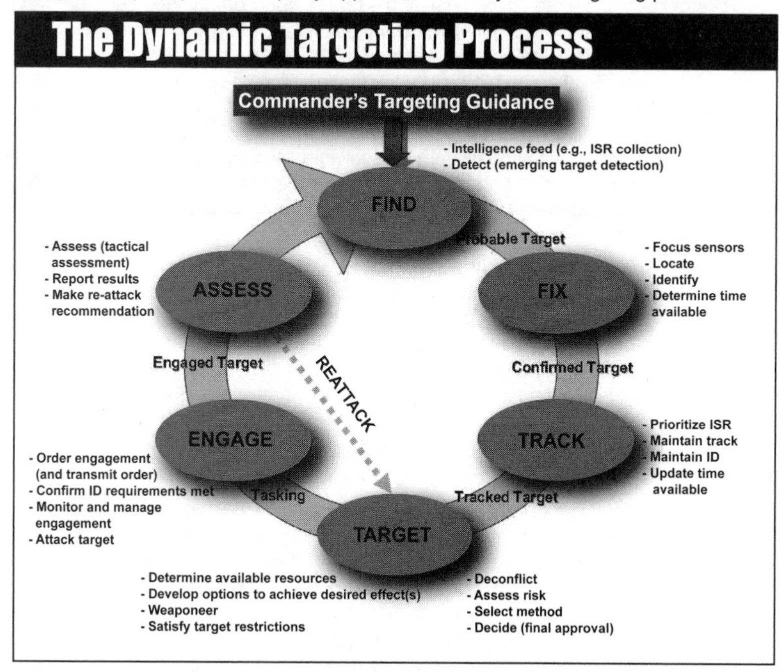

Ref: Adapted from AFFD3-60, Targeting, fig. 3-1 (2006). This graphic is not provided in Annex 3-60.

Also by definition, dynamic targeting occurs in a much more compressed timeline, requiring special consideration and attention for all personnel assigned to work the dynamic targeting process. The importance of dynamic targeting is further emphasized by joint targeting doctrine. While not the sole responsibility of the commander, Air Force forces (COMAFFOR), Airmen are heavily involved in the planning and execution of the dynamic targeting. The joint force commander (JFC) ultimately designates the responsibilities and authorities associated with the prosecution of dynamic targets and may often designate specific component responsibilities, based on location, capability, or target types.

Dynamic targeting is a term that applies to all targeting that is prosecuted outside of a given day's preplanned air tasking order (ATO) targets (i.e., the unplanned and unanticipated targets). It represents the targeting portion of the "execution" phase of effects-based approach to operations (EBAO). It is essential for commanders and air operations center (AOC) personnel to keep effects-based principles and the JFC's objectives in mind during dynamic targeting and ATO execution. It is easy for those caught up in the daily battle rhythm to become too focused on tactical-level details, losing sight of objectives, desired effects, or other aspects of commander's intent. When this happens, execution can devolve into blind target servicing, unguided by strategy, with little or no anticipation of enemy actions.

Dynamic targeting is different from deliberate targeting in terms of the timing of the steps in the process, but not different in the substance of the steps. Ultimately, dynamic targets are targets—as such, their nomination, development, execution, and assessment still takes place within the larger framework of the targeting and tasking cycles. Some are fleeting and require near-immediate prosecution if they are to be targeted. Such targets require a procedure that can be worked through promptly and that facilitates quick transition from receipt of intelligence ("trigger events") through targeting solution to action against the target. This compressed decision cycle is best handled through the specialized dynamic targeting sub-processes. Seen from the larger cycle's perspective, dynamic targeting takes place within phases five (execution planning and force execution) and six (assessment) of the targeting and air tasking cycles. The earlier phases serve to provide commanders' targeting guidance and determine concept of operations (CONOPS) for making the resources that may prosecute dynamic targets available. Ultimately, the JFC and COMAFFOR should make decisions about these targets based on critical and timely intelligence information and may likely require reallocation of resources that could impact ongoing deliberate plan execution.

The combat operations division (COD) is responsible for implementation of dynamic targeting for the ATO currently in execution. Successful dynamic targeting, however, requires a great deal of prior planning and coordination with other divisions within the AOC and with other components based on the type of target. If dynamic targeting is to be done correctly, planners should develop a plan that makes assets available to the COD prior to the start of execution. This can be done in a number of ways but the most common methods are:

- Preplanning target reference methods and coordination measures such as kill boxes and combat area entry points/routes for cruise missiles.
- Preplanning on-call or pre-positioned strike and intelligence, surveillance, and reconnaissance (ISR) packages (including tanker support) for rapid response to emerging targets (such as on-call electronic warfare, space, cyberspace operations, interdiction, or close air support missions available for tasking during ATO execution; missions on ground alert; and/or air-to-ground weapons loaded on aircraft performing defensive counterair missions).
- Using joint intelligence preparation of the operational environment (JIPOE) to determine the most probable areas where targets may emerge during execution.
- Diverting airborne assets assigned to lower priority targets to strike the recently identified target.
- Coordinating and synchronizing dynamic targeting operations by streamlining procedures.
- Developing procedures for rapid handover of the mission tasking to another component for mission execution, if the air component cannot attack an emerging target.

Divisions other than the COD have important roles to play in dynamic targeting. The strategy division (SD) should capture macro-level targeting guidance to include component priorities in the air operations directive (AOD). Many items in the AOD, like commander's intent, anticipated weapons available, ROE, acceptable risk levels, and elements of the ISR collection plan provide vital information needed by operators and targeteers to develop and implement effective and timely effects based responses. For instance, ROE are especially important to this form of time-compressed targeting. While the SD typically drafts ROE inputs with advice from the servicing judge advocate, all involved in planning and execution should clearly understand the ROE. Compliance with ROE is a shared responsibility between the COMAFFOR staff, subordinate command elements, and aircrews/operators. Due to the probable time-sensitive nature of targets prosecuted during execution, clear guidance should be developed to enable rapid prosecution. Planning personnel may need to convey the priority of the dynamic target planned for engagement in terms relative to the target planned for deliberate execution that may not be engaged due to the reprioritization. In that same light, the priority of the ISR asset that may provide assessment information on that target should also be addressed, especially if there may be a dynamic change to the ongoing joint integrated prioritized collection list (JIPCL) missions.

Liaison officers (LNOs) from coalition partners, other components, and other Services are essential during dynamic targeting. LNOs—particularly the special operations liaison element (SOLE), battlefield coordination detachment (BCD), and other government agencies—may be able to provide the COMAFFOR with additional options for dealing with emerging targets as well as provide locations and activities of friendly forces. LNOs work de-confliction issues and their forces may also assist friendly forces by finding, fixing, tracking, targeting, and assessing targets.

As stated earlier, dynamic targeting occurs in a much compressed timeline. Successful prosecution of a target may require that targeting be completed in minutes. To achieve this time compression, the COMAFFOR should consider implementing procedures that enable the phases of dynamic targeting to be performed simultaneously rather than sequentially. Ideally, one COD team should perform targeting of all dynamic targets. Creating separate teams may result in unwanted isolation, impede unity of effort, and inhibit the cross-flow of information.

Successful prosecution of targets during execution also requires well organized and well-rehearsed procedures. There is a need for sharing sensor data and targeting information, identifying suitable strike assets, obtaining mission approval, and rapidly deconflicting weapon employment. The reaction time between the sensor and shooter can be greatly accelerated if there are clearly articulated objectives, guidance, priorities, and intent for dynamic targeting before targets are even identified. The appropriate response for each target depends heavily on the level of conflict, the clarity of guidance to define the desired outcome, and ROE.

I. Dynamic Targeting & the Tasking Process

Ref: Annex 3-60, Targeting (14 Feb '17), pp. 45 to 47.

A. Categories of Targets

Dynamic targeting includes prosecution of several categories of targets:

JFC-designated time sensitive targets (TST)

Targets or target sets of such high importance to the accomplishment of the JFCs mission and objectives, or one that presents such a significant strategic or operational threat to friendly forces or allies, that the JFC dedicates intelligence collection and attack assets, or is willing to divert assets away from other targets in order to engage it.

Component critical targets (CCT)

Targets that are considered crucial for success of friendly component commanders' missions, but are not JFC-approved TSTs. Component commanders may nominate targets to the JFC for consideration as TSTs. If not approved as TSTs by the JFC, these component-critical targets may still require dynamic execution with cross-component coordination and assistance in a time-compressed fashion.

Target Changes

Targets that are scheduled to be struck on the ATO being executed but have changed status in some way (such as fire support coordination measures changes).

Other Targets

Other targets that emerge during execution that friendly commanders deem worthy of targeting, prosecution of which may not divert resources from higher-priority targets.

Each of the four categories of targets specified is prosecuted via the same dynamic targeting portion of the tasking process—they differ only in relative priority.

Emerging Terminology: JFC Critical Targets

By definition, "time-sensitive target" implies that creating effects against a target needs to happen quickly. Many targets are "time-sensitive" – they may be fleeting targets of opportunity (e.g., enemy leader leaving his compound) or pose an imminent, direct danger to friendly forces (e.g., enemy forces flanking friendly forces). "JFC-TST" is appropriate for such circumstances. However, the way the term is currently used, it also encompasses high priority targets that are not time-sensitive (e.g. enemy HHQ or communication node). Using "time-sensitive" in these cases may add confusion to the process and adversely affect decisions relating to whether or not to strike a target and how to strike it. In order to distinguish between high priority targets and time-sensitive targets, the term "JFC-Critical Target (JFC-CT)" has been introduced. While "JFC-TST" is still extant practice, targeteers must ensure they understand target prioritization and timing involved regardless of the title. Of note, JFC-TSTs and JFC-CTs can be targeted in deliberate and dynamic targeting cycles.

B. ROE, CID, PID & Target Validation

Rules of engagement (ROE), combat Identification (CID), positive identification (PID), and target validation all play important part roles in dynamic targeting.

Rules of Engagement (ROE)

ROE comprises the directives that delineate the circumstances and limitations under which US forces will conduct combat operations. They provide a framework that encompasses national policy goals, mission requirements, and the rule of law. All targeting decisions must be made in light of the applicable ROE.

Combat Identification (CID)

For prospective targets, there are essentially three levels of CID, which must be acquired prior to engagement that are relevant to AOC personnel and those tasked to carry out actions against them. At the first level, the track or entity is identified as friendly, foe, or neutral. At the next level, the prospective target's type of platform is identified. This may aid in determining the nature of tactical action required and assist in prioritizing the target. Finally, a third level entails determining the prospective target's intent (as by its track relative to friendly forces) when possible. This should further aid in establishing the prospective target's priority, and may sometimes entail reclassifying a target as a TST based on its potential threat to friendly forces. CID characterizations, when applied with combatant commander ROE, enable engagement decisions and the subsequent use or prohibition of lethal and nonlethal force. CID is used for force posturing, command and control, situational awareness, and strike/no-strike employment decisions. Effective CID not only reduces the likelihood of friendly fire incidents, but also enhances joint fire support by instilling confidence that a designated target is, in fact, as described.

Positive Identification (PID)

PID is conducted through observation and analysis of target characteristics including visual recognition, electronic support systems, non-cooperative target recognition techniques, identification friend or foe systems, or other physics-based identification techniques, and is informed by CID processes.

Target Validation

Target validation ensures that targets meet the objectives and criteria outlined by the commander's guidance and ensures compliance with the law of war and ROE. Target validation during dynamic targeting includes analysis of the situation to determine whether planned targets still contribute to objectives, whether targets are accurately located, and how planned actions will impact other friendly operations. The PID decision is crucial to having a valid target.

II. Dynamic Targeting Phases (F2T2EA)

Ref: Annex 3-60, Targeting (14 Feb '17), pp. 47 to 49.

Dynamic targeting consists of six distinct phases: find, fix, track, target, engage, and assess (F2T2EA). These are the same phases used to prosecute joint TSTs, as explained in the Multi-Service Tactics, Techniques, and Procedures for Dynamic Targeting *(AFTTP 3-2.3)*. This method referred to as F2T2EA or colloquially as the "kill chain."

F - Find

The find phase involves detection of an emerging target, which various aspects of its characterization will result in it being binned into one of the dynamic targeting categories listed above. The find phase requires clearly designated guidance from commanders, especially concerning target priorities, and the focused ISR collection plan based on JIPOE, to include named areas of interest and target areas of interest. Following this collection plan leads to detections, some of which may be emerging targets, that meet sufficient criteria (established by the AOC with commander's guidance) to be considered and developed as a target. The time sensitivity and importance of this target may be initially undetermined. Emerging targets usually require further ISR and analysis to develop and confirm.

Commanders should not task sensors without an idea of what they may collect. They should anticipate results, not request unfocused detection. The result of the find phase is a potential target that is nominated for further investigation and development in the fix phase.

F - Fix

The fix phase positively identifies an emerging target as worthy of engagement and determines its position and other data with sufficient fidelity to permit engagement. When the emerging target is detected, sensors are focused upon it to confirm its identity and precise location. This may require implementing a sensor network or diverting ISR assets from other uses to examine it. The COMAFFOR may have to make the decision on whether diversion of ISR resources from the established collection plan is merited, but this decision can often be made by COD personnel. Data correlation and fusion confirms, identifies, and locates the target, resulting in its classification in one of the four target categories listed above. Target location and other information should be refined enough to permit engagement in accordance with ROE. An estimation of the target's window of vulnerability frames the timeliness required for prosecution and may affect the prioritization of assets and the associated risk assessment.

If a target is detected by the aircraft or system that may engage it (for example, by an armed remotely piloted aircraft, or platform with an advanced targeting pod), this may result in the find and fix phases being completed near-simultaneously, without the need for additional ISR assets. It may also result in the target and engage phases being completed without a lengthy coordination and approval process. Battle management systems [i.e., airborne warning and control system (AWACS) and joint surveillance target attack radar system (JSTARS) aircraft] can often fix target locations precisely enough to permit engagement without the need for further ISR collection. Growth in sensor technology has permitted "non-traditional" sources of ISR to supplement the find, fix, and track phases. Integrating data from platforms other than those traditionally dedicated to intelligence collection, to include information gleaned from weapons systems or even munitions themselves, helps to build a common operating picture that commanders can use to shorten the F2T2EA cycle.

T - Track

The track phase takes a confirmed target and its location, maintaining a continuous track. Sensors should be coordinated to maintain situational awareness and track continuity on targets. Windows of vulnerability should be updated when warranted. This phase may require re-prioritization of ISR assets, just as the fix phase may, in order to maintain situational awareness. If track continuity is lost, it may be necessary to re-accomplish the fix phase—and possibly the find phase as well. The track phase results in track continuity and refining the target identification. This is maintained by appropriate sensors or sensor combinations, a sensor prioritization scheme (if required), and updates on the target's window of vulnerability (if required). The process may also be run partially "in reverse" in cases where an emerging target is detected and engaged. Once it becomes clear that it is a valid target, the sensors detecting it can examine recorded data to track the target back to its point of origin, such as a base camp. This could potentially identify threats or more lucrative targets. Such point of origin hunting has proven especially useful during stability and counterinsurgency operations such as those in Iraq and Afghanistan.

T - Target

The target phase takes an identified, classified, located, and prioritized target; determines the desired effect and targeting solution against it; and obtains required approval to engage. During this phase, COD personnel should review target restrictions, including collateral damage, ROE, LOW, the no strike list (NSL), the restricted target list (RTL), and fire support coordination measures (FSCM). In essence, the targeting and operational members of the COD must accomplish all facets of the "target validation" process. This phase also accomplishes effects validation, weaponeering/capabilities analysis, and collateral damage estimation (CDE) analysis. COD personnel match available strike and sensor assets against desired effects, then formulate engagement options. They also submit assessment requirements.

The selection of assets for a specific target may be based on many factors, such as the location and operational status of ISR and strike assets, support asset availability, weather conditions, ROE, target range, the number and type of missions in progress, available fuel and munitions, the adversary threat, and the accuracy of targeting acquisition data. This can be the lengthiest phase due to the large number of requirements that should be satisfied. In many cases, however, dynamic targeting can be accelerated if target phase actions can be initiated and/or completed in parallel with other phases.

E - Engage

In this phase, identification of the target as hostile is confirmed and engagement is ordered and transmitted to the pilot, aircrew, or operator of the selected weapon system. The engagement orders should be sent to, received by, and understood by the operator of the weapons system. The engagement should be monitored and managed by the engaging component (for the air component, by the AOC). The desired result of this phase is successful action against the target.

A - Assess

In this phase, predetermined assessment requests are measured against actions and desired effects on the target. ISR assets collect information about the engagement according to the collection plan (as modified during dynamic targeting) and attempt to determine whether desired effects and objectives were achieved. In cases of the most fleeting targets, quick assessment may be required in order to make expeditious re-attack recommendations.

Targeting

III. Dynamic Targeting Engagement Authority

The authority to engage should be delegated to the C2 node that has the best information or situational awareness to execute the mission and direct communications to the operators and crews of the weapon systems involved. If the COMAFFOR is delegated TST engagement authority by the JFC, that commander may delegate his engagement authority to a lower level (e.g., AOC director or chief of the combat operations division). The COMAFFOR has the authority to redirect those forces over which he has operational or tactical control. For all others, the affected component commander should approve all requests for redirection of apportioned air assets. Components execute the ATO as tasked and recommend changes to the AOC as appropriate, given emerging JFC and component requirements.

Functional commands like US Cyber Command (USCYBERCOM) and Joint Functional Component Command-Space (JFCC-Space) may have operational and tactical control of some functional capabilities. In such cases, coordinating authorities at the JFC and/or component level should be authorized to plan, coordinate, integrate, and execute their respective functional capabilities within the operational area. Coordination requirements associated with these functional capabilities may result in long-lead times that should be considered within the AOC planning and execution processes.

At the tactical level, engagement authority normally resides with the "shooter" (aircrew, system operator, etc.) for those planned events on the current tasking order being executed; this follows the tenet of decentralized execution. The fact that planned missions on an ATO have been approved for release by the COMAFFOR passes engagement authority to the "shooters" personally executing those ATO missions, who should adhere to all guidance included in the ATO [special instructions (SPINS), airspace control order (ACO), ROE, etc.]. In dynamic targeting situations, where the target is not specified in the ATO prior to takeoff or execution, engagement may require that the "shooter" be "cleared to target" from a C2 entity outside the AOC like JSTARS, AWACS, tactical air control party (TACP), and forward air controllers (ground or airborne) due to identification or other restrictions required prior to attack.

Engagement authority for those events that the AOC maintains control over may be passed to crews, via the Theater Air Control System (TACS), with required criteria to be met for weapon release, when appropriate. Engagement authority for certain sensitive targets may reside at a higher level than the JFC and should be passed appropriately through the component commander when the situation dictates.

Placing the appropriate level of battle space awareness at subordinate C2 nodes can streamline the C2 cycle and allow timely engagement during dynamic targeting. Decentralized C2 nodes will exchange target information (type, classification identification, location, etc.) through common data links (e.g., Link 16, UHF, wide area networks, etc.) with a fidelity that permits them to operate as a single, integrated C2 entity in order to effectively perform decentralized, coordinated execution of time-sensitive attacks.

IV. Dynamic Targeting Risks

Ref: Annex 3-60, Targeting (14 Feb '17), p. 52.

Understanding the level of acceptable risk is critical to successful targeting during execution. With compression of the decision cycle comes increased risk due to insufficient time for the more detailed coordination and deconfliction that takes place during deliberate targeting. Commanders should assess risk early, determine what constitutes acceptable risk, and communicate their intent. JFC guidance may stipulate acceptable risk when engaging TSTs, if not, then the COMAFFOR should seek to obtain it. When new targets are acquired, Airmen in the AOC and in the field should rely on commanders' guidance, ROE, and their own experience to assess acceptable risk.

Particular targets may be determined to be such a threat to the force or to mission accomplishment that the COMAFFOR may accept a higher level of risk in order to attack the target immediately upon detection. Items to be considered in the risk assessment include:

- Risk of potential friendly fire incidents, risk to non-combatants, and collateral damage potential.
- LOW and ROE compliance.
- Increased risk to attacking forces due to accelerated planning and coordination.
- Redundant attacks and wasting limited resources.
- Accepted use of non-optimum capabilities and potentially limited effects.
- Opportunity costs of diverting assets from their planned missions.

These considerations should be balanced against the risk of not attacking the target in time and thus risking mission failure, harm to friendly forces, or losing the opportunity to strike the target. More commonly, the risk associated with dynamic targeting involves the trade-off of diverting ISR and strike assets from already-scheduled missions to emerging targets. This should only be done when commanders' priority given to the new target exceeds that of the old. However, proper planning for on-call assets can mitigate much of this opportunity cost.

V. Changes and Limitations (Dynamic)

Ref: Annex 3-60, Targeting (14 Feb '17), p. 52.

The COD should be ready to respond with new targeting information in order to provide seamless operations when changes occur. These include:

Responding to Changes in Friendly Operations

For instance, if an aircraft that was tasked to prosecute a target has to abort for maintenance reasons, the COD should know the target's relative priority in order to provide appropriate targeting guidance. If the target is low priority, it may be best to place it on a subsequent day's ATO. If it is of higher priority, COD personnel may determine how best to direct or divert resources to prosecute it. COD personnel may have the best picture of what resources are available to prosecute it and what diverting resources may cost. Likewise, if an aircraft or package is diverted to prosecute a TST, the COD should identify the target(s) which may no longer be struck, as well as the new target which may be attacked. This information should be passed to the targeteers and collection managers to ensure coordinated collection and assessment on these new targets.

Responding to Changes in Weather

A target planner's actions may be similar to when he responds to changes in friendly operations. Further, changes in weather may require changes to the platforms and/or weapons required to engage a particular target. Target planners should ensure that the AOC weather specialty team is engaged.

Re-Targeting

If a target that was to be prosecuted is no longer a viable target for whatever reason, targeteers should have alternate targets to assign to a strike mission. Time is important because assets may already be airborne.

Responding to TSTs

When a TST is identified, the COD should decide the best time to engage it. COD targeteers are involved in these efforts and provide guidance to planners concerning the characteristics and vulnerability of the target. Targeteers should be familiar with possible targets so that quick assessments and guidance can be given before the window of opportunity to strike the TST is gone.

Dynamic targeting has two significant limitations compared with deliberate targeting: the lack of detailed capability analysis and increased threat exposure. Commanders and the COD should consider these limitations when deciding whether to prosecute a target using dynamic targeting methods.

Capability Analysis

Due to the reduced planning time available, targets prosecuted using dynamic targeting may be engaged with less consideration given to key employment issues such as fuse settings or axes of attack. In some cases, assets may be diverted to prosecute these targets with munitions that are not optimum for the given task. Since these considerations may carry increased risk of mission failure, collateral damage, or even harm to friendly troops, commanders should weigh the potential benefits gained by prosecuting the target quickly. COD personnel should work with their targeteers to ensure that proposed capability analysis solutions are sufficient for the given task.

Increased Threat

Denied environment targets are normally attacked by packages with dedicated support, such as electronic jamming and suppression of enemy air defense capabilities. The shortened dynamic targeting planning window may not allow for the same level of support, thereby exposing aircrews to greater risk. Time for target area threat analysis is also reduced, further increasing risk to aircrews and weapon survivability.

Chap 7

V. Targeting & the Air Tasking Cycle

Ref: Annex 3-60, Targeting (14 Feb '17), pp. 55 to 80.

The air tasking cycle is the COMAFFOR's process for effective and efficient employment of joint air capabilities. It is a methodical, iterative, and responsive process that translates operational level guidance into tactical level plans. The air tasking cycle promotes flexibility and versatility with a series of Air Tasking Orders (ATOs) and related products in progress at any time and by responding during execution to changes in the operational environment. The air tasking cycle consist of the following stages:

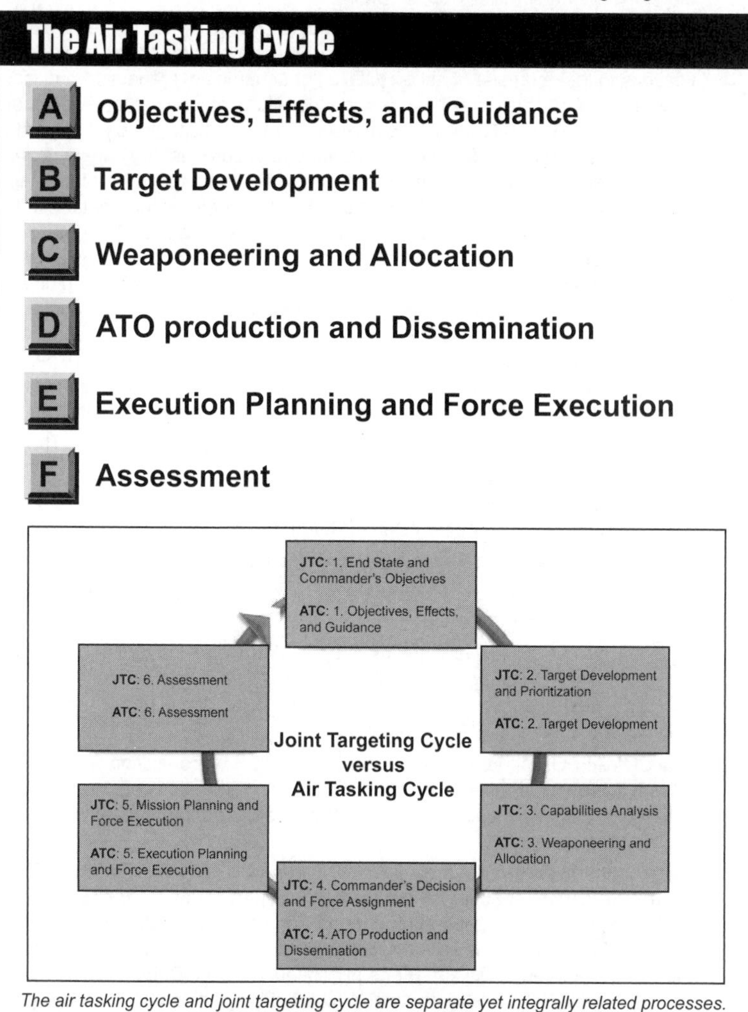

The air tasking cycle and joint targeting cycle are separate yet integrally related processes.

See pp. 6-23 to 6-28 for related discussion of the Joint Targeting Cycle.

I. Air Tasking Cycle Phases

The air tasking cycle consist of the following stages:

A. Objectives, Effects and Guidance

Purpose of the Phase

Clear understanding of the commander's objectives and guidance is essential for effective tasking and targeting. Objectives are the clearly defined, decisive, attainable, and measurable goals toward which every military operation should be directed. They provide focus for those at all stages of the tasking cycle and give targeting personnel the overarching purpose for their efforts. Guidance sets limits and boundaries on the objectives and how they are attained. It establishes constraints—things we must do—and restraints—things we must not do. Together, the two embody commander's intent for military operations.

This phase starts with joint force commander (JFC) guidance to the joint force components. The JFC consults with his component commanders, decides on modifications to their courses of action (COAs) or schemes of maneuver, and issues guidance and intent. This may occur through the efforts of the joint targeting coordination board (JTCB). The JTCB provides a forum in which all components can articulate strategies and priorities for future operations to ensure that they are synchronized and integrated. The JTCB normally facilitates and coordinates joint force targeting activities with the components' schemes of maneuver to ensure that the JFC's priorities are met. Accordingly, the commander, Air Force forces (COMAFFOR) should issue further guidance on the specific scheme of maneuver Additionally, the JFC should delegate authority to conduct execution planning, coordination, and deconfliction associated with joint air component tasking to the COMAFFOR and should ensure that this process is a joint effort. The COMAFFOR should possess a sufficient C2 infrastructure, adequate facilities, readily available joint planning expertise, and a mechanism for accomplishing targeting, weaponeering, and assessment. The air operations center (AOC) provides the COMAFFOR with these capabilities.

The JFC determines whether a JTCB will be held and defines its role. The JTCB should cover four broad topics:

- Assessment of campaign progress since the last meeting (usually the last 24 hours), with recommendations for future action.
- Broad guidance for the next 72 hours issued by the JFC.
- Major operations (schemes of maneuver) over the next 48 hours, briefed by each of the components.
- Macro-level review and guidance on joint maneuver and fires [including, especially, targeting and intelligence, surveillance, and reconnaissance (ISR) priorities] over the next 24 hours, to help guide joint dynamic targeting efforts for the upcoming execution period.

The COMAFFOR should prepare prior to the JTCB by consulting with senior component liaisons and the staff to determine what modifications are needed to the air scheme of maneuver and to determine the air apportionment recommendation for the JFC's approval.

Once battle rhythm starts, the apportionment period is usually 24 hours. The apportionment recommendation can be approved as part of the JTCB or separately after it. Once approved, the apportionment decision should be included in the ultimate product of this phase, the air operations directive (AOD). In deriving guidance that may be considered at the JTCB and published in the AOD, the COMAFFOR is supported by the AOC strategy divisions (SD)24 strategy plans and strategy guidance teams. The strategy guidance team is primarily responsible for producing the AOD. The SD should also ensure the cyber operations directive and space operations

Joint Air Tasking Cycle

Ref: JP 3-30, Command and Control of Joint Air Operations (Feb '14), fig. III-13, p. III-21.

Editor's note: See pp. 6-23 to 6-28 for related discussion of the Joint Air Tasking Cycle from JP 3-30.

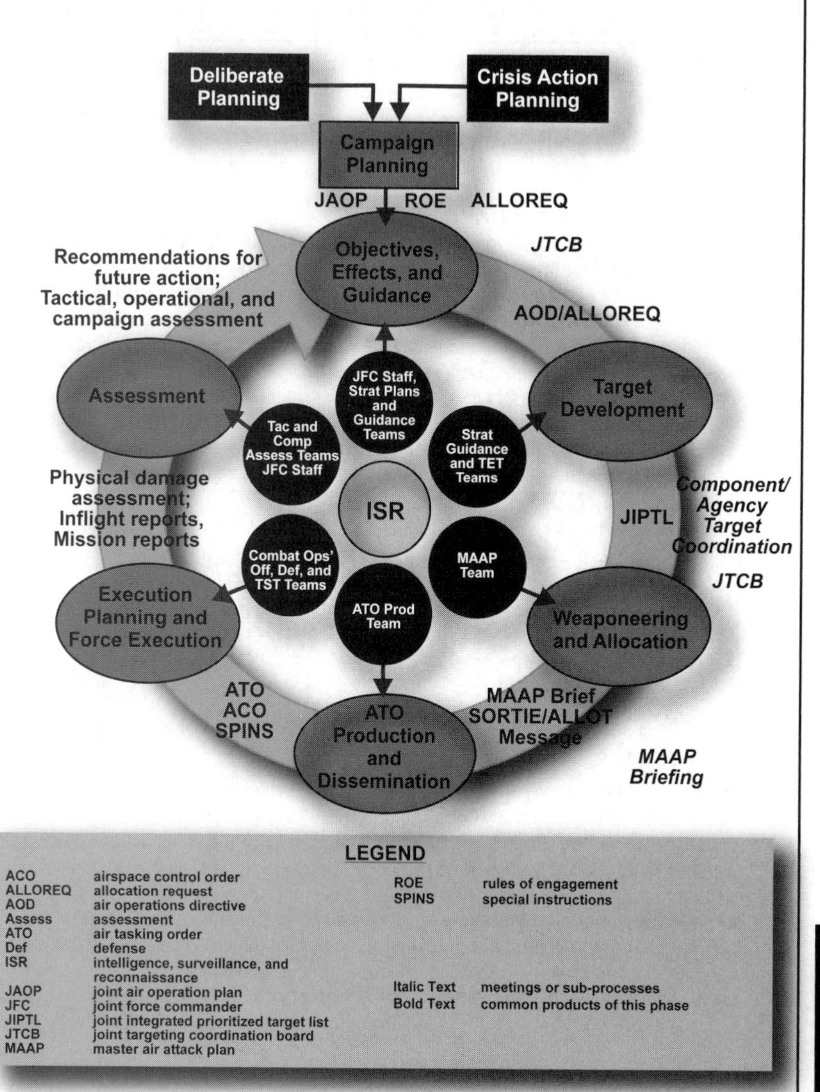

directive (SOD) are fully integrated and synchronized with the AOD produced by the AOC.

The objectives, effects, and guidance phase is also where effects and their accompanying measures of effectiveness (MOE) and measures of performance (MOP) are determined. Strategy guidance and strategy plans teams work closely with the CPD targeting effects team (TET), and the ISR division (ISRD) to determine effects that achieve the stated objectives, select appropriate measures and indicators for assessment, and determine ISR requirements to collect against the MOEs. Results of this effort may be published as lists of tasks or desired effects in the AOD.

Finally, considerations of the law of war (LOW) and rules of engagement (ROE) for the conflict may directly affect all phases of the tasking process (and thus targeting). Targeteers should understand and be able to apply the basic principles of these disciplines as they relate to targeting. See Appendix A for further discussion of LOW and ROE.

Products of the Phase

The AOC SD drafts the air operations directive (AOD) for COMAFFOR approval. In a normal battle rhythm, this is done on a daily basis. The AOD is the vehicle for the COMAFFOR to express his intent for a specific day and communicate the JFC's apportionment decision. Apportionment guidance should reflect prioritized operational objectives and relevant tactical tasks with approximate weights of effort for each objective. Specific weights of effort should be avoided due to the difficulty in precisely measuring effects of air, space, and information operations (IO), and to allow maximum flexibility in planning the application of airpower. However, the CPD can use these weights of effort, along with existing friendly force capabilities, to estimate the numbers of aimpoints by effect or objective to focus target development.

The JIPTL and JFACC/COMAFFOR Responsibilities

Just as joint doctrine describes the option for the JFACC to serve as the Airspace Control Authority and Area Air Defense Commander, the JFACC/COMAFFOR should be delegated authority to approve the JIPTL as the JFC's representative. Accomplished in full view and in coordination with all components, this arrangement yields efficiencies by locating the JIPTL approval process with the targeting expertise resident within the AOC and potentially eliminating the requirement for a JTCB. Alternatively, similar efficiencies can be gained by appointing the JFACC/COMAFFOR chair of the JTCB vice a member to the JFC staff.

The prioritized tasks in the AOD should be effects-based and reflect commander's guidance and intent. By crafting effects-based tasks for the AOD, target developers within the AOC's ISRD gain the flexibility to identify and nominate the most effective means to achieve the desired effects. Tasks that are not effects-based are often target-based, meaning that there is little flexibility in the selection of targets, and can lead to the inefficient use of scarce airpower resources. The AOD is the primary vehicle for communicating desired effects to target developers and others involved in targeting on a daily basis. Detailed, logical lists of effects-based tasks with appropriate measures and ISR collection requirements are a necessary part of the AOD.

The AOD should also be used to express the JFC's and COMAFFOR's guidance regarding what target categories (target sets) are time-sensitive, what the priority is among them, and what types of dynamic targeting would cause preplanned missions to be re-tasked. Categories of time sensitive targets (TST), high-value targets, and other objects of dynamic targeting should be presented in the context of the desired effects, and those desired effects prioritized against the desired effects for preplanned targets. This allows the COD to rapidly assess the value of preplanned targets against TST or emerging targets to determine whether or not to re-task air, space, or information assets. This guidance also reduces the possibility of all newly detected targets being struck. Just because a target can be engaged within the air tasking order (ATO) execution period does not mean that effort should be diverted from preplanned targets to engage it.

While daily guidance is critical to subsequent phases of the ongoing tasking cycle, the SD strategy plans team also works on longer-range planning, including study of branches and sequels. Conclusions drawn from this study should be disseminated throughout the AOC to assist in focusing later target development and intelligence collection efforts.

Finally, the AOD should include the COMAFFOR's guidance on which targets or target sets require immediate assessment feedback. ISR and PED assets are usually limited in number and the collection requirements for target development, joint intelligence preparation of the operational environment (JIPOE), indications and warnings, and other taskings may have a higher priority than combat assessment. Operations may be more efficient if assessment is focused on a select few high priority targets or sets.

B. Target Development

Purpose of the Phase

This is the phase in which the efforts of deliberate targeting relate specific targets to objectives, desired effects, and accompanying actions. Targeteers within the ISRD and the CPD TET take the effects determined during the objectives, effects, and guidance phase and analyze which targets should be struck (or otherwise affected) to accomplish them. Target development requires thorough examination of the adversary as a system of systems in order to understand where critical linkages and vulnerabilities lie. Critical elements are those elements of a target that enable the target to perform its primary function. Targeteers will determine which critical elements enable enemy capabilities and/or actions which are the focus of the commander's objectives and thus the source of the desired direct and cascading effects on the system. Critical linkages within a system often enable the functioning of several interrelated parts of the system, and so affecting them in the right way can disable several components, or even cause cascading system-wide failure. Vulnerable targets are those that can be attacked or otherwise affected. Thorough analysis should identify critical vulnerabilities, if they exist. These are elements of the adversary's system that are both critical and vulnerable. Analysis is made effective through access to the community of subject matter expertise and information regarding the functioning of systems that support adversary capabilities. This research may require expertise beyond that normally available on the COMAFFOR's planning staff. In such cases, reachback/federation entities may fill COMAFFOR staff shortfalls. It requires cooperation with other planning staffs and national interagency groups throughout the process. Target development involves five distinct functions, each discussed below:

- Target analysis
- Target vetting
- Target validation
- Target nomination
- Identification of intelligence gaps, collection and exploitation requirements

The purpose of these together is to relate target development to tasking. The target nomination part of the process, the component target nomination list (TNL) development, usually culminates in a target coordination meeting, held by the TET within the CPD (when the JFC delegates joint targeting coordination authority to the joint force air component commander (JFACC/COMAFFOR) with the assistance of the various joint components and multi-national liaison elements). The TET collates target

Target Development Phase (Overview)

Ref: Annex 3-60, Targeting (14 Feb '17), pp. 60 to 70. See following pages (pp. 7-38 to 7-41) for further detail and discussion of these functions.

Target development involves five distinct functions, each discussed below. The purpose of these together is to relate target development to tasking. The target nomination part of the process usually culminates in a target coordination meeting, held by the TET within the CPD, with the assistance of the various joint components and multi-national liaison elements. The TET collates target nominations from all sources. It works with the ISRD and other agencies to analyze targets. It screens all nominated targets to ensure they meet commander's intent and are relevant. It allocates and prioritizes the nominated targets based on the best potential to achieve desired effects and objectives and coordinates this target allocation to ensure other components' priorities and timing requirements are met. The product of this effort, when approved by the CFC or designated representative, is the JIPTL.

1. Target Analysis

Target analysis takes the desired effects determined during planning or the first phase of the tasking cycle and matches them to specific targets. This analysis looks at the importance of various potential targets as enablers of enemy capabilities, as critical elements within enemy systems, or as potential trigger points for desired enemy behavior changes. There are many means available to accomplish this. Two of the most common that have been used in the past are target system and critical node analysis.

Target system analysis (TSA), as its name implies, approaches targets and target sets as systems to determine vulnerabilities and exploitable weaknesses. Targeteers review how a functional target system works as a whole and analyze the interactions between components. TSA takes a system-of-systems approach to look at interdependencies and vulnerabilities between systems as well as intra-system dependencies.

2. Target Vetting

Target vetting assesses the accuracy of the supporting intelligence used to develop the target. Additionally, the vetting process results in the identification and documentation of collateral concerns associated with a specific target, as well as intelligence gain-loss concerns.

Refer to Annex 3-60, app. A for further discussion of targeting legal requirements.

3. Target Validation

Target validation ensures all vetted targets achieve the effects and objectives outlined in a commander's guidance and are coordinated and de-conflicted with agencies and activities that might present a conflict with the proposed action. It also determines whether a target remains a viable element of the target system. During the development effort, the targets may also require review and approval based on the sensitive target approval and review process, coordinated through the combatant commander to national authorities. The first part of validation asks such questions as:

- Does the target meet COMAFFOR or higher cdr's objectives, guidance, and intent?
- Is the target consistent with LOAC and ROE?
- Is the desired effect on the target consistent with the end state?
- Is the target politically or culturally "sensitive?"
- What will the effect of striking it be on public opinion (enemy, friendly, and neutral)?
- What are the risks and likely consequences of collateral damage?
- Is it feasible to attack this target? What is the risk?
- Is it feasible to attack the target at this time?
- What are the consequences of not attacking the target?
- Will attacking the target negatively affect friendly operations due to current or planned friendly exploitation of the target?

The second part of validation starts the coordination and integration of actions against the target with other operations. This continues after the ATO is produced and responsibility is assumed by the COD. Part of coordination is deconfliction, which is largely a checklist function. The checklist should be developed during JAOP development and be appropriate to the particular organization and conflict. Many offices and agencies must be coordinated with to prevent fratricide, collateral damage, or propaganda leverage for the enemy.

The first three stages result in what might be called "target allocation"—working interactively with other elements in the CAOC to determine which targets "make the cut" on the given day's ATO. This is not always an easy decision, especially in conflicts where resources are limited and/or the target lists are lengthy. Still, it is a vital part of what the CPD does. The final stage produces a list of validated target nominations that will be submitted to higher authority for approval on target nomination lists.

4. Target Nomination

Once all component, allied, and agency target nominations for a given ATO are received, the TET prioritizes the nominated targets and places them in a TNL based on the commander's objectives. The TET then presents the TNLs through the appropriate coordinating bodies representing the joint force components and other required agencies to ensure their requirements are supported, joint force priorities are met, and desired effects are achieved.

If targeting functions are delegated appropriately, the final deconfliction and coordination of components' nominations should be at a target coordination meeting run by the TET. Component representatives should be prepared to justify target selections, since not all targets may be engaged based on the JFC's apportionment decision and the COMAFFOR's allocation. If differences arise and cannot be resolved at the meeting, the issue should be coordinated at higher levels for resolution. The meeting should not generally address mating of specific weapons to targets, but it should consider all capabilities and initiate the planning and coordination needed for those options. Additionally, the meeting may address the availability of certain high demand weapons or munitions on a particular ATO. However, the availability of weapons or capability should not drive the nomination of targets—this is antithetical to an effects-based approach.

The result of coordination is the draft JIPTL, which is submitted to the JFC or designated representative for approval. Again, targets may be added to no-strike or restricted target lists as a result of this part of the process highlighting RTL targets (for possible approval) and sensitive target approval and review (STAR) targets.

5. Identification of intelligence gaps, collection and exploitation requirements

This stage attempts to answer the question, "How may we know that we have achieved the desired effects?" by establishing intelligence collection and exploitation requirements for each nominated target. This stage begins with target analysis and runs parallel to the other stages. The requirements should be articulated early in the tasking process to support target development and ultimately assessment. Targeteers should work closely with collection managers to ensure that target development, pre-strike, and post-strike requirements are integrated into the collection plan, along with any changes that occur throughout the tasking cycle. This intelligence support is also required to prepare for future targeting during execution (e.g., to pre-task real time ISR assets) and to support post-strike assessment of success. It should be noted that first-order effects of nonlethal operations are often subtle; in various instances may be of short duration for enabling purposes only or require days to months for the effect(s) to resolve, if at all, and may have effects that relate to the broader context of the target system (e.g., only visible at the operational or strategic level). Further, assessment of second- and third-order effects can be even more difficult. For these reasons nonlethal pre- and post-strike collection requirements are critical for ensuring a cohesive means exists to assess the intended effects. The product of this stage may be a joint integrated prioritized collection list (JIPCL).

nominations from all sources. It works with the ISRD and other agencies to analyze targets. It screens all nominated targets to ensure they meet commander's intent and are relevant. It allocates and prioritizes the nominated targets based on the best potential to achieve desired effects and objectives and coordinates to ensure other components' priorities and timing requirements are met. The product of this effort, when approved by the JFC or designated representative, is the JIPTL.

Target development influences and ultimately leads to target nominations and development of the JIPTL, joint target list (JTL), restricted target list (RTL), and no strike list (NSL). In combination with each component TNL, the JIPTL is ultimately created. As noted, all the phases of the tasking process are intertwined. Target development efforts can frequently force refinement of desired effects or even objectives, especially if weaponeering and allocation efforts indicate that a particular targeting avenue of approach is impractical. Target development efforts also frequently reach forward to influence weaponeering and allocation choices, dynamic targeting during execution, and the assessment process. The results of detailed target development are often stored in target system studies, individual target folders and targeting databases that can be studied by all levels of command and used in future target development efforts. Additionally, when detailed targeting development data are not available (i.e., a non-Joint Strategic Capability Planning directed planning effort), targeting and planning staffs should leverage the intelligence community functional target systems studies, models and simulations, experts to support target development efforts.

1. Target Analysis

Target analysis takes the desired effects determined during planning or the first phase of the tasking cycle and matches them to specific targets. This analysis looks at the importance of various potential targets as enablers of enemy capabilities, as critical elements within enemy systems, or as potential trigger points for desired enemy behavior changes. There are many means available to accomplish this through the application of capabilities across the spectrum of targeting (i.e., influence operations, physical attack, cyberspace attack, etc.). Two of the most common that have been used in the past are target system and system of system analysis.

Target system analysis (TSA) approaches targets and target sets as systems to determine vulnerabilities and exploitable weaknesses. Targeteers review how a functional target system works as a whole and analyze the interactions between components. TSA takes a system-of-systems approach to look at interdependencies and vulnerabilities between systems as well as intra-system dependencies in order to maximize the effectiveness of target development. Ideally, TSA production begins in peacetime, before the commencement of conflict, and is accomplished with federated support and "reachback."

As part of a comprehensive system-of-systems analysis (SOSA) approach, TSA focuses on one or more of the many functional target systems identified by the Defense Intelligence Agency (DIA). These include infrastructure targets across an entire region or nation (i.e., electrical power or petroleum, oil, and lubricants (POL) production), or non-infrastructure systems such as financial networks. SOSA seeks to find nodes common to more than one system, focusing on the interactions and interrelationships between system elements, in order to determine their degree and points of interdependence and to discern linkages between their functions. The ultimate goal of TSA is to find critical nodes and vulnerabilities that, if disrupted or affected in a specific manner, create effects that achieve the commander's objectives.

The analysis performed in target development proceeds through successively greater levels of detail, flowing from the macro (broad scope) level to the micro (narrowly focused) level. This winnowing approach is essential to preserve the linkage between desired effects and objectives and the specific actions that are taken against particular targets. It determines the necessary type, breadth, and duration of action that should be exerted on each target to generate effects that are consistent with the commander's objectives.

Targets for consideration come from a variety of sources. Many are developed pre-conflict and confirmed during planning. These may or may not come from a theater JTL maintained in peacetime. Many more are suggested during joint air operations plan (JAOP) development or by the SD as the air component's strategy evolve during a conflict. Many are derived by the AOC's targeteers themselves, as target analysis suggests the means of achieving desired effects.

Many targets are nominated by space and cyberspace support elements and other joint force components in the form of a TNL in order to achieve that component's desired effects. Upon dissemination of the AOD, and based on JFC guidance, components begin to develop their nominations for inclusion in the next ATO. Some targets may be suggested by government agencies outside the DOD or by foreign governments. The product of target analysis is a list of proposed target nominations designed to achieve the effects determined in earlier stages of planning (such as JAOP development or the objectives, effect, and guidance stage of the tasking cycle), which may then be validated. Other products may include creation of or additions to no-strike or restricted target lists (see "products of the phase," below).

Target research within the tasking cycle often entails studying previously unidentified or unlocated targets. Responsibility for the research lies primarily, but not solely, with the targets and tactical assessment (TGT/TA) team of the ISRD, which uses federated and reachback support to ensure that the AOC obtains, analyzes, and disseminates the information needed for further target development. Integration of full spectrum targeting capabilities is a critical part of identifying targeting opportunities and creating the appropriate lethal and nonlethal effects.

Determining the status of previously struck targets, enemy recovery and recuperation efforts, and changes in enemy tactics, processes, and strategy is a function of the TGT/TA team of the ISRD. This information is critical in validating the effectiveness of friendly action. It helps shape ongoing target development within the tasking cycle by showing where re-strikes or other further action may be required. It is also crucial to the SD's efforts to identify needed changes in the overall campaign strategy.

2. Target Vetting

Target vetting assesses the accuracy of the supporting intelligence used to develop the target. Additionally, the vetting process results in the identification and documentation of collateral concerns associated with a specific target, as well as intelligence gain-loss concerns.

3. Target Validation

Target validation ensures all vetted targets are compliant with LOW and ROE. Validation also ensures targets achieve the effects and objectives outlined in commander's guidance and are coordinated and de-conflicted with agencies and activities that might present a conflict with the proposed action. It also determines whether a target remains a viable element of the target system. During the development effort, the targets may also require review and approval based on the sensitive target approval and review process, coordinated through the combatant commander to national authorities. This phase is done by targeteers within the CPD TET, in consultation with the strategy plans team within the SD and other experts and agencies, as required.

See following page (p. 7-40) for further discussion of target validation (questions and coordination/integration requirements).

Target Validation (Questions/Integration)

Ref: Annex 3-60, Targeting (14 Feb '17), pp. 63 to 64.

The first part of validation asks such questions as:

- Does the target meet COMAFFOR or higher commanders' objectives, guidance, and intent?
- Is the target consistent with LOW and ROE?
- Is the desired effect on the target consistent with the end state?
- Is the target politically or culturally sensitive?
- What may the effect of striking it be on public opinion (enemy, friendly, and neutral)?
- What are the risks and likely consequences of collateral damage?
- Is it feasible to attack this target? What is the risk?
- Is it feasible to attack the target at this time?
- What are the consequences of not attacking the target?
- May attacking the target negatively affect friendly operations due to current or planned friendly exploitation of the target?

The second part of validation starts the coordination and integration of actions against the target with other operations. This continues after the ATO is produced and responsibility is assumed by the COD. Part of coordination is de-confliction. Many offices and agencies must be coordinated with to prevent friendly fire incidents, collateral damage, or propaganda leverage for the enemy. Some examples of where coordination and integration are required:

- **Special operations forces (SOF)**. The joint force special operations component commander (JFSOCC) must deconflict joint special operations with the JFC and the other component commanders to avoid friendly fire incidents. This is best done at a COMAFFOR targeting coordination meeting held as part of the TET's function. The AOC should work through the special operations liaison element (SOLE) for deconfliction.
- **Land forces**. AOC personnel should work through the BCD (and Marine liaison element, when appropriate) and the air support operations center (ASOC) to ensure that air component targeting is coordinated and integrated with land component operations. Careful crafting and placement of fire support coordination measures (FSCM) facilitate this.
- **Maritime forces**. AOC personnel maintain close liaison with the maritime component through the naval and amphibious liaison element (NALE) and provide air, space, and cyberspace support, as required.
- **Search and rescue (SAR)**. SAR personnel must deconflict with current targeting operations and other ongoing operations to ensure the safety of any SAR operations.
- **Space, cyberspace, and information operations**. Space, cyberspace, and information operations should be cognizant of both intended and unintended effects created by the targeting process and ensure that these effects support the JFC's objectives and strategies.
- **Other government agencies**. Targeting personnel should be aware of agency involvement and should work closely with the JFC's national intelligence support team (NIST).

4. Target Nomination

Once all component, allied, and agency target nominations for a given ATO are received, the TET prioritizes the nominated targets and places them in a TNL based on the commander's objectives. The TET then presents the TNLs through the appropriate coordinating bodies representing the joint force components and other required agencies to ensure their requirements are supported, joint force priorities are met, and desired effects are achieved.

If targeting functions are delegated appropriately, the final deconfliction and coordination of components' nominations should be at a target coordination meeting run by the TET. Component representatives should be prepared to justify target selections, since not all targets may be engaged based on the JFC's apportionment decision and the COMAFFOR's allocation. If differences arise and cannot be resolved at the meeting, the issue should be coordinated at higher levels for resolution. The meeting should not generally address mating of specific weapons to targets, but it should consider all capabilities and initiate the planning and coordination needed for those options. Additionally, the meeting may address the availability of certain high demand weapons or munitions on a particular ATO. However, the availability of weapons or capability should not drive the nomination of targets—this is antithetical to an effects-based approach.

The result of coordination is the draft JIPTL, which is submitted to the JFC or designated representative for approval. Again, targets may be added to no-strike or restricted target lists as a result of this part of the process highlighting RTL targets (for possible approval) and sensitive target approval and review (STAR) targets.

5. Identification of intelligence gaps, collection and exploitation requirements

Identifying collection and exploitation requirements through assessment is critical to targeting efforts. This stage attempts to answer the question, "How may we know that we have achieved the desired effects?" by establishing intelligence collection and exploitation requirements for each nominated target. This stage begins with target analysis and runs parallel to the other stages. The requirements should be articulated early in the tasking process to support target development and ultimately assessment. Targeteers should work closely with collection managers to ensure that target development, pre-strike, and post-strike requirements are integrated into the collection plan, along with any changes that occur throughout the tasking cycle. This intelligence support is also required to prepare for future targeting during execution (e.g., to pre-task real time ISR assets) and to support post-strike assessment of success. It should be noted that first-order effects of nonlethal operations are often subtle; in various instances may be of short duration for enabling purposes only or require days to months for the effect(s) to resolve, if at all, and may have effects that relate to the broader context of the target system (e.g., only visible at the operational or strategic level). Further, assessment of second- and third-order effects can be even more difficult. For these reasons nonlethal pre- and post-strike collection requirements are critical for ensuring a cohesive means exists to assess the intended effects. The product of this stage may be a joint integrated prioritized collection list (JIPCL).

Target List Development

Various target lists are created for use by the JFC to ensure the accuracy of target intelligence and validity of deliberate targeting in relation to guidance and LOW. These JFC managed lists include the JTL, RTL, and the NSL. The daily joint integrated prioritized target list (JIPTL), is created for use by the COMAFFOR to support the desired effects to be achieved on the corresponding ATO. Responsive and verifiable procedures should be in place for additions or deletions to any of the lists. However, commanders should be aware of the larger impact to effects based planning when individual targets are removed from the JIPTL or restrictions are applied. The removal or servicing restriction of one seemingly isolated target on a JIPTL may cause an entire target set grouping to become invalid thus requiring the identification of a different grouping of targets within the same or across one or more additional/ alternate target sets to create the same effect.

Before a nomination becomes a target, it is a candidate target that is developed, vetted, and validated. The candidate target list (CTL) is a list of selected target development nominations (TDN) submitted to the JFC for inclusion in the joint targeting process that are considered to create an effect that is consistent with the commander's objectives. The JTF staff, joint forces subordinate to the JFC, supporting unified commands, and components all submit TDNs to the JFC for inclusion on the CTL.

The second step of Phase 2 (Target Development) begins with the TDNs on the CTL being vetted and validated, and the JFC determining on which list the target should be placed. JTL is a consolidated list of targets upon which there are no restrictions placed and are considered to have military significance in the joint force commander's operational area. Essentially, the JTL is a compilation of all known, vetted, and validated targets that may be selected by any component for any type of action; exploitation or attack, lethal or nonlethal, air, ground, space, cyberspace, or other execution methods. The air component, as with other components, may develop target nominations for inclusion on the JTL via the CTL process.

JTF components select targets from the JTL to compile their respective TNLs and forward them to the JFC. Even in a mature theater unanticipated conflicts may not have a JTL from which components may select their TNLs. In this case, as we saw in Afghanistan, components will nominate targets for engagement without reference to a standing list. The TNLs are then combined, validated, and prioritized to form a draft JIPTL that is submitted to the JTCB for finalization. At each successive level throughout the life cycle of a target, a validation process occurs that checks targets against the NSL, RTL, ROE, current intelligence, commander's guidance, etc. Component commanders request the JFC (or the JFC's appointed representative) review and approve RTL targets nominated to the JIPTL that exceed the specified restrictions before execution. During operations, the JFC may delegate the authority to create the draft JIPTL to the COMAFFOR. If given this authority, the COMAFFOR's TET should execute the function of draft JIPTL creation.

Draft Joint Integrated Prioritized Target List (JIPTL)

The draft JIPTL is formed from consolidating and prioritizing the component TNLs based on prioritized JFC objectives. Those compiling the JIPTL consider the estimated available force capabilities and their ability to affect the targets on the list. The list usually contains more targets than can be serviced by the resources available. Thus, a draft JIPTL "cut line" is usually established. This "cut line" should reflect which targets should most likely be serviced for that ATO cycle, as well as the joint space tasking order (JSTO) and cyber tasking order (CTO) cycles. It should be clearly understood that the "cut line" simply reflects an estimate of the line above which targets are expected to be serviced by available resources, in priority order, and does not guarantee that a specific target will be attacked. Other variables like TSTs, changes in JFC priorities, emerging crisis, and changing resource availability may have an impact on target servicing. The JFC may also prohibit or restrict joint force attacks on specific targets or objects based on military risk, LOW, ROE, or other considerations. Targeting restrictions fall into two categories, no strike (sometimes called prohibited) and restricted.

See following page (p. 7-44) for further discussion.

No Strike List (NSL)

The NSL is a list of objects or entities characterized as protected from the effects of military operations under international law or the ROE. Attacking these targets may violate the LOW (e.g., cultural and religious sites, embassies belonging to countries not party to the conflict, hospitals, and civilian schools), interfere with friendly relations with other nations, indigenous populations, or governments; or breach national guidance and ROE that stipulates authorized targets/target systems (e.g., national guidance to not damage the nation's economic infrastructure). The NSL is compiled independent of, and in parallel to, the CTL. It is important to note, however, that entities from the CTL may be moved to the NSL if, as a result of additional target development, it is determined that attacking them may violate the LOW and/or guidance. Conversely, targets placed on a NSL may be removed and become subject to military action if their status as a protected object or entity has changed. It is critical to include the relevant staff judge advocate (SJA) in all aspects of target development and target list management. For example, religious and medical structures that function as weapons storage or barracks facilities may lose their protected status and may be legally attacked. However, not all situations create an automatic revocation of protection. For instance, the placement of an anti-aircraft artillery (AAA) piece on a medical facility, though an action in violation of LOW, does not result in the loss of protection; but neither does the protection status negate the legal authority to attack the AAA. The situation requires special handling by planners and attackers to determine whether the AAA must be attacked and to ensure minimal effects upon the hospital when attacked, to include the appropriate collateral damage estimation (CDE) review and approval.

See following page (p. 7-44) for further discussion.

Restricted Target List (RTL)

A restricted target is a valid target that has specific restrictions placed on the actions authorized against it due to operational considerations. Actions that exceed specified restrictions are prohibited until coordinated and approved by the establishing HQ. Attacking restricted targets may interfere with projected friendly operations. This list also includes restrictions on targets directed by higher authorities. The targets on the RTL are nominated by elements of the joint force, approved by the JFC, and include restricted targets directed by higher authorities. Targets may have certain specific restrictions associated with them that should be clearly documented in the RTL, such as do not strike during daytime or strike only with a certain weapon. Some targets may require special precautions, such as chemical, biological, or nuclear facilities, or targets in close proximity to no-strike targets. When targets are restricted from lethal attacks, targeteers should consider nonlethal capabilities as a means to achieve desired effects or support the objectives.

The previous section identifies key linkages between the joint targeting process and the air tasking cycle. Both elements should synchronize in every aspect of the process to ensure that the air component is adhering to the JFC's guidance and objectives with regards to targeting.

Products of the Phase

See following page (p. 7-44) for an overview and listing of target development phase products.

Target Development Phase Products

Ref: Annex 3-60, Targeting (14 Feb '17), pp. 68 to 69.

Joint Integrated Prioritized Target List (JIPTL)

The JIPCL is a prioritized list of intelligence collection and exploitation requirements needed to support indications and warning, analysis, and future target development efforts and to measure whether desired effects and objectives are being achieved. Requirements and priorities are derived from the recommendations of components in conjunction with their proposed operations supporting the JFC's objectives and guidance. An approved JIPCL is a product of answering information gaps as well as the collection and exploitation requirements stage of target development. The ISRD has primary responsibility within the AOC for the JIPCL, although considerable consultation with the SD OAT is required.

Joint Integrated Prioritized Collection List (JIPCL)

The JIPCL is a prioritized list of intelligence collection and exploitation requirements needed to support indications and warning, analysis, and future target development efforts and to measure whether desired effects and objectives are being achieved. Requirements and priorities are derived from the recommendations of components in conjunction with their proposed operations supporting the CFC's objectives and guidance. An approved JIPCL may be a product of answering information gaps as well as the collection and exploitation requirements stage of target development. The ISRD has primary responsibility within the CAOC for the JIPCL, although considerable consultation with the SD OAT is required.

No Strike List (NSL)

The NSL is a list of objects or entities characterized as protected from the effects of military operations under international law and/or rules of engagement. Attacking these may violate LOW—interfere with friendly relations with indigenous personnel or governments or breach ROE. Combatant commanders (CCDRs) and JFCs determine which targets are included on the NSL based upon inputs from components, supporting unified commands, or higher authorities. Targets on this list require national-level approval to strike. Targets on the NSL can only be moved to the RTL or JIPTL with national-level approval.

Restricted Target List (RTL)

The RTL is a list of targets that have specific restrictions imposed upon them. Some actions on restricted targets are prohibited until coordinated and approved by the establishing headquarters. Targets are restricted because certain types of actions against them may have negative political, cultural, or propaganda implications, or may interfere with projected friendly operations. The RTL is nominated by elements of the joint force and approved by the JFC. This list also includes restricted targets directed by higher authorities. Actions taken by an opponent may remove a target from the RTL.

Target System Analysis

Target System Analysis which provides an all-source examination of potential target systems to determine relevance to stated objectives, military importance, and priority of attack.

Electronic Target Folders (ETF)

Electronic target folders (ETF) developed to intermediate level. Depending on the level of intermediate development, ETFs will contain data on the target characterization, significance, location, type, function, expectation, elements, collateral damage considerations, intelligence gain/loss, and facility graphics (see CJCSI 3370 for complete details on ETF content at basic and intermediate levels to include graphics types).

C. Weaponeering and Allocation

Purpose of the Phase

Weaponeering is the process of determining the quantity of a specific type of lethal or nonlethal means required to create a desired effect on a given target. Allocation, in the broadest sense, is the distribution of limited resources among competing requirements for employment. There are two aspects relevant to the air tasking cycle: allocation of targets and allocation of forces. Weaponeering and allocation function together to produce the master air attack plan (MAAP). These efforts commence before the JIPTL is approved and continue past MAAP production into execution planning. They are integral to all aspects of targeting.

Weaponeering considers such things as the desired effects against the target (both direct weapons effects and indirect desired outcomes the second and third order effects), target vulnerability, delivery accuracy, damage criteria, and weapon reliability. Targeteers quantify the expected results of lethal and nonlethal capabilities employment against prioritized targets to produce desired effects. It results in probable outcomes given many replications of an event. It does not predict the outcome of every munitions delivery, but represents statistical averages based on modeling, weapons tests, and real-world experience. With modern weapons, however, the probabilities of accurate delivery and of achieving intended direct effects are high and steadily increasing. Weaponeering is normally done by TGT/TA team prior to TET using validated data and methodologies automated by the Joint Technical Coordinating Group for Munitions Effectiveness and the Defense Threat Reduction Agency, as well as appropriate data and methodologies for specialized/emerging capabilities associated with space and cyberspace capabilities. Weaponeering for space (non-terrestrial) and cyberspace targets is conducted by the Joint Space Operations Center (JSpOC) and 624th Operations Center (OC), through their parent combatant commands respectively, using applicable tools and methods. The final weaponeering solution is chosen by the MAAP Team. The output of the air tasking cycle weaponeering planning process is a recommendation of the quantity, type, and mix of lethal and nonlethal weapons needed to achieve desired effects while avoiding unacceptable collateral damage. All approved targets are weaponeered to include at least the following:

- Target identification and description
- Recommended aim points/joint desired point of impact (JDPI) and nonlethal reference points (NLRP)
- Desired scope, level(s) and duration of damage, destruction, degradation, denial, disruption, deterrence, suppression, corruption, usurpation, neutralization, delaying, influence, exploitation, or other planned effects
- Weapon system and munitions recommendations
- Fuzing requirements (if required)
- Probability of achieving desired direct effect(s)
- Target area terrain, weather, and threat considerations for the operational environment, including its physical, electromagnetic, and information (including cyberspace) components
- Collateral damage considerations
- Collateral effects

Precautions must be taken to avoid or minimize civilian casualties and damage to civilian infrastructure, and nonlethal collateral effects to civilian property which may also inadvertently affect civilian property outside the area of operations. The danger of collateral damage and effects varies with the type of target, terrain, weapons used, weather, the proximity of civilians and their structures, and the level of integration or shared communication infrastructures among the military, civil, government, private, and corporate environments.

According to LOW, incidental damage to civilian objects must not be excessive in relation to the expected military advantage to be gained. Collateral damage criteria were established on this foundational principle.

Collateral damage methodologies are aids to the decision-maker when approving targets for military action. They provide logical and repeatable methods to ensure due diligence in limiting civilian suffering while enabling the commander to assess risk in the accomplishment of military objectives. Collateral damage estimates are not designed to limit military action, but to mitigate, to the best of our ability, the unintended consequences of that military action. Military objectives are limited to those objects which, by their nature, location, purpose, or use make an effective contribution to the adversary's military action. Only those targets whose total or partial destruction, capture, or neutralization, in the circumstances ruling at the time, offer a definite military advantage may be attacked.

If an attack is directed against dual-use objects that might be legitimate military targets but also serve a legitimate civilian need (e.g., electrical power or telecommunications), then this factor should be carefully balanced against the military benefits when making a weapons selection, as must reconstruction and stabilization considerations following the end of hostilities. Thus, those conducting weaponeering should always keep commander's objectives and the end state in mind, as should those in other AOC teams and divisions who review weaponeering solutions and the MAAP. This includes the non-AOC weaponeering and attack planning processes for nonlethal operations. The methodologies and data used for weapon effectiveness estimation are also capable of producing estimations of collateral damage risk to noncombatants and non-targeted facilities. Established ROE and LOW also address collateral damage concerns. Targeteers must comply with Joint Chiefs of Staff (JCS) CD estimation directives and instructions. For example, it may sometimes be necessary to strike a target more precisely than would otherwise be necessary in order to avoid collateral damage. Certain levels of collateral damage estimation require expertise that lies outside of the COMAFFOR's or even the JFC's control and should be coordinated through the TGT/TA Team via federated and reachback relationships. External organizations should also comply with the same strict guidance on CDE that is imposed under ROE, LOW, and current CJCS instructions.

It is critical to stress that all estimates generated during this phase are situation-specific, reflecting the pairing of a particular capability against a particular target, under a particular condition of employment. As such, users of this information should be cautioned against assuming that the estimated effectiveness of a force capability under one set of circumstances is broadly applicable to other circumstances. Relatively minor targeting variations may have an exaggerated impact on effects estimates. It is equally important to stress that these estimates of performance are not designed to take into account considerations outside of the realm of weapon-target interaction (e.g., they do not address whether or not the delivery system may survive to reach the target.).

Targeteers should know the capabilities and availability of lethal and nonlethal platforms, weapons, and fuses. They should also be familiar with the standard conventional load platforms in their theater and delivery tactics. Weaponeering results may only be useful if the employment parameters assumed in weaponeering match those used in combat. Targeteers should work closely with the operations and logistics staff to obtain required information. As a rule of thumb, theater component targeting branches should request a copy of the time-phased force and deployment data (TPFDD) to obtain units' expected input options selected from the employed automated weaponeering programs, and to provide realistic planning data. Targeteers should be in constant coordination with space and cyberspace officers, and other special access programs throughout the process for capabilities not available via TPFDD and weaponeering tool synchronization. Weaponeering should also take into account the availability of the various weapons being considered. Certain high value weapons, such as those capable of deep penetration or other special effects, are normally limited in number and should only be used against those targets that both require the weapon for successful attack and are of sufficiently high priority to warrant the expenditure of the resource. Finally, some weapons, particularly certain capabilities, require long lead times in planning, deployment, and approval, which means that such capabilities should be thought about early and included at the beginning of the JOPPA process.

The weaponeering phase of the planning process is also where lethal and nonlethal effects may be planned against targets. Coordination with the information operations/non-kinetic operations team (IO/NKOT) is critical during this phase to ensure all operations (space, cyberspace, information, EW, etc.) are deconflicted, appropriately resourced, and phased over the battle space. There are a variety of tools available to planners to attempt to summarize and quantify the assessed impact of nonlethal operations. Since these techniques and capabilities are not fully normalized in most AOCs, it may be necessary to leverage the assistance of specialized teams in the DOD and academic communities.

Allocation is the translation of the air apportionment decision into the total number of sorties or missions by weapon system type available for each objective or task. It falls under the CPD MAAP team, which takes the final prioritized list of weaponeered targets and allocates airpower by melding available capabilities and resources, and weaponeering recommendations. The result is a translation of the total weight of air effort into the total number or sorties or missions required to achieve desired effects.

Prior to the TET target coordination meeting, the MAAP team determines how many aimpoints can be serviced on the given ATO day. The TET then reviews the lists of nominated targets and determines which "make the cut" on that day's proposed JIPTL. The TET should work closely with the SD and the MAAP Team to ensure that the prioritized list ties into the JAOP and AOD appropriately. The SD should ensure that the TET understands how effects and objectives are prioritized, how they are to be achieved over time, and that it has a macro-level idea of the number of targets associated with each objective. The TET then collects target nominations from other sources and works a daily allocation of targets that have been planned against the effects and objectives to build the daily JIPTL. Approaching JIPTL construction in this way helps avoid an ad hoc, target-servicing approach.

Each air capable joint force component submits an allocation request (ALLOREQ) message to the COMAFFOR (timed to coincide with the beginning of the MAAP part of the tasking process, usually not later than 36 hours prior to the start of a given ATO day). ALLOREQs contain requests for air and space component support and information on sorties from other components not required for organic component support that are available for COMAFFOR tasking. The MAAP team works with the TET to take the approved JIPTL (to include weapon restrictions, timing issues, and other restraints) and inputs from the component liaisons, the AMD (especially concerning tanker availability), and others to produce the MAAP. They determine an overall sortie flow for the ATO period and determine how that flow should be divided into packages—discrete sets of missions and sorties designed to complement each other or provide required support (for example, tankers and electronic warfare assets packaged with the strike assets supported). They also determine required times over target or times on station. Packages are arranged in sequence and used to determine a timeline and resource requirements for the ATO period. Each package should be de-conflicted in time, space, and effect.

Part of the allocation and MAAP portions of the tasking process is the creation of an ISR collection and assessment plan. Early planning for assessments is critical to ensure that target status can be quickly determined to meet restrike recommendation criteria. Theater ISR collection assets should be carefully orchestrated to ensure optimal coverage of the operational environment. Collection assets should be positioned not only to provide assessment of targets planned for attack, but should be able to detect and collect on emerging targets and be flexible enough to collect against them as well. At the same time, ISR collection assets should continue to monitor the operational environment in order to help discern whether desired effects are being created and whether the enemy is adapting his courses of actions (COAs) to our actions. The collection assessment plan cannot be made in a vacuum and should be closely coordinated with all other planning efforts.

The AOC should establish procedures to ensure that the organizations nominating targets receive continuous feedback on the status of their nominations throughout the tasking cycle. For example, not all targets nominated may be approved for the draft JIPTL, nor may all targets on the approved JIPTL be included on the ATO. There should be a feedback mechanism to ensure that targets not attacked, for any reason, are reported to the nominating authority for consideration on future TNLs.

Products of the Phase

See facing page for an overview and listing of weaponeering and allocation phase products.

D. Air Tasking Order (ATO) Production and Dissemination

Purpose of the Phase

Accomplished by the CPD ATO production team, this phase finalizes the ATO and associated orders, produces them, and disseminates them to combat units. It is based on commander's guidance (especially the AOD), the MAAP, and component requirements. Airspace control and air defense instructions should be provided in sufficient detail to allow components to plan and execute all missions listed in the ATO. These are usually captured in the airspace control order (ACO) and the day's special instructions (SPINS). Instructions contained in the SPINS and the ACO are updated as frequently as required. The ATO, ACO, and SPINS provide operational and tactical direction at appropriate levels of detail. The level of detail should be very explicit when forces operate from different bases and multi-component and/or composite missions are tasked. By contrast, less detail is required when missions are tasked to a single component or base. Components may submit critical changes to target requests and asset availability during this phase of the cycle. Parallel IRC processes may also result in the production of functional specific task orders like the cyber tasking order (CTO) and joint space tasking order (JSTO), as based upon applicable functional guidance like the Cyber Control Order (CCO) and SOD.

This stage of the process is where targeting instructions are communicated from the operational level to the tactical level (i.e., weapons standard conventional loads, weapon pairing with target and JDPI's, time on target, and fuse settings). It is imperative that targeting instructions include the desired objective of the mission. The mission commander is the final decision-maker prior to execution and must understand the desired effect to be achieved. Concurrent with the ATO, the AOC should make available relevant target materials that may assist tactical units in their mission planning efforts.

Products of the Phase

See following page (p. 7-50) for an overview and listing of air tasking order production and dissemination phase products.

Weaponeering & Allocation Phase Products

Ref: Annex 3-60, Targeting (14 Feb '17), pp. 74 to 75.

Master Air Attack Plan (MAAP)

The MAAP is the COMAFFOR's time-phased air component scheme of maneuver for a given ATO period, synthesizing commander's guidance, desired effects, supported components' schemes of maneuver, friendly capabilities, and likely enemy COAs, and allocating friendly resources against approved targets. The MAAP is developed by CPD's MAAP team and usually presented in the form of a decision briefing for the COMAFFOR. This product is critical for the targeting personnel to provide information to the collection managers in developing their collection and assessment planning.

Weaponeering solution determines the quantity of lethal or nonlethal weapons required to achieve an effect on the target, considering target vulnerability, weapons characteristics and effects, and delivery parameters. Weaponeering identifies the whole range of engagement options that may affect the target and highlights and/or selects the most appropriate engagement capabilities commensurate with desired effects for each relevant phase of the campaign.

Joint Desired Point of Impact (JDPI)

Joint Desired Point of Impact (JDPI) and associated graphic comprise the mensurated, three-dimensional, geophysical coordinates that identify the aimpoint for kinetic weapon employment.

Nonlethal Reference Point (NLRP)

Nonlethal Reference Point (NLRP) comprises the virtual location for the employment of nonlethal capabilities.

Collateral Damage Estimation (CDE)

Collateral Damage Estimation (CDE) and associated graphic establish the potential given the specific weapon-target pairing to create unintentional or incidental injury or damage to persons or objects that would not be lawful military targets in the circumstances ruling at the time. The CDE may result in specific constraints on weapons and delivery parameters.

Sensitive Target Approval and Review (STAR)

Sensitive Target Approval and Review (STAR) package conveys the request of the commander to garner approval from higher headquarters for the authority to strike a target that is considered to have a high CDE or will create significant political or media interest.

Air Tasking Order (ATO) Phase Products

Ref: Annex 3-60, Targeting (14 Feb '17), pp. 76 to 77.

Air Tasking Order (ATO)

The ATO is a medium used to task and disseminate to components, subordinate units, and command and control agencies projected sorties, capabilities and/or forces to targets and specific missions. It normally provides specific instructions to include call signs, targets, controlling agencies, etc., as well as general instructions. The ATO may subsume the ACO and SPINS or published as separate orders.

Special Instructions (SPINS)

SPINS are a set of instructions that provide information not otherwise available in the ATO, but are necessary for its implementation. This may include such information as commander's guidance (often including the AOD itself), the C2 battle management plan, combat search and rescue procedures, the communications plan, and general instructions for inter- and intratheater airlift.

Rules of Engagement (ROE)

ROE are directives issued by competent military authority that delineate the circumstances and limitations under which United States forces will initiate and/or continue combat engagement with other forces encountered. They should be published separately, versus being buried in the SPINS or another document.

Airspace Coordination Order (ACO)

The ACO provides direction to integrate, coordinate, and deconflict the use of airspace within the operational area. (Note: this does not imply any level of command authority over air assets.)

Reconnaissance, Surveillance, and Target Acquisition (RSTA) Annex

The reconnaissance, surveillance, and target acquisition (RSTA) annex is produced during this stage by the ISRD. The RSTA annex is the ISR supplement to the ATO. It contains detailed tasking of intelligence collection sensors and processing, exploitation, and dissemination (PED) nodes and provides specific guidance to tasked ISR assets, including ISR platforms, sensors, and PED.

The **finalized JIPTL cutline** associated with the ATO is fed back into the target development process for situational awareness on status of targets to be serviced in order to accurately produce the follow-on JIPTL.

As the ATO is finalized, the targeting staff will continue to update and/or refine targeting products in accordance with the coordination activities in developing the MAAP. Guidance may preclude a particular weaponeering solution or risk assessment may require combined kinetic and non-kinetic solutions to create the desired effect(s).These refinements will be documented within the ETF and specific products modified (e.g., JDPI, CDE, etc.) accordingly.

E. Execution Planning and Force Execution

Purpose of the Phase

Execution planning includes the preparation necessary for combat units to accomplish the decentralized execution of the ATO. Force execution refers to the 24-hour period an ATO is executed by combat units, which generally includes 12 hours immediately prior to the start of a given day's execution period. The AOC aids both, preparing input for, supporting, and monitoring execution. The COMAFFOR, as the Air Force's warfighting commander, directs execution of Air Force capabilities. If a JFACC is appointed, that commander directs execution of air component capabilities and forces made available for joint or combined operations. It is normal, of course, for the COMAFFOR to also be the JFACC. Inherent in this is the authority to redirect joint or combined air assets made available for tasking. Under the Air Force tenet of centralized control and decentralized execution, unit commanders are given the freedom and flexibility to plan missions and delivery tactics as long as they fall within timing requirements, ROE, and intent of effects. The COMAFFOR coordinates redirection of sorties that were previously allocated for support of component operations with affected component commanders. For targeting, this is the application of all previous steps of targeting and monitoring the execution in preparation for assessment. During execution, the AOC is the central agency for revising the tasking of air forces, the JSpOC is the central agency for revising the tasking of Air Force space forces, and the 624th OC is the central agency for revising the tasking of Air Force cyberspace forces. They are also responsible for coordinating and deconflicting any changes with appropriate agencies or components. These operations centers may or may not have authority to re-direct use of other capabilities supporting theater efforts, depending upon the asset.

Due to operational environment dynamics, the COMAFFOR may be required to make changes to planned operations during execution. The AOC should be flexible and responsive to changes required during execution of the ATO. Forces not allocated for joint or combined operations, but included on the ATO for coordination purposes, can be redirected only with the approval of the respective component or allied commanders. During execution, the COMAFFOR is also responsible for retargeting air assets to respond to emerging targets or changing priorities. The COMAFFOR may delegate the authority to re-direct missions made available for higher priority targets to C2 mission commanders as necessary. The AOC should be notified of all redirected missions. This can have significant impact on the ISR and collection planning efforts and require significant oversight by targeting personnel within the AOC.

The COD supervises the detailed execution of the ATO. Targeteers monitor ATO execution and recommend alternate targets when necessary. Normally, targeting changes are needed due to adverse weather, assessment requirements, or modification of priorities. The ability to quickly recommend good alternate targets is very important to the flexibility of airpower. Combat operations targeteers should be aware of all significant information on the current ATO to include targets, desired effects and objectives, guidance, and ROEs, and weaponeering and collateral damage estimates.

The rational use of force relies on the capability to achieve positive identification (PID) and geolocation of adversary entities as a precursor to taking action against them. Conducting CID of all operational environment entities is thus a critical enabling capability in any use, or potential use, of military force. Identifying adversary or enemy entities is essential, of course, but so is identifying friendly and neutral entities. Friendly force tracking (FFT) is a core function of combat identification (CID). FFT is the process of fixing, observing, and reporting the location and movement of friendly forces. The purpose of FFT is to provide commander's enhanced situational awareness and to reduce friendly fire incidents.

Targeting products produced in the previous phases become the primary means for imparting targeting information to the unit level in their preparation for and execution of force employment. This includes coordinating and deconflicting changes to targeting information with tasked units.

F. Assessment

Assessment is a continuous process that measures the overall effectiveness of employing joint force capabilities during military operations. It is also the determination of the progress made toward accomplishing a task, creating a condition, or achieving an objective. It helps answer basic questions such as:

- "Are we doing things right?"
- "Are we doing the right things?"
- "Are we measuring the right things?"

Assessment, as related to Annex 3-60 Targeting, is covered in further detail on pp. 7-53 to 7-58. See pp. 5-31 to 5-40 for a broader discussion of assessment from Annex 3-0, Operations and Planning (4 Nov 16), as related to the common operations framework.

VI. Targeting Assessment

Ref: Annex 3-60, Targeting (14 Feb '17), pp. 80 to 87.

Editor's note: See pp. 5-31 to 5-40 for a broader discussion of assessment from Annex 3-0, Operations and Planning (4 Nov 16), as related to the common operations framework.

Assessment is a continuous process that measures the overall effectiveness of employing joint force capabilities during military operations. It is also the determination of the progress made toward accomplishing a task, creating a condition, or achieving an objective. It helps answer basic questions such as:

- "Are we doing things right?"
- "Are we doing the right things?"
- "Are we measuring the right things?"

The first question addresses the performance of planned air operations by assessing the completion of tasks. The second question addresses the level at which the commander's desired effects are being observed in the operational environment and prompts examination of the links between performance and effects. The third question addresses the process of assessment itself and the importance of understanding how we choose to measure the links between performance, cause, and effect. When determined properly, the answers to these questions should provide the commander with valid information upon which to base decisions about strategy.

In an effects-based construct, it is not possible to think about actions and effects without considering how accomplishment of those effects should be measured. Effects and objectives should always be measurable and planning for them should always include means of measurement and evaluation. Assessment is not a separate phase of the air tasking—or any other—cycle, as descriptions and graphics often imply for the sake of conceptual clarity. Rather, it is interwoven throughout the planning and execution phase and is inseparable and integral component of the effects-based approach to conflict. Planning for assessment begins prior to commencement of operations and continues well after operations are over. It is a central part of an effects-based approach to conflict assessment that occurs at the strategic, operational, and tactical levels. From an Air Force perspective, assessment is conducted at unit level with intelligence and operational personnel identifying estimated level of mission success with supporting data (e.g., mission reports [MISREP], weapon system video [WSV], etc.) and at the operational level by AOC, JSpOC, and 624th OC personnel, who may leverage other organizations for reachback support. Each lower level feeds the levels above it and provides a basis for broader-based evaluation of progress. Products from each level provide the foundation for strategic level assessments that include target system and overall campaign assessment.

Any comprehensive view of assessment should tie evaluation of progress at the tactical level to all other levels of war, up to and including the national strategic level. The proper focus of assessment conducted by the air component should be on the operational level of war. An effective assessment construct should also support commanders' objectives at all levels, support commanders' decision cycles in real time, and provide the basis for analysis. To accomplish these things, an effective assessment construct should address the entire spectrum of operations and all levels of war, permit component validation of assessment elements, focus on effects, standardize federation, utilize intelligence specialties effectively, and integrate analysis efforts to the maximum extent possible.

I. Levels of Assessment

Assessors perform many types of assessment across the strategic, operational, and tactical levels to inform a wide array of decisions. These levels are distinct yet inter-related.

A. Strategic-Level Assessment

Strategic-level assessment addresses issues at the joint force (e.g., winning a particular conflict) and national levels (e.g., enduring security concerns and interests). It involves a wide array of methodologies, participants, and inputs. The President and SECDEF rely on progress reports produced by the CCDR or other relevant JFC, so assessment at their levels often shapes the nation's, or even the world's, perception of progress in an operation.

B. Operational-Level Assessment

Operational-level assessment begins to evaluate complex indirect effects, track progress toward operational and strategic objectives, and make recommendations for strategy adjustments and future action extending beyond tactical re-attack. Assessment at this level often entails evaluation of COA success, assessment of the progress of overall strategy, and joint force vulnerability assessment. These are commonly performed by joint force component commanders (e.g., JFACC) and the JFC and their staffs.

C. Combat Assessment (CA)*

Combat assessment (CA) is defined in JP 3-60 as the determination of the overall effectiveness of force employment during military operations. CA is composed of three major components: (a) battle damage assessment; (b) munitions effectiveness assessment; and (c) reattack recommendation. CA typically focuses on task accomplishment and specific engagements. The results of tactical tasks, measured by MOPs, are often physical in nature, but also can reflect the impact on specific functions and systems. CA may include assessing progress by phase lines; destruction of enemy forces; control of key terrain, people, or resources; and security or reconstruction tasks. Assessment of results at the tactical level helps commanders determine operational and strategic progress, so JFCs should have a comprehensive, integrated assessment plan that links assessment activities and measures at all levels. From the Air Force perspective, these would include but not be limited to: in-flight reporting, weapon system video (WSV), mission reports (MISREPs), full motion video (FMV) and cyberspace ISR activities.

CA determines the results of weapons engagement (with both lethal and nonlethal capabilities), and thus is an important component of joint fires and the joint targeting process. To conduct CA, it is important to fully understand the linkages between the targets and the JFC's objectives, guidance, and desired effects. CA includes the three related elements: battle damage assessment, munitions effectiveness assessment, and reattack recommendations or future targeting.

See following pages (pp. 7-56 to 7-57) for an overview and further discussion of the three related CA elements: battle damage assessment, munitions effectiveness assessment, and reattack recommendations or future targeting.

** With a broader concern for assessing operational, campaign level results, Air Force Annex 3-0 uses the term "Tactical Assessment" over "CA" because it is more broadly applicable and descriptively accurate: Not all operations (and hence not all assessments at the tactical level) involve combat. The name should apply to all tactical-level evaluation. The terms, however, are functionally equivalent for most purposes.*

II. Measures and Indicators

Ref: Annex 3-60, Targeting (14 Feb '17), pp. 82 to 83.

At all levels of assessment, planners should choose criteria that describe or establish when actions have been accomplished, desired effects created, and objectives achieved. These criteria are called "measures and indicators." There are two common types of measures:

Measures of Performance (MOP)

A criterion used to assess friendly actions that are tied to measuring task accomplishment. An example of this would be five offensive cyberspace operations performed, 100 combat sorties flown, and 98% ordnance delivered effectively.

Measures of Effect (MOE)

A criterion used to assess changes in system behavior, capability, or operational environment that is tied to measuring the attainment of an end state, achievement of an objective, or creation of an effect. An example would be to prevent the enemy's weapons factory from delivering weapons to the enemy for at least 48 hours.

Measures and Indicators

Measures and indicators are selected MOEs and MOPs established during planning. When selecting assessment measures, planners should identify the essential elements of information required to collect against them and provide guidance in the collection plan and JIPCL if special ISR resources are needed. These measures should be refined or amended during the tasking cycle, as the tactical situation or the status of the target changes. Selection of assessment measures is an iterative, ongoing effort.

To be useful as a gauge of effectiveness, a measure, whether a MOP or MOE, should be meaningful, reliable, and either observable or capable of being reliably inferred. Meaningful means it should be tied, explicitly and logically, to objectives at all levels. Reliable means it should accurately express the intended effect. If quantitative measures are used, they should be relevant. It is not sufficient to choose, for example, "fifty percent of enemy armor attrited" as an MOE without understanding why that measure is relevant to objectives. Observable means that existing ISR collection methods can measure it with the required precision to detect the intended change.

MOEs and MOPs may be quantitative or qualitative. Sometimes subjective measures, independent of other empirical measures, determine whether indirect effects and the objectives they lead to are being accomplished. Qualitative means primarily that judgment should be made in the absence of meaningful quantitative measures. Military personnel tend to be less comfortable with these rather than with more empirical, quantitative, measures, since they are generally trained to regard their profession as more of a science than an art, but often the numbers themselves involved in quantitative measures can deceive. Seemingly "scientific" quantitative measures are often poorer representations of what should happen in the operational environment than more qualitative measures, like "enemy armor units A, B, and C not offering larger than platoon sized resistance to forces closing on Phase Line X until at least day Y." Such a measure may be much more relevant to the friendly scheme of maneuver, be easier to collect against, and be easier for commanders to act upon. It is often easier, especially at the higher levels of assessment, to choose qualitative measures that are logically tied to objectives. Quantitative measures, on the other hand, can, through their very seeming certainty, take on a life of their own, leading to actions that do not contribute to accomplishing objectives or the end state.

Combat Assessment (CA)

Ref: Annex 3-60, Targeting (14 Feb '17), pp. 85 to 87.

CA determines the results of weapons engagement (with both lethal and nonlethal capabilities), and thus is an important component of joint fires and the joint targeting process. To conduct CA, it is important to fully understand the linkages between the targets and the JFC's objectives, guidance, and desired effects. CA includes the three related elements: battle damage assessment, munitions effectiveness assessment, and reattack recommendations or future targeting.

Battle Damage Assessment (BDA)

The purpose of battle damage assessment (BDA) is to compare post-execution results with the projected results generated during target development. Comprehensive BDA requires a coordinated and integrated effort between joint force intelligence and operations functions. Traditionally, BDA is a phased process. It begins with aimpoint-level evaluations of primary damage mechanisms and effect upon the targeted elements of a given target type (facility, individual, virtual, equipment, or organization). These assessments are aggregated and form the basis of system-level assessments. BDA is defined in three phases:

- **Phase 1 BDA**: This is the Initial Target Assessment reporting on physical damage assessment (PDA) and or change assessment with initial functional damage assessment (FDA) of the target. This BDA level phase is often derived from single source reporting. Typical timelines associated with this phase are 1-2 hours after information becomes available (e.g. sortie debrief, WSV review, Initial Imagery Report). It also provides initial inputs for a Restrike Recommendation.
- **Phase 2 BDA**: This is the Supplemental Target Assessment report on the physical, change assessment, and functional damage assessment of the target. This report is a detailed Physical Damage Assessment (PDA), Functional Damage Assessment FDA, and change assessment normally based on multi-source reporting. Phase 2 BDA reporting is provided when there is a significant change to the Phase 1 reporting to include the multi-source verification and change to the confidence level of the initial reporting.
- **Phase 3 BDA**: This is the Target System Assessment (TSA) and represents the aggregate of previous phase reporting. This assessment is normally produced by national-level intelligence agencies working closely with the Joint Task Force assessment teams (J2, J3, & J5). It represents an in-depth target system functional damage assessment with respect to a target system (collection of related facilities/entities) and provides commanders with high level assessments that help determine future weights of effort for future planning and execution. Reporting for this phase is normally provided 24 hours after information becomes available.

For additional information on the BDA process, refer to Defense Intelligence Agency (DIA) publications DI-2820-4-03, Battle Damage Assessment Quick Guide; DI 2800-2-YR, Critical Elements of Selected Generic Installations (Critical Elements Handbook); and JP 3-60, Appendix D, The Targeting Assessment Process.

Munitions Effectiveness Assessment (MEA)

MEA evaluates whether the selected weapon or munition functioned as intended. It examines the munitions' known parameters, the delivery tactics used, and the interaction between the munition and the delivery platform. MEA is fed back into the planning process to validate or adjust weaponeering and platform selections. It is also the form of assessment with the highest potential return on investment in terms of weapons and tactics development, because the data it generates is fed into the JMEM revision process, resulting in more accurate future capability analysis. MEA is combined operations and intelligence function.

Estimated Damage Assessment (EDA)

EDA is a type of physical damage assessment and is the process of anticipating damage using the probability of weapon effectiveness to support Estimated Assessments and allows the commander to accept risk in the absence of other information. Many times during execution, it is not possible to wait on ISR verification of strike results without inordinately delaying presentation of assessments to decision makers. EDA is an evolving technique of using Service documented munitions effectiveness (e.g., reliability, accuracy, effects, etc.), MISREPs, and other data to predict weapons effectiveness on targets and target systems as place holders for the probabilities of success in absence of reported BDA; a process facilitated by the precision and reliability of modern weapon systems. For instance, depending on the target type, size, number of weapons employed, and associated probability of damage, a prediction can be made of the target's continued level of operational capability. This information is also used to weigh the need for additional collection in lieu of inherent reporting from the weapon(s), aircraft, or aircrew to provide an assessed prediction of the level of physical and functional damage inflicted on selected targets and target systems. Essentially, the prediction becomes more accurate as additional information is received and incorporated, if the additional accuracy is needed. Due to EDA's requirements for empirical data, its use should be limited to weapons that have Air Force certified data and/or are contained in JMEM. How and when EDA is used should be determined during deliberate planning but should also be reviewed prior to each ATO execution. In general, it is appropriate for all but high-priority targets, but considerations for schemes of maneuver and strategic implications must always be considered. Normally, the COMAFFOR will provide guidance as to what level of risk he/she is willing to accept for a given target/target set when authorizing assessments based on EDA.

Reattack Recommendations and Future Targeting

Future target nominations and reattack recommendations merge the picture of what was done (BDA) with how it was done (MEA) and compares the result with predetermined MOEs that were developed at the start of the joint targeting cycle. The purposes of this phase in the process are to determine degree of success in achieving objectives and to formulate any required follow-up actions, or to indicate readiness to move on to new tasks in the path to achieving overall JFC objectives. Both operations and intelligence should work closely to present each target considered for restrike recommendation with the best and most current available information. Analysts may also discover that other targets in the system/network are now logical follow-on targets, or that the commander's objectives have now been met in regard to certain target(s), and that it is appropriate to recommend an end to further targeting within that target system or network. From the Airman's perspective, this element of Tactical Assessment occurs at the operational level. AOC planners are an integral part of providing the information to accomplish this for the COMAFFOR. Reattack recommendations should be consistent with JFC objectives and guidance.

Assessment is an inherently joint force process. It relies upon intelligence and operational data from multiple levels. As such, organizations and individuals who may conduct assessment require access to the intelligence analyses of those who developed the targets and the operational information from the ATO which executes against those targets. Both joint and national agencies often provide federated subject matter expertise to support all phases of BDA and other assessments. See Appendix B for an expanded discussion on federated support for targeting and assessment.

Assessment products are standardized but can be tailored in accordance with the level and type of assessment. For more on tactical assessment refer to JP 5-0, Joint Operation Planning, Appendix D; JP 3-60, Appendix D; AFTTP 3-2.87; and AFI13-1AOCV3. For more on combat assessment refer to JP 5-0, Appendix D; JP 3-60, Appendix D; CJCSM 3162.01, CJCSI 3370.01; and DI-28209-2-03, Commander's Handbook for Joint Battle Damage Assessment.

III. Targeting and Legal Considerations

Ref: Annex 3-60, Targeting (14 Feb '17), app. A.

Legal considerations and international legal obligations directly affect all phases of targeting. Those involved in targeting should have a thorough understanding of these obligations and be able to apply them during the targeting analysis.

App A. briefly discusses the following legal considerations impacting targeting:

- Basic principles of law of war (LOW).
- LOW considerations concerning personnel, objects and places.
- Rules of engagement (ROE) considerations.
- The role of judge advocate general (JAG) in targeting.

Targeting must adhere to the LOW and all applicable ROE. It is the policy of the Department of Defense to comply with the law of war during all armed conflicts and other military operations regardless of how such conflicts and operations are characterized. The law of war is that part of international law that regulates the conduct of armed hostilities. The law of war encompasses all international law for the conduct of hostilities binding on the United States or its individual citizens, including treaties and international agreements to which the United States is a party, and applicable customary international law. Military necessity does not provide authorization or justification for acts that are otherwise prohibited by the LOW. Instead, military necessity must be applied in conjunction with other LOW principles.

Appendix A is not all encompassing and is no substitute for legal advice from the appropriate Staff Judge Advocate (SJA). Constant coordination between planners, operators and JAGs is essential. The legal framework for the functional capability being employed (e.g., kinetic, space, cyberspace, etc.) depends on the nature of the activities to be conducted. Commanders, planners, operators, and targeteers must understand the relevant legal framework in order to comply with the laws and policies, the application of which may be challenging given the nature of nonlethal operations (e.g., ubiquity of cyberspace operations, regional effect of information operations (IO), etc.) and the often geographic orientation of domestic and international law.

The **law of war (LOW)** rests on five fundamental principles that are inherent to all targeting decisions: military necessity, unnecessary suffering, proportionality, distinction (discrimination), and honor (chivalry).

Rules of engagement (ROE) are directives issued by competent military authority to delineate the circumstances and limitations under which air, ground, and naval forces may initiate and/or continue combat engagement with other forces encountered. Essentially, ROE are rules for a particular operation that govern the use of force to reflect the will of the civilian and military leadership. ROE constrain the actions of US military forces to ensure their actions are consistent with domestic and international law, national policy, and objectives. Although ROE are not law, they are authoritative restrictions issued at the appropriate level of command to control the use of force. ROE are based upon domestic and international law, history, strategy, political concerns, and a vast wealth of operational wisdom, experience, and knowledge provided by military commanders and operators.

The **JAG** assists the planners and operators with reviewing targets for compliance with applicable LOW/ROE restrictions (including collateral damage and other CCDR restrictions) prior to mission execution. Legal advice and counsel is necessary to the development, interpretation, modification, and proper implementation of the ROE. JAGs and their support staff should be trained, operationally oriented, and readily accessible to assist planners and operators with international legal considerations and ROE or related issues.

I. Combat Support

Ref: Annex 4-0, Combat Support (21 Dec '15).

The Air Force defines combat support (CS) as the foundational and crosscutting capability to field, base, protect, support, and sustain Air Force forces across the range of military operations. The nation's ability to project and sustain airpower depends on effective CS.

A Readied Force A Prepared Operational Environment A Positioned Force	**CORE EFFECTS**	An Employed Force A Sustained Force A Reconstituted Force
Readying the Force Preparing the Operational Environment Positioning the Force	**CORE PROCESSES**	Employing the Force Sustaining and Recovering the Force Reconstituting the Force
Field Forces Base Forces Posture Responsive Forces	**CORE CAPABILITIES**	Protect Forces Generate the Mission Support the Mission, Forces, and Infrastructure Sustain the Mission, Forces, and Infrastructure

FUNCTIONAL COMMUNITIES

Acquisition AFOSI Airfield Operations Analyses, Assessments, and Lessons Learned Chaplain Corps Civil Engineer Communications/Information Contracting	Distribution Force Support Financial Mgmt/Comptroller Health Services Historian Judge Advocate Logistics Planning Maintenance Materiel Management	Munitions Mgmt Postal Services Public Affairs Safety Science/Technology Security Forces Test and Evaluation Weather Services

CS enables airpower through the integration of its functional communities to provide the core effects, core processes, and core capabilities required to execute the Air Force mission. The integration of these functional communities ensures Air Force forces are ready, postured, equipped, employed, and sustained at the right place and time to support the joint force.

Refer to SMFLS4: Sustainment & Multifunctional Logistics SMARTbook (Guide to Logistics, Personnel Services, & Health Services Support). Includes ATP 4-94 Theater Sustainment Command (Jun '13), ATP 4-93 Sustainment Brigade (Aug '13), ATP 4-90 Brigade Support Battalion (Aug '14), Sustainment Planning, JP 4-0 Joint Logistics (Oct '13), ATP 3-35 Army Deployment and Redeployment (Mar '15), and more than a dozen new/updated Army sustainment references.

I. Combat Support Construct

Ref: Annex 4-0, Combat Support (21 Dec '15), pp. 4-6.

Combat Support Principles

The foundation of combat support (CS) is a ready force, properly sized, organized, trained, and integrated. The structure comes from diverse functional communities that train and are equipped to provide a wide variety of capabilities. CS derives its capabilities from three overarching principles:

- CS enables operations in peacetime and wartime with effects supporting US national interests at any time or place across the range of military operations. CS includes the essential capabilities, functions, activities, and tasks necessary to employ all Air Force elements of air, space, and cyberspace forces at home station or while deployed.
- CS provides essential support while minimizing the forward footprint and maximizing reachback, thus increasing effectiveness and responsiveness. This essential support ensures the Air Force can quickly respond to a mission with a right-sized force, and with maximum effectiveness worldwide.
- CS provides the ability to transition swiftly from home station to a deployed environment and between operational requirements. CS planners should carefully examine requirements at deployed locations while operations continue at home station.

Core effects, the end result of combat support (CS), are produced from the core processes. Core capabilities are then used within the core processes to produce the effects necessary to achieve mission objectives. The core capabilities are formed by the employment of functional communities in a synergistic manner. The functional communities are those areas where Airmen who perform CS duties operate. This construct represents an Air Force-wide enterprise; some elements can be deployed forward in direct support of a contingency, while other elements can provide additional support to forward forces through reachback.

CS Core Effects

CS core effects are the products provided to a commander, Air Force forces (COMAFFOR), as outcomes of the CS core processes. The six CS core effects are:

- **Readied Forces**. Mission ready forces.
- **Prepared Operational Environment**. An environment conducive to mission execution.
- **Positioned Forces**. The right types and amounts of forces and materiel at the right places and times to meet mission objectives.
- **Employed Forces**. Forces, infrastructure, and materiel meeting mission requirements.
- **Sustained Forces**. Forces and materiel conducting persistent operations.
- **Reconstituted Forces**. A recovered force readied for operations.

CS Core Processes

The CS core processes are the standardized, overarching set of macro procedures that use core capabilities to produce CS effects. These macro procedures are the primary means of arranging CS practices due to their cyclical nature. The six CS core processes are:

- **Readying the Force**. Organizing, training, and equipping a fit force to provide mission capability.

- **Preparing the Operational Environment**. Analyzing, planning, and posturing forces, infrastructure (built and natural), and materiel for rapid employment.
- **Positioning the Force**. Deploying, receiving, and integrating forces and materiel at the point of employment.
- **Employing the Force.** Generating the mission, providing right-sized support, and ensuring timely regeneration of forces and materiel.
- **Sustaining and Recovering the Force**. Maintaining effective levels of forces, materiel support, including the physical plant, and infrastructure capability for ongoing operations. Recovering forces, materiel support, and infrastructure damaged from attack, accident, or other incident.
- **Reconstituting the Force**. Reset or redeployment of forces and materiel, ensuring airpower can be reapplied to meet operational needs.

CS Core Capabilities

The CS core capabilities result from the proper employment and integration of the functional communities. These capabilities form the structure of the remainder of this document. The CS core capabilities enable the Air Force to:

- **Field Forces**. Providing fully prepared CS forces to enable a COMAFFOR to meet the joint force commander's requirements. It includes organizing, acquiring, and tailoring forces to produce a responsive, sustainable, and survivable force.
- **Base Forces**. Establishing, sustaining, recovering, and closing airbases and operating locations (OLs). Providing enduring bases, installations, and OLs with the assets, programs, and services necessary to support and project airpower. For more information, see the discussion on Engineer Execution.
- **Posture Responsive Forces**. Assessing, structuring, scheduling, and processing force capabilities to support mission requirements. It also includes executing a dynamic positioning strategy to maximize CS responsiveness and speed of employment.
- **Protect Forces**. Providing an integrated all-hazards approach for force protection to detect threats and hazards to the Air Force and its mission. Applying measures to deter, pre-empt, negate, or mitigate the identified threats and hazards based on an acceptable level of risk. Actions required to protect forces specifically against hostile action include detecting, identifying, and defeating penetrative or standoff threats to personnel and resources; assessing OLs for threats and available support from host civil and military agencies; disseminating information and warning personnel; and protecting infrastructure.
- **Generate the Mission**. Preparing, configuring, launching, recovering, and regenerating weapon systems and payloads. It also includes conducting security cooperation engagements with partner nations as required in support of the combatant commander's (CCDR) theater campaign plan.
- **Support the Mission, Forces, and Infrastructure**. Supplying, distributing, and maintaining goods, services, and infrastructure throughout the operational area.
- **Sustain the Mission, Forces, and Infrastructure**. Ensuring CS is maintained for the duration of operations, maximizing the use of reachback, to include the industrial base, when needed.

CS Functional Communities

CS functional communities are fundamental to effective airpower. Each makes unique contributions to the overall mission.

Refer to Annex 4-0, Appendix "Functional Communities", for more information.

II. Command Relationships

A combatant commander (CCDR) exercises combatant command authority (COCOM) and directive authority for logistics (DAFL). The CCDR exercises these authorities over assigned and, if provided by the Secretary of Defense, attached Air Force forces (AFFOR) through the commander, Air Force forces (COMAFFOR). Air Force command and control (C2) structures for combat support (CS) are designed to enable a COMAFFOR to execute the Service's Title 10, United States Code (U.S.C.) responsibility for logistical support while also supporting the CCDR's exercise of DAFL.

When an Air Force major command (MAJCOM) is also the Service component to a CCDR (component MAJCOM, or C-MAJCOM), the C-MAJCOM organizes and employs forces to accomplish assigned missions. C-MAJCOMs provide the first echelon of reachback support to forces in the CCDR's area of responsibility. A numbered Air Force (NAF), if designated as a component NAF (C-NAF), provides the senior Air Force warfighting echelon and the organizational combat support planning expertise. The C-NAF staff plans the C2 architecture for operations. Regardless of the source of support or the support C2 structure, the Service component is responsible for ensuring essential support for all assigned and attached Air Force personnel within a joint force. Air Force commanders should be prepared to accept single-Service responsibility for joint common use items.

The C2 of CS operations produces a fully integrated CS capability extending from the lowest levels of capability (i.e., base and below) to the highest levels of resource allocation (headquarters Air Force) and operational planning (Air Force component, joint force, and above). Commanders and decision-makers have an immediate need for capabilities that capture, transmit, and share data about the status of current operations, courses of action, future plans, and predictive analyses. At each level there should also be a common set of dynamic and tailorable reporting and tracking tools.

Roles and Responsibilities

Major CS responsibilities for the COMAFFOR and AFFOR staff include:

- Develop supporting plans to meet CCDR mission requirements.
- Coordinate planning activities and requirements with force providers.
- Coordinate with commanders' staffs at all appropriate levels to identify employment locations.
- Plan and coordinate communications and information support.
- Plan and coordinate force protection support.
- Plan, coordinate, and provide materiel distribution.
- Plan and coordinate maintenance and munitions support.
- Plan, coordinate, and provide emergency services.
- Establish and identify manpower and equipment requirements.
- Identify host-nation support requirements.
- Ensure legality of all aspects of operations.
- Develop expeditionary site plans for approved employment locations.
- Manage allocated war reserve materiel.
- Ensure efficient use of physical plant to ensure available facilities and infrastructure to support in-garrison operations.
- Identify initial material capability gaps and provide input to acquire or modify new or existing weapon systems.
- Plan and execute operations security in support of military operations, activities, plans, training, exercises, and capabilities.

II. Engineering Operations

Ref: Annex 3-34, Engineer Operations (15 Aug '17).

Air Force civil engineer forces establish, operate, sustain, and protect installations as power projection platforms that enable Air Force and other supported commanders core capabilities through engineering and emergency response services across the full mission spectrum. Air Force forces generally operate from fixed bases, yet are mobile enough to project combat airpower worldwide from both enduring and non-enduring bases. To support this concept, Air Force civil engineers organize, train, and equip to rapidly respond as part of the air expeditionary task force performing comparable functions in peacetime and during contingencies. Engineers enable combat forces to operate across the range of military operations.

Engineer operations are increasingly being conducted in joint, interagency, and multinational environments. Operations require operating and maintaining both built and natural environments, including bases, facilities, infrastructure, and lines of communication for sustainment. Air Force civil engineers possess distinctive skills to provide engineering support throughout all phases of military operations, including airbase opening, establishing operating locations, receiving and bedding down forces; and sustaining, recovering, and closing operating locations.

I. Engineering: The Airman's Perspective

Air Force civil engineers prepare for war during peacetime, train as organic units, and deploy fully capable of rapidly establishing airbases to support the projection of airpower.

More than aircraft, missiles, or weapons, airpower is the coordinated activities of the weapon system, the weapon support system, and the basing system. The weapon system comprises the delivery vehicle, weapon, and operator. The weapon support system directly supports the weapon system. The basing system includes the infrastructure, people, materiel, and information needed to sustain operations for both the weapon and the weapon support system. Examples of expeditionary basing include bare bases, main operating bases, joint operating bases, forward operating locations, combat outposts, and cooperative security locations. There are differences in how the Services view expeditionary bases. The Air Force views an expeditionary base as an airfield, described as an area prepared to accommodate (including buildings, installations, and equipment), landing and takeoff of aircraft. The Army refers to these types of bases collectively as base camps: an evolving military facility that supports military operations of a deployed unit and provides the necessary support and services for sustained operations. Regardless of Service component lead, expeditionary bases vary in purpose and size and are built using different standards based on factors such as the mission, anticipated life span, and expected population. At the heart of the basing system is the installation itself.

All Services provide capability to establish and maintain bases. However, each Service maintains core engineering capabilities stemming from its traditional roles to meet specific operational needs. Air Force civil engineers have expertise in providing close support to conventional and unconventional forces while maneuvering; similar to Army and Marine engineer close support to ground forces. Furthermore, Air Force civil engineers enable airpower operations through rapid repair of damaged airfields, or construction of expeditionary landing surfaces; employing obstacles to deny enemy freedom of maneuver through denial of potential enemy airfields; and constructing protective structures to counter the effects of direct and indirect weapons through

II. Engineer Functions

Ref: Annex 3-34, Engineer Operations (15 Aug '17), pp, 6 to 7.

In joint and Air Force operations, engineering functions are categories of related engineering capabilities and activities that are grouped together to help commanders integrate, synchronize, and direct engineering operations. These functions fall into three basic groups: general engineering, combat engineering, and geospatial engineering:

Engineer Functions

General Engineering

Combat Engineering

Geospatial Engineering

A. General Engineering

General engineering consists of those engineer capabilities and activities that provide infrastructure and modify, maintain, or protect the physical environment. Examples include construction, repair, maintenance, and operation of infrastructure, facilities, lines of communication, and bases; airfield damage repair (ADR), terrain modification and repair, and selected explosive hazard activities. General engineering provides the means to develop installations to project airpower. It can occur under combat conditions but differs from combat engineering in that it is not in support of maneuver of forces. General engineering focuses on rapidly responding to establish, sustain, and recover airbases, conducting ADR as needed. These types of activities are usually required during initial stages of major operations when base infrastructure is unavailable or inadequate to support the commander, Air Force forces (COMAFFOR) in achieving the joint force commander's objectives. Engineering tasks are time consuming, requiring centralized planning and control to effectively manage limited resources. Commanders may employ a combination of military engineers, civil service, contractors, multinational engineers, and host nation personnel to fulfill engineering requirements. Although the nature of some tasks or the threat of violence in an operational area may require military engineers, once the area begins to stabilize the tasks can be performed using multiple available resources. For more detailed information on general engineering capabilities, see Appendix B.

B. Combat Engineering

Combat engineering is defined as those engineering capabilities and activities that provide close support to the maneuver of land combat forces. It consists of mobility, countermobility, and survivability operations. The primary difference between combat engineering and general engineering is combat engineering's requirement for close support to land combat forces and its focus on mobility/maneuver versus supporting base and mission operations from fixed locations. This should not be confused with "engineering under combat conditions."

Although Air Force civil engineers are not specifically organized, trained, and equipped to conduct combat engineering, their inherent skills are used to conduct tasks to support the installation and are in close support of Air Force forces maneuvering within the operational environment. For example, engineers support mobility operations by removal and demolition of obstructions on captured airfields. Engineers can also enhance mobility by establishing and recovering airfields and forward operating bases, providing the COMAFFOR with additional options of maneuver and flexibility. In support of countermobility, engineers emplace obstacles to achieve standoff, and work with Security Forces personnel to create obstacles to funnel enemy forces into firing zones that support integrated defense. To enhance base survivability, engineers build aircraft revetments; construct fighting positions and watch towers; reinforce overhead protection of key facilities; harden critical infrastructure; recover aircraft; and provide fire emergency services, explosive ordnance disposal, and emergency management capabilities.

C. Geospatial Engineering

Geospatial engineering contributes to a clear understanding of the physical environment by performing tasks that provide information and services to enhance effective use of the operational environment. Examples of geospatial data, information, and services include: terrain analyses and visualization, digitized terrain products, nonstandard tailored maps, precision survey, data management, baseline survey data, and force beddown analysis. Installation geospatial data enable commanders to make informed decisions during installation planning throughout all phases of operations.

Engineer Function of Installations Support

Installations support is the ability to provide installation assets and services necessary to support military forces. This includes activities necessary for effective real property life cycle management and installation services, the two elements of installations support. Installations support focuses on managing real property facilities and infrastructure in the US and at enduring and non-enduring bases in other geographic combatant commanders' areas of responsibility while providing protection, safety, security, and sustainability for personnel and mission-critical built and natural assets. Installations support activities may take place before, after, or concurrent with general engineering activities. During contingencies, engineers can expect to perform general engineering, geospatial engineering, and installations support activities, depending on the type of base required and whether existing facilities are already in place to support the mission, population, and expected mission duration. These concepts apply across the range of military operations.

Real Property Life Cycle Management

Real property Life Cycle Management is defined as the ability to plan and provide for the acquisition, operation, sustainment, recapitalization, realignment, and disposal of real property assets to meet the requirements of the force.

Engineer Function of Logistics Services

In joint and Air Force operations, logistics service is divided into water and ice service, contingency base services, and hygiene services When Air Force forces are deployed in support of major operations where Air Force beddown and sustainment support is unavailable, the supported command, Service, or agency provides logistical sustainment requirements.

expeditionary hardening. Similarly, Army and Marine engineers train extensively on combat engineering functions to provide close support to ground forces. They focus on eliminating obstacles to maneuver, employing obstacles to deny enemy freedom of maneuver, and constructing protective structures to counter effects of direct and indirect weapons.

Engineer capabilities are a significant force multiplier for a commander of Air Force forces, joint force commander, and combatant commander to meet mission objectives. These capabilities are grouped into categories: Engineering, Installations Support, and Logistics Services to achieve objectives at the strategic, operational, and tactical warfare levels. From the Air Force engineer's perspective, installation support and logistics services categories are at the same level as engineering.

III. Command and Organization

Once the President or the Secretary of Defense (SecDef) authorizes implementation of a specific plan, deployment of Prime Base Engineer Emergency Force (Beef) and RED HORSE forces is executed through the joint deployment system and conducted using guidance issued by Headquarters Air Force. During deployments, civil engineer forces are part of an Air Expeditionary Task Force (AETF), commanded by a commander, Air Force forces (COMAFFOR), and follow command relationships affecting all Air Force forces.

This is the case in engagement, cooperation, and deterrence operations; homeland operations; crisis response and limited contingency operations; and major operations and campaigns. To preserve unity of command, the joint force commander (JFC) will usually delegate operational control (OPCON) of Air Force forces to the COMAFFOR. Through the AETF structure, the COMAFFOR has both OPCON (as delegated by the JFC) and administrative control, as established by the Air Force, of Air Force civil engineers assigned or attached to the JFC.

The civil engineer organization consists of a total force mix of regular Air Force, Air Force Reserve, Air National Guard, and civilians. Air Force engineers are assigned or attached to organizations performing installation support, construction projects, and emergency response. Engineers provide training activities that support base and homeland operations, major commands and numbered Air Forces along with their subordinate wings. The civil engineer governance structure provides standardized guidance, training, equipment, and procedures through a corporate process.

To support the COMAFFOR, civil engineers are deployed as Prime BEEF or RED HORSE forces. Both train as organic units and remain fully prepared to rapidly deploy as full unit type codes (UTC) or tailored force packages. UTCs are ideally suited to provide the right skills at the right time. Prime BEEF organizations rapidly respond worldwide to provide a wide range of engineer support required to establish, operate, sustain recover, and reconstitute installations. RED HORSE is a self-sufficient, mobile heavy construction unit capable of rapid response and independent operations in a Level 1 threat environment. They provide heavy repair and construction capability that exceed Prime BEEF capacities.

The organization of Prime BEEF and RED HORSE forces under a single engineering commander has been proven as an alternative operational concept to support COMAFFOR requirements. Under this concept, limited theater-wide engineering forces can effectively be leveraged to prioritize and mass engineering capabilities, either light or heavy, at the right time and place to meet the demands. This concept first emerged in 2009 with the establishment of the Expeditionary Prime BEEF Group and evolved in 2012 into the Expeditionary Civil Engineer Group which placed both Prime BEEF and RED HORSE forces under a single engineering commander with an empowered staff.

Maintaining unit integrity should be considered when building deliberate plans. Unit integrity enables engineers who train together to also deploy together, providing the COMAFFOR an integrated mission-ready team.

III. Medical Operations

Ref: Annex 4-02, Medical Operations (29 Sept '15).

Air Force medical forces provide the joint force with several distinct capabilities. These include health services support, en route casualty support, and health care to eligible beneficiaries. Air Force medical forces assist in sustaining the performance, health and fitness of every Airman in-garrison and while deployed within the Continental United States (CONUS) or overseas (OCONUS) in support of global operations. The Air Force is increasingly called upon to deliver medical capabilities throughout the range of military operations. Diverse medical missions may consist of civil-military operations, global health engagement, or humanitarian assistance/disaster relief as part of joint or multinational operations.

I. Airman's Perspective

First and foremost, Air Force medical operations are focused on life-saving expeditionary medical support. They ensure rapid casualty stabilization, treatment, staging, and evacuation to definitive care, all while maintaining the standard of care and providing patient safety en route. The En Route Casualty Care System (ERCCS) is a versatile and flexible component of military medicine. Personnel are able to determine the required medical capability and designate the optimal mix of assets based on mission demands. Home station medical operations are postured at military and civilian facilities to provide health care while maintaining advanced clinical skills currency.

Air Force medical forces are made up of regular, Air Force Reserve, Air National Guard, Air Force civilians and contractors. Though regular, Reserve, and Guard medical forces are interoperable, each has distinctive mission areas.

Command and organization of medical forces consist of both the internal command and operational relationships within the Air Force Medical Service (AFMS) and the external command relationships for home station and expeditionary operations.

Medical capability objectives are designed to support the force. These objectives are to: promote and sustain a healthy and fit warfighter; prevent illness and injury; restore health (to include expeditionary health care); and optimize and sustain human performance.

Expeditionary Operations Planning Considerations tie together expeditionary operations, home station operations, medical stability operations, and other medical support operations worldwide. These considerations include Medical Forces Employment, Medical Forces Support in Joint Operations, the En Route Casualty Care System and Medical Logistics.

The global security environment is constantly evolving. No single nation can address every challenge and priority alone. With this in mind, the USAF actively partners with allies to further US and partner nation mutual interests in air, space, and cyberspace. Air Force medical forces may be required to support stability operations, build partnerships, and improve partner capacity. They leverage military health engagement and training opportunities to enhance military capabilities necessary to achieve objectives for all operations.

Responsive medical assistance within our borders is also vital to effective support of the Service as a whole. Air Force medical personnel support homeland operations as effectively as they do in an expeditionary environment.

II. Air Force Medical Forces Employment

Ref: Annex 4-02, Medical Operations (29 Sept '15), pp. 4, 19 to 21

Fundamentals of Air Force Medical Forces

Air Force medical forces provide a combat support (CS) functional capability. They provide the force health protection capability of CS. Likewise, medical forces are by design not self-sustaining; they depend upon CS capabilities for security and infrastructure support. They are an integral part of forces employed to open, establish, and operate airbases.

Air Force medical forces leverage speed, range, and flexibility by using hub and spoke operations to quickly form and maneuver customized medical capabilities to forward bases. Air mobility forces move cargo and personnel through one or more en route staging bases (the spokes) to arrive at a main operations base (the hub) within a theater. Before placing medical teams at airlift hubs, medical planners should consider the feasibility of the airlift web supporting routine hub operations and non-routine spoke requirements. Hub and spoke operations are further detailed in Medical Forces Support in Joint Operations.

Medical planners are integrated into the commander, Air Force forces' A-staff and the air operations center. They plan en route casualty care and aeromedical evacuation (AE) missions. Centralized control over Air Force medical, AE, and airlift forces is essential. It enables seamless stabilization and worldwide evacuation of casualties or patients from forward airfields to definitive care hospitals. Decentralized execution provides flexibility for en route medical support and local health services.

A key component of expeditionary and operational planning is the employment of Air Force medical forces. Medical forces deploy in capability-based modules that are flexible and tailored to each contingency operation to provide the appropriate level of medical support to an operational area.

When opening an airbase, medical teams assess the potential health impact of a bed-down location, base configuration, and provide advice on mitigating health hazards. These teams also provide initial medical support to the fielded forces from the initial commencement date. Based on existing theater and local medical capabilities, they provide input into additional medical capabilities required to support the projected population at risk. Continual health surveillance and assessment of operational, disease, and environmental exposures and risks are part of force health protection. This continual surveillance and assessment is essential for optimal health outcomes and operational performance.

Upon establishing the airbase, medical personnel and materiel assets continue to flow into the operational area and basing locations to expand medical support capabilities. Medical forces remain focused on threats and countermeasures to sustain and optimize warfighter performance. Medical forces establish a theater health care system using the following assets:

- Initial Aeromedical Evacuation (AE) and patient movement item assets to evacuate casualties.
- Theater En Route Casualty Care System tied into the theater AE system.
- A theater contingency and disaster casualty management plan to integrate theater, host nation, and coalition medical services; expeditionary medicine platforms; and the AE network.

During the operations at an expeditionary location, Air Force medical capabilities are planned based on the Air Force population at risk (PAR) and access to available AE organic or contract aircraft. In a joint deployment, additional resources may be required to care for sister-Service personnel. Air Force Medical Personnel also work closely with line of the Air Force personnel to monitor operational threats and provide risk management data for maximum operational effectiveness. The theater health care system is tied together with a robust network of local, host nation, joint, and coalition medical force capabilities linked by air, ground, and naval evacuation platforms.

Employment Tailoring

Medical force employment tailoring includes PAR support, rapid incremental employment, combat support (CS) force module employment, hub and spoke employment, or flexible tasking of an in-place force. Each has benefits as well as risks to be weighed for the operation at hand. In practice, when considered at the theater level, medical forces use these employment methods to optimize a theater medical system characterized by speed, responsiveness, flexibility, and agility. The goal is to strike a balance in devising a medical operations plan that exploits the capabilities yet limits the risks that come with a light and lean system of capabilities. The plan should be designed to maximize the commander's capability to stabilize, treat, stage, and evacuate casualties and patients from points of injury to definitive care on a worldwide scale.

Rapid Incremental Employment

Air Force medical forces possess the ability to insert forces into forward areas with a team tailored to the specific operational mission. Tailored forces may include preventive medicine, primary care, trauma surgery, intensive care, and connectivity to the AE system. Within this dynamic window of rapid deployment, combat and support forces compete for limited airlift into new airfields based on priority. This priority is not always an "all or nothing" decision for the deployment of combat support forces. Rather, the decision may be a balanced response to increase combat support capability as the airfield opens and begins operations, or as requirements change. During the period of medical vulnerability, en route critical care capabilities Tactical Critical Care Evacuation Team, Critical Care Air Transport Team, etc.) are able to expedite evacuation (while continuing active resuscitation and treatment) of casualties from initial forward resuscitative care teams. When deploying, medical forces strive to ensure health protection capability arrives as early as the warfighters and minimizes the demand on limited airlift resources.

During the periods of opening and closing airbases, Air Force forces are at a high risk of injury or illness due to non-combat vulnerabilities such as poor food, water, or sanitation and industrial or occupational accidents. The use of tailored medical forces allows a tiered approach to flowing medical capabilities in or out to match changing medical support requirements, mission or threat scenarios, availability of airlift, or the PAR. Flowing in only essential medical capability on the first available aircraft provides the necessary force health protection yet maximizes the limited airlift available for competing priorities. Additional medical capability flows in to meet requirements as operations dictate and airlift becomes available. When Air Force forces redeploy, medical force capabilities decrease incrementally as the PAR decreases and the threat allows.

Employment as Part of a CS Force Module

During the planning stage leading up to an operation, force module elements are linked together in planning systems so they may be rapidly identified and tasked to deploy. The figure titled Air Force Medical Force Module Capabilities depicts medical force capabilities integrated into each air expeditionary task force force module with specific capability types and quantities based on the PAR in the force module and the typical force health protection threats found at most airbases. Medical forces in CS force modules are those required to provide direct support to an expeditionary unit conducting operations from one airbase.

III. Air Force Medical Forces Objectives

Ref: Annex 4-02, Medical Operations (29 Sept '15), pp. 11 to 17.

Sustain a Healthy and Fit Force

To sustain a healthy and fit force is the first of four objectives of Air Force medical operations. A fit and healthy force increases the Air Force's capability to withstand the physical and mental rigors associated with combat and other military operations. The ability to remain healthy and fit despite exposure to numerous health threats is a force multiplier at home station and in deployed settings.

Fit and healthy Airmen can deploy on short notice and operate effectively in austere environments. Early identification and intervention of health conditions that could otherwise prohibit Airmen from being fully ready to deploy, increases the commander, Air Force forces' (COMAFFOR's) ability to mass forces.

Prevent Illness and Injury

To prevent illness and injury is the second of four objectives of Air Force medical operations. Illness and injury prevention is the framework by which Air Force leaders and individuals optimize health readiness and protect Airmen. The force health protection goal is to prevent illness and injury from the physical and mental stress caused by environmental, occupational, operational, and warfare, to include chemical, biological, radiological, and nuclear (CBRN) threats. Air Force medical personnel recognize and prepare for emerging man-made and natural threats. They make reasonable efforts to identify and protect our forces from emerging infectious diseases, as well as potential genomic/proteonomic, directed energy, and other new technologies. Casualty prevention is a continuous process conducted throughout pre-deployment, deployment, and post-deployment phases. Illness and injury prevention requires the full commitment of commanders, leaders, and individuals.

Restore Health

To restore health is the third of four objectives of Air Force medical operations. Medical forces use combined processes to rapidly restore each Airman to a combat ready status or arrange for the appropriate rehabilitative services. Restoring health requires a continuum of medical capabilities that includes first responders, forward resuscitative care (FRC), en route care, theater hospitalization, and definitive care.

Optimize Human Performance

To optimize human performance is the last of four objectives of Air Force medical operations. Personnel are the most important and valuable resource for the Air Force. Accordingly, Air Force Medical Service (AFMS) focuses on human performance in addition to health care as a primary means of supporting the COMAFFOR. Given the prerequisite need for health, addressing human performance requires achievement of the AFMS effects of "a healthy and fit force" and "prevent illness and injury"—two key objectives of force health protection.

The AFMS becomes a force multiplier by focusing on human performance in addition to health care as the primary means of supporting Air Force and joint forces. Air Force medical personnel work to sustain the performance of Airmen, whether in the face of enemy conflict, environmental threats and stressors, or advancing age. Any activity that supports or encourages improvement in physical, mental, or emotional health and fitness contributes to sustaining human performance. Additionally, Air Force medical personnel develop risk mitigation approaches. They employ approved countermeasures to help Airmen maintain performance (or minimize performance degradations) during warfare or upon exposure to environmental threats such as climatic extremes, g-forces, fatigue, weapons effects, prolonged mental or physical stress, witnessing or participating in violent acts, etc.

IV. Force Protection

Ref: Annex 3-10, Force Protection (17 Apr '17).

The 21st Century has, thus far, been characterized by a significant shift in Air Force responsibilities and an increased exposure of its resources to worldwide threats. This point is underscored by the terrorist attacks of 11 September 2001 and ongoing operations worldwide. Today, potential opponents are less predictable, leveraging the increased availability of both high and low technology weapons, including weapons of mass destruction. The Air Force's ability to project US airpower requires protection from these threats at home, in transit, and abroad.

Due to the increased lethality of international and domestic threats, it is imperative the Air Force take strong measures to protect our personnel and installations around the world. How the Air Force protects its forces is critical to global engagement. An air expeditionary task force poised to respond to global taskings within hours should establish the capability to fully protect its forces.

Force Protection Defined

Joint doctrine defines FP as "[p]reventive measures taken to mitigate hostile actions against Department of Defense personnel (to include family members), resources, facilities, and critical information" (Joint Publication 3-0, Joint Operations). FP is a fundamental principle of all military operations as a way to ensure the survivability of a commander's forces.

A comparison of the joint definition with the North Atlantic Treaty Organization (NATO), definition is instructive. NATO doctrine explains that "[t]he operational environment may have no discernable 'front-lines' or 'rear area' and an adversary may be expected to target Allied vulnerabilities anywhere with a wide range of capabilities." Consequently, NATO defines FP as "[m]easures and means to minimize the vulnerability of personnel, facilities, materiel, operations, and activities from threats and hazards in order to preserve freedom of action and operational effectiveness thereby contributing to mission success."

Commanders at all levels should have an effective force protection program. Commanders are responsible for protecting their people and the warfighting resources necessary to perform any military operation. We are obligated by the moral necessity of protecting our Airmen to ensure force protection (FP) is a part of Air Force culture.

Understanding and using FP doctrine will help ensure the successful protection of our people and resources.

FP supports **combat support**, and its supporting capability of "Protect the Force." Protecting Air Force personnel and resources is critical to the Service's ability to perform its mission.

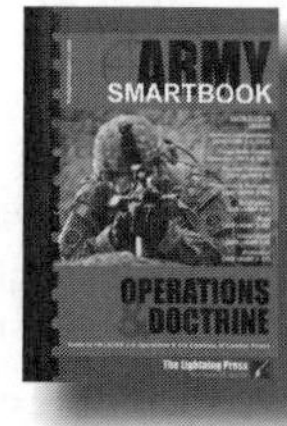

Protection is the preservation of the effectiveness and survivability of mission-related military and nonmilitary personnel, equipment, facilities, information, and infrastructure deployed or located within or outside the boundaries of a given operational area (JP 3-0). Refer to AODS6: The Army Operations & Doctrine SMARTbook for an entire chapter on protection to include the protection warfighting function, protection supporting tasks, and protection planning.

I. Command Responsibilities for Force Protection

Ref: Annex 3-10, Force Protection (17 Apr '17), pp. 4 to 7.

Centralized control and decentralized execution of force protection measures and resources are essential to protect forces against threats worldwide. Force protection (FP) is a task for every commander at every level. Clarity of command responsibilities for FP is essential for a comprehensive, unambiguous, and integrated response. Integration of all aspects of FP, including interoperability with civilian command and control systems, should enable commanders to react quickly to threats. The legal basis for commanders' FP authority has been greatly clarified in recent years, and it is essential that commanders understand their responsibilities and jurisdictions. Discussion of FP command responsibilities begins above the Air Force organizations in a joint force because of the top-down guidance that permeates the military in support of FP.

The Role of the Geographic Combatant Commander

Force protection is not exclusively a Service responsibility. According to both the Unified Command Plan and Joint Publication (JP) 1, Doctrine for the Armed Forces of the United States, geographic combatant commanders (GCC) have the overall requirement to establish and implement FP in their areas of responsibility (AORs). GCCs exercise authority for force protection over all Department of Defense (DOD) personnel (including their dependents) assigned, attached, transiting through, or training in the GCC's AOR, except for those for whom the Department of State Chief of Mission (COM) retains security responsibility. Examples of the latter include air attachés and Marine Corps embassy security group personnel. Additionally, GCCs develop and maintain memoranda of agreement with COMs that delineate security responsibility for DOD personnel based on whether the COM or the GCC is in the best position to provide FP. This is referred to as "proximity." Examples of this include US military personnel attending a foreign nation's defense college or Air Force personnel supporting military cargo aircraft at an international airport. Although the GCC is ultimately responsible, the GCC can work with the US Embassy to assume FP support duties to include intelligence sharing and threat warning.

Tactical Control Authority for Force Protection

GCCs have the authority to enforce appropriate FP measures to ensure the protection of all DOD elements and personnel subject to their control within their geographic AORs. This includes personnel on temporary duty, with the exception of DOD personnel for whom the COMs have security responsibility. This authority includes tactical control (TACON) for FP over military personnel within a GCC's AOR.

Further, TACON for FP authorizes the GCC to change, modify, prescribe, and enforce FP measures for covered forces. This relationship includes the authority to inspect and assess security requirements, and submit budget requests to parent organizations to fund identified corrections. The GCC may also direct immediate force protection condition measures (including temporary relocation and departure) when in his or her judgment such measures must be accomplished without delay to ensure the safety of the DOD personnel involved. Persons subject to TACON for FP of a GCC include regular and Reserve Component personnel (including National Guard personnel in a title 10 status) in the AOR.

A commander with TACON for FP can be different from the commander with mission responsibility. For example, 18th Air Force (18 AF) (Air Forces Transportation [AFTRANS]) has strategic airlift assets forward deployed in the United States Central Command AOR. Although the aircraft are staged in the Middle East, the commander with mission responsibility is the 18 AF (AFTRANS) commander (CC) and is responsible for securing these assets during mission execution. The 18 AF (AFTRANS)/CC has determined that Phoenix Ravens, specially trained Security Forces who travel with the aircraft, are required to support these missions. Therefore, Phoenix Ravens are forward deployed with these assets to secure the aircraft on missions. The protection of these aircraft and their personnel at their beddown location, however, remains an installation commander responsibility and not an 18 AF (AFTRANS) responsibility.

Although GCCs may delegate authority to accomplish the FP mission, they may not absolve themselves of the responsibility for the accomplishment of those missions. Authority is never absolute; the extent of authority is specified by the establishing authority, directives, and law.

Force Protection in US Northern Command

In most theaters, the senior DOD member serves as the combatant commander and assumes FP responsibilities. In US Northern Command's (USNORTHCOM's) AOR, where the Secretary of Defense and other senior DOD officials outrank the USNORTHCOM commander, the combatant commander maintains responsibility for FP. While this is a unique situation for USNORTHCOM, the principle is the same–there must be a commander responsible for the protection of DOD assets in the USNORTHCOM AOR to ensure unity of effort, and that commander is the commander, USNORTHCOM. The Title 10, United States Code, requirements of the military departments to support USNORTHCOM are the same as in any other theater, including supporting the USNORTHCOM FP mission.

USNORTHCOM executes a comprehensive all-hazards approach to provide an appropriate level of safety and security for the DOD elements (to include the Reserve components, DOD civilians, family members, and contractors supporting DOD at DOD facilities or installations), resources, infrastructure, information, and equipment from the threat spectrum to assure mission success. The authorities of commanders in the USNORTHCOM AOR are similar to those of commanders in other AORs.

The Airman's Perspective (Force Protection)

"Airminded" Force Protection (FP). Airmen normally think of airpower and the application of force from a functional rather than geographical perspective. Airmen do not divide up the battlefield into areas of operation as do surface forces. Airmen typically approach battle in terms of the effects they create on the adversary, rather than on the nature and location of specific targets. This approach normally leads to more inclusive and comprehensive perspectives that favor strategic solutions over tactical ones. This perspective extends to the Service's views on FP. Unlike surface forces that differentiate between security and protection, the Air Force's holistic approach is better suited to accomplishing its missions.

How Air Force forces are commanded and organized to execute FP responsibilities is influenced by this Airman's perspective. Because of the unique strategic nature of airpower operations, Airmen have developed a distinct perspective that guides how they think about the domains within which they conduct operational warfighting. General Henry "Hap" Arnold referred to this Airman's perspective as "airmindedness."12 This airmindedness reflects the range, speed, and capabilities of air, space, and cyberspace forces, as well as threats and survival imperatives unique to Airmen. Airmen have a common understanding of airpower operations, the threats and hazards to those operations, and the role they play in defending them. The Airman's perspective is an approach that shapes the conduct of operations and training to maximize operational effectiveness. It is built and developed through shared culture, ethos, values, and experience and makes a major contribution to an agile, flexible, expeditionary Air Force able to protect its force regardless of place or circumstance. Airmen should use their Airmen's perspective to drive how FP is applied in support of airpower operations.

II. Force Protection and Command Relationships in a Joint Environment

Since protecting the force is an overarching mission responsibility inherent in the command of all military operations, joint force commanders (JFCs) should consider force protection (FP) in the same fashion that they consider other aspects of military operations, such as movement and maneuver; intelligence, surveillance, and reconnaissance; employing firepower; sustaining operations in a chemical, biological, radiological, and nuclear environment; environmental conditions; and providing command and control during the execution of operations across the range of military operations. The geographic combatant commander (GCC) or a subordinate joint task force commander can delineate the force protection measures for all Department of Defense (DOD) personnel not under the responsibility of the Department of State. If a JFC designates command of an installation to a specific Service component commander, that commander has FP responsibility over all personnel on that installation, regardless of Service or status. For instance, if an Air Force commander is given FP responsibility for an installation, it is his or her responsibility to coordinate FP operations with commanders in adjoining or surrounding geographic regions; this includes intelligence sharing and deconfliction of operations that span the seams between operational areas.

The Service authority of administrative control (ADCON) is used to support various measures of FP, but is not the appropriate term to describe where the responsibility for implementation lies. For example, each Service may have ADCON responsibility to equip its personnel deploying to a hostile environment with appropriate body armor, but the requirement to wear that armor, and under what circumstances, is the responsibility of the commander on the ground at the deployed location, as these are operational, not administrative, decisions. As the JFC normally delegates operational control to the commander, Air Force forces (COMAFFOR) for all Air Force forces assigned or attached, the COMAFFOR normally exercises tactical control (TACON) for FP over those forces. TACON for FP over Air Force forces also resides with the joint commander of another Service who has Air Force forces attached with specification of TACON for a given responsibility.

Commander, Air Force Forces

In any operation in which the Air Force presents forces to a JFC, there will be a designated COMAFFOR who serves as the commander of US Air Force forces assigned or attached to the US Air Force component. The COMAFFOR, with the air expeditionary task force, presents the JFC a task-organized, integrated package with the proper balance of force sustainment and force protection elements. This applies on installations when the JFC has designated an Air Force officer as the base commander, i.e., when the Air Force is the primary occupant of a base.

Commanders at appropriate subordinate echelons (such as wing, group, and squadron level) retain ultimate responsibility for protecting people and property subject to their control and have the authority to enforce security measures. To this end, those commanders should ensure FP standards are met and make it an imperative to have an effective force protection program. These commanders face three major FP challenges: planning for FP integration and support as tasked in applicable operational plans, training for FP, and providing FP for those interests within their influence. These commanders have the added responsibility of accomplishing FP planning for the units identified to deploy to their location during contingency operations. Commanders should designate a member of their staffs as the integrator of FP subject matter experts to establish guidance for, program for, and manage FP requirements for their organizations.

(AFOPS2) Index